Communications in Computer and Information Science 2378

Series Editors

Gang Li⃝, *School of Information Technology, Deakin University, Burwood, VIC, Australia*

Joaquim Filipe⃝, *Polytechnic Institute of Setúbal, Setúbal, Portugal*

Zhiwei Xu, *Chinese Academy of Sciences, Beijing, China*

Rationale

The CCIS series is devoted to the publication of proceedings of computer science conferences. Its aim is to efficiently disseminate original research results in informatics in printed and electronic form. While the focus is on publication of peer-reviewed full papers presenting mature work, inclusion of reviewed short papers reporting on work in progress is welcome, too. Besides globally relevant meetings with internationally representative program committees guaranteeing a strict peer-reviewing and paper selection process, conferences run by societies or of high regional or national relevance are also considered for publication.

Topics

The topical scope of CCIS spans the entire spectrum of informatics ranging from foundational topics in the theory of computing to information and communications science and technology and a broad variety of interdisciplinary application fields.

Information for Volume Editors and Authors

Publication in CCIS is free of charge. No royalties are paid, however, we offer registered conference participants temporary free access to the online version of the conference proceedings on SpringerLink (http://link.springer.com) by means of an http referrer from the conference website and/or a number of complimentary printed copies, as specified in the official acceptance email of the event.

CCIS proceedings can be published in time for distribution at conferences or as post-proceedings, and delivered in the form of printed books and/or electronically as USBs and/or e-content licenses for accessing proceedings at SpringerLink. Furthermore, CCIS proceedings are included in the CCIS electronic book series hosted in the SpringerLink digital library at http://link.springer.com/bookseries/7899. Conferences publishing in CCIS are allowed to use Online Conference Service (OCS) for managing the whole proceedings lifecycle (from submission and reviewing to preparing for publication) free of charge.

Publication process

The language of publication is exclusively English. Authors publishing in CCIS have to sign the Springer CCIS copyright transfer form, however, they are free to use their material published in CCIS for substantially changed, more elaborate subsequent publications elsewhere. For the preparation of the camera-ready papers/files, authors have to strictly adhere to the Springer CCIS Authors' Instructions and are strongly encouraged to use the CCIS LaTeX style files or templates.

Abstracting/Indexing

CCIS is abstracted/indexed in DBLP, Google Scholar, EI-Compendex, Mathematical Reviews, SCImago, Scopus. CCIS volumes are also submitted for the inclusion in ISI Proceedings.

How to start

To start the evaluation of your proposal for inclusion in the CCIS series, please send an e-mail to ccis@springer.com.

Abolhassan Razminia · Dinh Hoa Nguyen
Editors

Intelligent Technology for Future Transportation

First International Symposium, ITFT 2024
Helsinki, Finland, October 19–21, 2024
Proceedings

Editors
Abolhassan Razminia ⓘ
Persian Gulf University
Bushehr, Iran

Dinh Hoa Nguyen ⓘ
Kyushu University
Fukuoka, Japan

ISSN 1865-0929　　　　　ISSN 1865-0937 (electronic)
Communications in Computer and Information Science
ISBN 978-3-031-84147-7　　ISBN 978-3-031-84148-4 (eBook)
https://doi.org/10.1007/978-3-031-84148-4

© The Editor(s) (if applicable) and The Author(s), under exclusive license to Springer Nature Switzerland AG 2025

This work is subject to copyright. All rights are solely and exclusively licensed by the Publisher, whether the whole or part of the material is concerned, specifically the rights of translation, reprinting, reuse of illustrations, recitation, broadcasting, reproduction on microfilms or in any other physical way, and transmission or information storage and retrieval, electronic adaptation, computer software, or by similar or dissimilar methodology now known or hereafter developed.

The use of general descriptive names, registered names, trademarks, service marks, etc. in this publication does not imply, even in the absence of a specific statement, that such names are exempt from the relevant protective laws and regulations and therefore free for general use.

The publisher, the authors and the editors are safe to assume that the advice and information in this book are believed to be true and accurate at the date of publication. Neither the publisher nor the authors or the editors give a warranty, expressed or implied, with respect to the material contained herein or for any errors or omissions that may have been made. The publisher remains neutral with regard to jurisdictional claims in published maps and institutional affiliations.

This Springer imprint is published by the registered company Springer Nature Switzerland AG
The registered company address is: Gewerbestrasse 11, 6330 Cham, Switzerland

If disposing of this product, please recycle the paper.

Preface

The 2nd International Symposium on Intelligent Technology for Future Transportation (ITFT 2024) was held in Helsinki, Finland on October 19–21, 2024. This symposium serves as a distinguished forum for scholars, researchers, and practitioners around the world to share and discuss innovative solutions, emerging trends, and the many challenges facing the development of intelligent transportation technologies.

As transportation systems around the world undergo rapid evolution, ITFT aims to share new ideas and new technologies amongst the professionals, industrialists, and researchers from the research areas of intelligent educational technologies. The symposium featured diverse presentations and discussions that highlighted breakthroughs in autonomous driving, smart transportation infrastructure, real-time traffic management, and sustainable transportation solutions.

All papers included in the ITFT 2024 conference proceedings reflect a broad spectrum of topics, encompassing both theoretical and applied research, and they provide valuable insights into current trends and future directions in intelligent transportation. A total of 74 manuscripts were submitted to ITFT 2024, each manuscript undergoing a meticulous single-blind peer-review process, involving no less than three reviewers. After careful scrutiny, we finally accepted 32 papers. We believe these papers will inspire your research endeavors and collectively steer you toward continuous advancements in these crucial fields.

We would like to thank all the organizing committee, reviewers, authors, and participants whose contributions made ITFT 2024 a tremendous success. We believe that with our joint efforts, research and application in the fields of intelligent transportation technologies will achieve more remarkable results.

We look forward to continued advancements in these exciting fields and to future ITFT symposiums that will continue to drive progress and shape the future of transportation.

December 2024

Abolhassan Razminia
Dinh Hoa Nguyen

Organization

Conference Chairs

Zaili Yang — Liverpool John Moores University, UK
Yu Zhang — University of South Florida, USA

Program Committee Chairs

Emrah Demir — Cardiff University, UK
Abolhassan Razminia — Persian Gulf University, Iran

Program Committee Co-chair

Fahimeh Farahnakian — University of Turku, Finland

Technical Program Committee Chairs

Dinh Hoa Nguyen — Kyushu University, Japan
Zhiwei Wang — Nanyang Technological University, Singapore
Wai Lok Woo — Northumbria University, UK

Publicity Chair

Zhenyu Zhao — Nanyang Technological University, Singapore

Local Chair

Mingyang Zhang — Aalto University, Finland

Technical Program Committee

Mahdi Abbasi	Bu-Ali Sina University, Iran
Agostinho Agra	University of Aveiro, Portugal
Waqas Ahmed	Universiti Kuala Lumpur, Malaysia
Aranzazu Berbey Alvarez	Universidad Tecnológica de Panamá, Panama
Nasrine Damouche	Université de Toulouse, France
Abolfazl Dehghanmongabadi	Shahrood University of Technology, Iran
Zigang Deng	Southwest Jiaotong University, China
Ehab Diab	University of Saskatchewan, Canada
Sameer-Ud-Din	National University of Sciences & Technology, Pakistan
Boban Djordjevic	KTH Royal Institute of Technology, Sweden
Ni Dong	Southwest Jiaotong University, China
Sunil Kumar Dube	Enphase Energy Inc., USA
Javier Faulin	Public University of Navarre, Spain
Hassen Fourati	Grenoble Alpes University, France
Pak Lun Fung	University of Helsinki, Finland
Stéphane Galland	Belfort-Montbéliard University of Technology, France
Christian Velasco-Gallego	Nebrija University, Spain
Álvaro Paricio García	Universidad de Alcalá, Spain
Marinella Silvana Giunta	Università Mediterranea di Reggio Calabria, Italy
Abdul Karim Gizzini	Institut Mines-Télécom, France
Arkadiusz Gola	Lublin University of Technology, Poland
Paweł Gora	University of Warsaw, Poland
Maciej Grzenda	Warsaw University of Technology, Poland
Ali Hajbabaie	North Carolina State University, USA
Sadeque Hamdan	University of Kent, UK
Zachary Hamida	Polytechnique Montréal, Canada
Md Mokammel Haque	Chittagong University of Engineering and Technology, Bangladesh
Tarek Hasan	University of Central Florida, USA
Sihong He	University of Connecticut, USA
Brayan González Hernández	Sapienza Università di Roma, Italy
Tran The Hoang	University of Auckland, New Zealand
Weiwei Jiang	Beijing University of Posts and Telecommunications, China
Golam Kabir	University of Regina, Canada
Arkadiusz Kampczyk	AGH University of Science and Technology, Poland
Mumtaz Karatas	Wright State University, USA

Mohamed Khan Afthab Ahamed Khan	UCSI University, Malaysia
Harald Kitzmann	University of Tartu Narva College, Estonia
Mladen Krstić	University of Belgrade, Serbia
Rafal Kucharski	Jagiellonian University, Poland
Yifu Lan	Aalto University, Finland
Chien-Sing Lee	Sunway University, Malaysia
Xin Li	Dalian Maritime University, China
Xi Lin	Shanghai Jiao Tong University, China
Xiaoli Liu	University of Helsinki, Finland
Jane Weizhen Lu	City University of Hong Kong, China
Elżbieta Macioszek	Silesian University of Technology, Poland
Nguyen Anh Minh Mai	Valeo, France
Edgar Emanuel Gonźalez Malla	Universitat Politècnica de València, Spain
Mohamed Amine Masmoudi	Rabat Business School, Morocco
Saleh Mobayen	National Yunlin University of Science & Technology, Taiwan
Faiz Ul Muram	Linnaeus University, Sweden
Husam Muslim	Japan Automobile Research Institution, Japan
Khoa Nguyen	Carleton University, Canada
Thi Thuy Hanh Nguyen	Vietnam National University, Vietnam
Mona Faraji Niri	University of Warwick, UK
Ankit R. Patel	University of Minho, Portugal
Narong Pleerux	Burapha University, Thailand
Waishan Qiu	McKinsey & Company (Shanghai), China
Ikjot Saini	University of Windsor, Canada
Bdereddin Abdul Samad	Higher Institute of Sciences and Technique Al Zahra, Libya
Hadi Sarvari	Hong Kong Polytechnic University, China
Qing Shen	University of Washington, USA
Haotian Shi	University of Wisconsin Madison, USA
Natalya Shramenko	Hochschule Karlsruhe, Germany
Dragan Simić	University of Novi Sad, Serbia
Željko Stević	University of East Sarajevo, Bosnia and Herzegovina
Burak Taşci	Firat University, Turkey
Burcu Tekeş	Başkent University, Turkey
Ashim Kumar Thapa	George Mason University, USA
Irfan Ullah	Dalian Maritime University, China
Shi'an Wang	University of Texas El Paso, USA
Yacan Wang	Beijing Jiaotong University, China
Zhiwei Wang	Southwest Jiaotong University, China

Pak Kin Wong	University of Macau, China
Lingxiao Wu	Hong Kong Polytechnic University, China
Yushu Yu	Beijing Institute of Technology, China
Noor Zaman	Taylor's University, Malaysia
Jun Zheng	Southwest Jiaotong University, China

Contents

Intelligent Vehicle Technology and Applications

The Comparative Analysis of Car-Pooling Algorithms for Ride-Sharing Systems 3
   Julien Baudru and Hugues Bersini

Acceptance of Automated Vehicles in Logistics and Beyond: A Survey-Based Investigation 25
   Guglielmo Papagni, Setareh Zafari, Johann Schrammel, and Manfred Tscheligi

Modeling Autonomous Delivery Robots Under Framework of Automated Driving System Using Dynamic Transport Assignment: A State-of-the-Art 38
   Yousuf Dinar, Carsten Gertz, Jacqueline Bianca Maaß, and Elena Queck

Virtual Variable Baseline Stereo Vision Algorithm 50
   Chia-Chiun Kuo and Liang-Kuang Chen

A Conceptual Framework to Operational Design Domain (ODD)-Based Scenario Generation for Technical Evaluation Obstacle Detection in Automated Shuttle Bus 63
   Kun Gao, Ulrike Weinrich, Thomas Riemer, and Hans-Christian Reuss

Precise Driver's Drowsiness Detection Using a Combination of Proven Methods with a Single Layer Neural Network 74
   Ghazal Abdolbaghi and Alireza Yazdizadeh

FusionSis: Analysis and Evaluation Framework for Fusion Safety State of Connected and Automated Vehicles Under Cyber Attacks 83
   Bowen Zheng, Shichun Yang, Weifeng Gong, Haoran Guang, Yi Shi, Mingjie Chen, and Yaoguang Cao

Deep Learning-Based Monocular Depth Estimation Method for Forklift Collision Avoidance 98
   Dong-Ju Kim, Chang-Yeop Lee, Hyo-Jin Kim, and Young-Joo Suh

Enhancing Multi-user Experience: Optimizing Explanation Timing Through Game Theory 106
   Akhila Bairy and Martin Fränzle

User-Intimate and Trustworthy Ontology-Based Requirement Engineering
Methodology: A Case Study on Connected and Autonomous Electrical
Vehicles 118
    *Alper Kanak, Ali Serdar Atalay, Oğuzhan Herkiloğlu,
Ahu Ece Hartavi Karcı, Elif Toy Aziziaghdam, and Salih Ergün*

Intelligent Transportation Management and Traffic Flow Analysis

AI and Leadership in Startup Innovation and Disruption: The Case
of Mobility as a Service (MaaS) 137
    Michele Vincenti

Research on Traffic Analysis Zones Division Model for New Roads Based
on Complex Network Theory 147
    Zhiyong Wen, Xiaoxiong Weng, and Bangquan Xie

Forecasting Vehicle Mobility on Various Segments of the Urban Transport
Network in the Fuzzy Paradigm 162
    Ramin Rzayev and Emil Ahmadov

The Emerging ICT for Exposure Reduction in Urban Mobility: Transport
Models for Risk Cycle 173
    Francesco Russo, Antonio Comi, and Corrado Rindone

Concept of a Virtual Test Field for Inland Waterway Transport 188
    *Jason Sutanto, Christian Hürten, Maximilian Jarofka,
Frédéric E. Kracht, and Dieter Schramm*

Critical Characterization of Three-Phase Traffic Flow in Severe Condition 200
    Bo Song, YongSheng Qian, JunWei Zeng, and Xu Wei

Design and Implementation of Traffic Congestion Relief Strategies Based
on Multi-objective Optimization Algorithms 209
    Jun Zhao

Wide-Area Ship Movement Prediction Using Random Forests 220
    *Tanja Vähämäki, Farshad Farahnakian, Paavo Nevalainen,
and Jukka Heikkonen*

Characteristics of Heterogeneous Traffic Flow Involving Different
Intelligent Level Autonomous Vehicles 246
    Xuan Wang, Junwei Zeng, Yongsheng Qian, and Xu Wei

Exploring the Potential Application of Ramp Metering Systems to Improve the Performances of Roundabout Corridors 260
Lorenzo Brocchini and Antonio Pratelli

The Research on Customer Demand of Asia-Europe Liner Shipping Companies Based on Kano Model 271
Yiyang Chen

The Death and Life of Free-Floating Car Sharing in China: Case Study of Chengdu and Changchun 282
Hongjie Wang, Xia Luo, Qiming Su, and Hongqing Bao

Intelligent Transportation Infrastructure and Sustainable Transportation

Research on Coordinated Control of Multiple Energy Storage Systems Considering No-Load Voltage Differences 295
Shi Xiao, Jingwen Zheng, Luqing Jiang, Yajie Zhao, Bin Li, Zhihong Zhong, and Hu Sun

Intelligent Transportation Systems: Enabling Sustainable Transportation and Efficient Traffic Management—A Review 311
Roberto D. Rosario, Arjel Alvarez, Ronnel C. Quinto, and Mark de Guzman

A Green Intelligent Transport Model for Urban Mobility 324
Gerald B. Imbugwa, Tom Gilb, and Manuel Mazzara

RPC Coordinated Control Strategy with Battery and Flywheel Energy Storage 340
Shi Xiao, Zhiqiang Zhang, Peijin Yang, Bin Li, Zhihong Zhong, and Hu Sun

Experimental Virtual-Reality Assessment of a Cycling Environment Using a One-Boundary Drift-Diffusion Model 360
Kaori Nakamura, Shun Su, Yusak Susilo, and Daisuke Fukuda

Reinforcement Learning Based Smart Charging for Electric Vehicle Fleet 375
Biao Xu, Aivars Rubenis, and Chao Long

Materials Recycling and Transportation Infrastructure 384
Christian Paglia

Dynamic Characteristics Analysis of High-Speed Trains Considering
Wheel Polygonal Wear Under Track Irregularity Excitation 396
   *Xin Wang, Hongzhang Yu, Lin Zhou, Junyi Mu, Imdad Ullah Khan,
and Chunrong Hua*

Modeling Interdependencies in Intelligent Traffic Systems and Sustainable
Urban Development Using Graph Neural Networks 406
   Qian Cao, Jing Li, and Paolo Trucco

Towards a Green Future: Case Analysis and Technological Prospects
for Sustainable Transportation Decarbonization 416
   Shu Liu, Shanshan Shi, and Chen Fang

Author Index 427

Intelligent Vehicle Technology and Applications

The Comparative Analysis of Car-Pooling Algorithms for Ride-Sharing Systems

Julien Baudru[1,2(✉)] and Hugues Bersini[1,2]

[1] IRIDIA, Université Libre de Bruxelles, Brussels, Belgium
julien.baudru@ulb.be
[2] FARI, AI Institut for the Common Good, Brussels, Belgium

Abstract. In this paper, we present a comparative analysis of five methods for constructing ride-sharing pools of users, focusing on their efficiency in terms of execution time, the percentage of user requests fulfilled, the distance of the detour made by the driver and the waiting time of the passenger. Furthermore, we introduce a model able to simulate user demand, based on car usage data across different time intervals in Belgium. Then, we use the proposed model as a basis for evaluating the performance of the five methods and their variants: OD Similarity, OD Clustering, OD Time Alignment, Trip Similarity, and Trip Buffering.

Keywords: Car-Pooling · Trip-Matching · Ride-Sharing · Sustainable Mobility · Smart Transportation · Road Networks

1 Introduction

For several decades, the use of private vehicles has exploded, leading to an increase in traffic congestion, pollution and accidents. Various solutions already exist to reduce the reliance on private cars, we can first think of public transport, for instance, as a promising alternative. However, since not all cities around the world have a well-developed public transport network, ride-sharing seems to be the most viable alternative for users in terms of economy, ecology and comfort [4]. In addition, according to [20], Peer-to-Peer (P2P) ride-sharing is the most flexible and lowest-cost mobility alternative.

Reducing the number of vehicles on the road appears to be essential from an economical and ecological point of view. Indeed, in Europe Union (EU), inefficient urban mobility, and road congestion in particular, costs the EU around 119 billion dollars a year (2020) [15]. In the United States of America (USA), the road congestion is estimated to be responsible for a cost of around 121 billion dollars a year (2019) [23]. From an ecological point of view, in EU, between 2014 and 2017, CO_2 emissions from road transport rose by 45 million tonnes, or 5% [19]. In addition, the use of personal vehicles is a major source of urban air pollution in Brussels, Stuttgart and Milan [28]. In USA, in 2021, CO_2 emissions from transportation in the United States totaled 1700 million tonnes, the most

© The Author(s), under exclusive license to Springer Nature Switzerland AG 2025
A. Razminia and D. H. Nguyen (Eds.): ITFT 2024, CCIS 2378, pp. 3–24, 2025.
https://doi.org/10.1007/978-3-031-84148-4_1

from any sector of the economy. And in 2019, 40 million tonnes of greenhouse gases, about 2% of all transportation-related emissions, were emitted because of traffic congestion [24]. As outlined in the aforementioned information, it is clear that solving the challenges posed by ride-sharing, or car-pooling, has become a major mobility issue in most regions.

Ride-pooling can be described as a shared transport system in which multiple users take a common route, and therefore vehicle, to reach their different, or common, destination [12]. This transport system is based on the shared use of private vehicles or the shared use of vehicles provided by Mobility as a Service (MaaS) companies. This system differs from the services of Transportation Network Companies (TNCs), such as Uber and Lyft, which offer a low-cost alternative to taxis [20]. P2P rides-haring seek to gain the advantages of TNCs while reducing their negative impact on the environment. The most suitable definition for the study presented in this paper is ride-sharing through the use of private vehicles. An extended definition provided by [22] is given in Text 1.

Private cars are utilized by various households or organizations in either a centralized (one large, open group) or decentralized system (several small, closed groups). The vehicle is owned by one member of the carpool group or can be jointly owned by several group members.

Text 1: Definition of ride-sharing.

As accurately stated in [9], passengers and drivers might have multiple varying objectives and preferences that must be optimised and satisfied but often these preferences are conflicting. For example, drivers aim to optimize their financial gains, minimize daily service hours, and express preferences for specific service areas, whereas passengers seek to minimize travel time, waiting time and cost and may prioritize a comfortable journey with pleasant personal space. However, according to [9], offering a car sharing service that optimises all the above objectives while simultaneously satisfying all preferences may not be feasible. This is why in the following of this paper we will deal with only two preferences and one objective. These preferences concern the maximum detours travelled by the driver and the maximum waiting time of the passenger and the objective is to maximise the number of satisfied passenger requests.

In this paper, we compare different methods for creating groups of users interested in carpooling. We evaluate these methods based on their execution time, the percentage of user requests satisfied, the distance of the detour made by the driver and the waiting time of the passenger. In addition, we propose a model that simulates user demand based on car usage data across various time periods for a day in Belgium. This approach provide insights into user behavior patterns, allowing a robust comparison of the OD Similarity, OD Clustering, OD Time Alignment, Trip Similarity, and Trip Buffering methods. We start by defining the different mathematical elements used, secondly we present two similarity functions, and then we detail different methods to generate pools of users. Next, we introduce a model to simulate user requests in car-pooling and we compare the results obtained by the different methods tested on this model and on randomized experiments with the two different similarity functions when possible, and finally propose a series of future improvements for further research.

2 Similar Works

To the knowledge of the authors, only a few reviews exist in the literature on the ride-sharing problem. One of the most notable is provided by [25], in this review the authors propose a classification of the various existing systems, distinguishing between dynamic and static systems. They then subdivide the category of dynamic systems into three parts: centralised, decentralised and hybrid systems. For each of the static and dynamic systems, they propose a distinction between systems that use heuristics and those that do not. Also, they compare the advantages and disadvantages of each system. Another notable review is that provided by [20], in their article they compare a series of flexible, dynamic, ride-sharing systems on the basis of five criteria, the use of flexible paths, the use of multi-hop, the possibility of having multiple rides and whether the solution is optimal. They note that few systems meet all the criteria at the same time, and that the execution times of the systems presented are too long to be used in large-scale real-life applications. The authors then propose two novel approaches to increase the performance of a ride-sharing system, one of which uses the Ellipsoid Spatiotemporal Accessibility Method (ESTAM), an idea similar was proposed in [2], to find the optimal meeting point between a passenger and a driver.

There exists a diverse array of strategies for user matching methods in car-pooling, all aimed at optimizing passenger-driver pairing for efficient transportation solutions. Traditional approaches often rely on heuristic algorithms, such as nearest neighbor, greedy algorithms or evolutionary algorithms, which prioritize proximity and availability. For instance in [8], they present two heuristic algorithms based on greedy method and the time-space network for the case of one origin to many destinations and many origins to one destination in the context of dynamic taxi-pooling problem. In [11], the authors demonstrate the effectiveness of evolutionary algorithms in minimizing total trip costs for distributing passengers traveling from a common origin to different destinations in multiple taxis. Another popular method to solve this problem is to use an operational search approach. In [3], the authors have developed an approximation algorithm for assigning cars to requests while aiming to minimize costs. Their algorithm guarantees solutions with at most 2.5 times the optimal cost, and experiments show that it often achieves a better ratio, around 1.2, on synthetic data.

Additionally, collaborative filtering methods, inspired by recommender systems, consider user preferences and historical data to improve matching accuracy. Among others, the model proposed by [9], MaMoP, uses social reasoning and evolutionary algorithms to simultaneously optimise ride-sharing solutions and take account of user preferences. In [26], they integrate user personality preferences into a matching model for passengers in ride-sharing systems. They modify the stable roommates problem algorithm for one-on-one passenger matching and consider factors such as personality and steadiness. In [1], the authors propose an algorithm to improve the matching optimization, taking into account the gender, age, professional status and the social tendencies of the participants. In addition, the proposed algorithm subdivides the unmatched segments of the path of the drivers, generating new trip requests to create additional matches using these unmatched segments. In [29], they present a model in which riders are matched based on a specific set of human characteristics using machine learning techniques. After trip completion, they record the user feedback and compute two main characteristics that are most important to riders. The registered and the computed characteristics are fed to a classification module, which later predicts the two main characteristics for new riders. And finally in [7], the authors propose a recommender system for carpooling services that leverages on learning-to-rank techniques to automatically derive the personalised ranking model of each user from the history of her choices (i.e., the type of accepted or rejected shared rides). Then, the system builds the list of recommended rides in order to maximise the success rate of the offered matches.

Nevertheless, there appears to be a lack of practical evaluations of existing algorithms for generating user pools in the context of ride-sharing. Consequently, this article aims to address this gap in the current literature.

3 Problem Formulation

In this section, we describe the theoretical background from which the pooling methods presented further on have been developed.

First, it is worth noting that the ride-sharing issue is classified under the classic Dial-a-Ride Problem (DARP) [21], known for its NP-hard complexity [27]. Within DARP, passengers request rides from designated origins o_p to specific destinations d_p. Therefore, we define the directed weighted graph $G = \{V, E\}$ with $V = \{v_1, v_2, ..., v_n / v_i \in \cap G\}$, i.e. the set of road intersections. The edges of $G = \{V, E\}$ are define as $E = \{arc(i,j)/i \in V, j \in V\}$, i.e. the set of roads between these intersections. For each $arc(i,j)$, a non-negative travel cost $\delta(i,j)$ is associated, which corresponds to the distance of the road between intersections i and j, each $arc(i,j)$ also have a travel speed $\sigma_u(i,j)$ depending on the user u. We denote by $p_u(i,j)$ the subset of V containing the sequence of nodes $\{v_1, v_2, ..., v_n\}$ from the arcs included in path of user u to travel from the source i to the destination j. The travel time for user u to complete path $p_u(i,j)$, $\tau_u(i,j)$, is given by the Eq. 1.

$$\tau_u(i,j) = \sum_{v,v' \in p_u(i,j)} \left(\frac{\delta(v,v')}{\sigma_u(v,v')} \right) \tag{1}$$

In this review, we distinguish between two types of user u, drivers u_d and passengers u_p. Let U be the set of users, P the set of pools, P_i the pool containing one user u_d and multiple users u_p, and n_p the number of passengers in the pool P_i. In each of the methods presented, we aim to create pools consisting of a single driver u_d accompanied by at least one passenger u_p, the number of passengers n_{u_p} not exceeding the maximum number of seats available in the vehicle $v_{capacity}$.

4 Similarity Functions

In this section, we describe the different similarity functions that will be used in the rest of this article to create user pools thanks to the different methods presented in Sect. 5. These two functions are deployed to determine how similar two users are according to distance or time criteria. One uses only the origin and destination, while the other uses the path taken by the users. In the following, all computations of distance or time between two points are based on the shortest path found by the Dijsktra [10] algorithm.

4.1 Distance Similarity

A common metric used to quantify the similarity between two points is the Euclidean distance. Thus, the Euclidean distances between the origins and the destinations are given by $\delta(o_i, o_j)$ and $\delta(d_i, d_j)$ respectively. By combining these two distances, we obtain the similarity function between two users i and j given by the Eq. 2.

$$sim(i,j) = \exp\left(-\alpha \cdot \left(\frac{\delta(o_i, o_j)}{\gamma_o} + \frac{\delta(d_i, d_j)}{\gamma_d}\right)\right) \quad (2)$$

Where α is a scaling factor to adjust the importance of the distance and γ_o and γ_d are scaling parameters that control the spread of the similarity function. The larger the value of α, the more emphasis is placed on the proximity of both origins and destinations.

This similarity function can be modified to take into account the average origin and destination of a group of users, a pool, enabling it to be used iteratively in one of the methods presented later. This function is defined in the same way as before, but with the addition of the current pool's average origin o_{pool} and average destination d_{pool}. Where γ_{pool_o} and γ_{pool_d} are scaling parameters that control the spread of the similarity function. This modified similarity function is given by Eq. 3.

$$sim(i,j) = \exp\left(-\alpha \cdot \left(\frac{\delta(o_i, o_j)}{\gamma_o} + \frac{\delta(d_i, d_j)}{\gamma_d} + \frac{\delta(o_{pool}, o_j)}{\gamma_{pool_o}} + \frac{\delta(d_{pool}, d_j)}{\gamma_{pool_d}}\right)\right) \quad (3)$$

4.2 Trip Similarity

Another way of creating user pools is to define the similarity between users on the basis of their paths rather than just their origin and destination. For this we use the similarity function defined by Eq. 4, where the length of a path is given by n and the node v_u^i is the ith node on the path of the user u.

$$sim(p_{u_d}, p_{u_p}) = \frac{\sum_{i=0}^{n} psim(v_{u_d}^i, v_{u_d}^i)}{n} \quad (4)$$

In [16] the authors use the spatio-temporal similarity measure between two nodes in the graph define by the Eq. 5.

$$psim(v_{u_d}, v_{u_d}) = \exp\left(\frac{w_1 \times \ln\left(\frac{1}{1+\delta(v_{u_d}, v_{u_d})}\right) + w_2 \times \ln\left(\frac{1}{1+\Delta\tau(v_{u_d}, v_{u_d})}\right)}{w_1 + w_2}\right) \quad (5)$$

Where $\delta(v_{u_d}, v_{u_d})$ is the spatial Euclidean distance of two points, $\Delta\tau(v_{u_d}, v_{u_d})$ is the absolute difference of the points in time and $w1$ and $w2$ are the weight given to the distance and time factors. Thus, when $w_1 = 1$ and $w_2 = 0$ this similarity function only takes into account the distance separating the pairs of points on the two paths. Conversely, when $w_1 = 0$ and $w_2 = 1$, the function only takes into account the travel time separating the pairs of points on the two paths.

5 Pooling Methods

In this section, we present and describe five methods to create user groups, pools, for ride-sharing. Although these methods are different, they have a number of common characteristics. The Trip Similarity and OD Time Alignment methods both use the time criterion, the Trip Similarity and Trip Buffering methods both use the paths of the users $p_u(o, d)$, the OD Similarity and OD Clustering methods are based on the origin and destination points of the users (o, d), and finally all the methods except OD Time Alignment use the notion of distance δ.

5.1 OD Similarity

This method consists of matching users who have similar origin and destination locations. The objective is to find pairs of users i and j whose origins o_i and o_j are close to each other and whose destinations d_i and d_j are also close to each other. To create pools of users with a maximum size of $v_{capacity}$ and ensuring that each pool contains at least one driver u_d, we use the following steps:

1. Select a driver u_d from the set of available users and assign it to a pool.
2. For each remaining user u_p:
   (a) Calculate the similarity between the origin and destination of u_p and the average origin and average destination of the current pool using the Eq. 3.
   (b) If the similarity exceeds a certain *threshold*, add u_p to the current pool.
   (c) Repeat steps 2 and 3 until the pool reaches its maximum size $v_{capacity}$.
3. Remove all users in the current pool from the set of available users.
4. Repeat steps 2–4 until all users are assigned to a pool or stop the algorithm if no more allocations are possible.

5.2 OD Clustering

This method consists of creating user groups by creating clusters based on the distance similarity of the origin o and destination d points of users using the Eq. 2. First, we create a graph where the nodes are the users who have sent their requests. We define this complete directed weighted graph as $G_u = \{V, E\}$ with $V = \{u_1, u_2, ..., u_n / u_i \in \cap U\}$, i.e. the set of user in U. The edges of $G_u = \{V, E\}$ are define as $E = \{sim(i,j)/i \in V, j \in V\}$, i.e. the similarity values between each users. Then we use the Louvain algorithm [5] to create the clusters, pools, based on the weights of the edges. The final pool construction is subject to two constraints applied to the clusters found.

$$n_{u_p} \leq v_{capacity} \quad (6)$$

$$n_{u_d} = 1 \quad (7)$$

The constraint 6 ensures that the size of each cluster should be at most $v_{capacity}$ and the constraint 7 ensures that each cluster contains at least one driver user.

5.3 OD Time Alignment

This method consists of optimizing the formation of the pools by matching users with similar departure and arrival times. This approach minimizes waiting times for users and ensures efficient utilization of vehicles. Let $\Delta\tau(o_{u_p}, o_{u_d})$ denote the difference between the departure times of a passenger u_p and a drive u_d. Similarly, let $\Delta\tau(d_{u_p}, d_{u_d})$ denote the difference between the arrival times of a passenger u_p and a drive u_d. A time threshold $\tau_{\text{threshold}}$ is defined to determine whether the departure and arrival times of a passenger and a driver are sufficiently aligned for carpooling. Mathematically, the conditions for time alignment is expressed by the Eqs. 8 and 9.

$$\Delta\tau(o_{u_p}, o_{u_d}) \leq \tau_{\text{threshold}_o} \quad (8)$$

$$\Delta\tau(d_{u_p}, d_{u_d}) \leq \tau_{\text{threshold}_d} \quad (9)$$

Then, to create the pools, we assign a number of $v_{capacity}$ passengers u_p to each driver u_d, based on a ranking of the most similar departure and arrival times.

5.4 Trip Similarity

This method consists of comparing different points on the path of two users from a spatial and temporal point of view. Pools are created based on the groups of users with the largest combined trip similarities. To define each individual trip i based on the origin o and destination d we use Dijsktra [10] to find the shortest path, $p_{u_i}(o, d)$. The similarity between two paths is given by the Eq. 4. Then we create pools of users based on this similarity measures. Let P_{ij} be the pool containing users u_i and u_j, and n_{ij} the total number of users in the pool P_{ij}. For each pair of users u_i and u_j with $u_i, u_j \in U$ and $i \neq j$:

1. Compute the similarity between the trips of u_i and u_j using Eq. 4.
2. If $sim(p_{u_d}, p_{u_p})$ is greater than a certain $threshold$, then add u_i and u_j to the same pool if there is exactly one driver user u_d in the pool P_{ij} and if $n_{ij} \leq v_{capacity}$.

Finally, instead of comparing each of the trip points of two users, we compare their origins and destinations as well as N randomly selected trip points. This choice of implementation greatly reduces the computation time while preserving the characteristics of the method.

5.5 Trip Buffering

This method consists of creating pools based on the users that the driver meets during his trip. Each driver has a buffer of a given distance δ_{buff} or time τ_{buff}. If the origin o_p of an user u_p is in the buffer, the driver u_d will make a detour to pick him up and add it to his pool, otherwise he ignores it and continues on his way to his destination. Let v_d be the current position of u_d in G. Mathematically, the fact of taking a passenger can be expressed by the binary variable p_{u_p} given in the Eq. 10.

$$p_{u_d} = \begin{cases} 1, & \text{if } \exists \text{ user } u_p \text{ such that } \delta(v_d, o_p) \leq \delta_{buff} \text{ or } \tau(v_d, o_p) \leq \tau_{buff} \\ 0, & \text{otherwise} \end{cases} \quad (10)$$

To ensure correctness of the method, two constraints must be satisfied. The first Constraint 11 ensures that the length of the detour is less than the maximum detour length $\delta_{detourMAX}$ tolerated by u_d and the second Constraint 13 ensures that there are still seats available in the vehicle. Let n_{u_p} be the current number of user u_p in the vehicle of user u_d. If the variant of this method using detour time rather than distance is selected, then the Constraint 11 is replaced by the Constraint 12.

$$\delta_{detour} \leq \delta_{detourMAX} \cdot p_{u_d} \qquad (11)$$

$$\tau_{detour} \leq \tau_{detourMAX} \cdot p_{u_d} \qquad (12)$$

$$n_{u_p} \leq v_{capacity} \cdot p_{u_d} \qquad (13)$$

For this method, there is no need to specify a constraint on the number of drivers in the pool u_d, as the pools are constructed from these drivers. It is also important to specify that the detour constraint, 11 or 12 depending on the preference, is also defined for the destination point of the passenger according to the destination point of the driver. This avoids taking on passengers who are on the same route but whose destination is not at all similar to the destination of the driver.

As shown, this method can be modified to use travel time instead of distance as presented above, so the distance buffer becomes a time buffer, the detour distance becomes the time taken by that detour and the maximum detour distance becomes the maximum detour time tolerated by the driver.

6 Experimental Setting

In this section, the various results obtained are presented. All experiments were carried out on a Windows 11 OS equipped with an 8-core AMD Rizen 7 5800X processor with a frequency of 3.80 GHz and 32 GB of RAM. For the sake of quick prototyping the algorithms have been written in Python 3.10.11. The road network, the graph, is stored in the form of a dictionary of dictionaries thanks to the NetworkX library [14].

6.1 Road Network

Table 1 shows the properties of the real road network G used in the two experiments, the randomized and the simulation. These data come from information available on Open Street Map thanks to the Python library OSMnx [6], each node is a road intersection and each intersection is linked by roads, the edges. These edges carry information such as the distance of the road, the speed limit, the transit time, the type of road and its name.

Table 1. Brussels road network specifications

Network	Nodes	Edges	Max deg	Avg. deg
Brussels	18547	40890	12	4.41

6.2 Modeling User Requests

To evaluate the different methods presented, we simulated the user requests by varying their occurrence according to the hour of the day. Figure 1 shows the number of requests as a function of time; this model is based on real data collected by [18] in Belgium, shown in Fig. 2. It should be noted that the proposed model only takes into account the number of requests at each time of day, without taking into account the real origin and destination points of users, which are chosen randomly from the road network G.

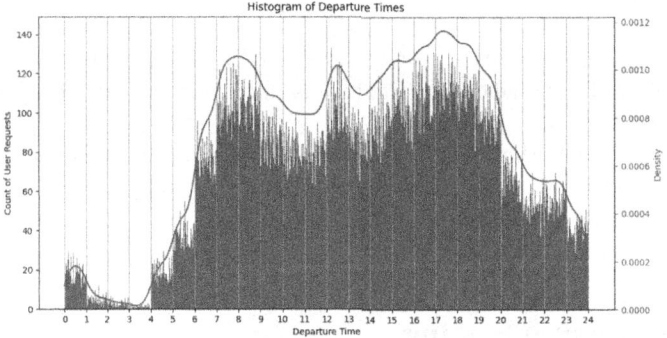

Fig. 1. Simulation data generated by our model.

To create this model we used a combination of uniform random function. The construction of the proposed model can be described as follow: Let N represent the total number of user in the simulation and d_h be the proportion of users at time h with h in $0, 1, ..., 23$. The d_h values have been experimentally defined to best match the real data from [18]. Let $U(h, h+1)$ represent a uniform distribution over the interval $[h, h+1)$, and let n_h represent the number of users in hour h, which is calculated as $n_h = N \times d_h$. The model generates n_h random numbers from $U(h, h+1)$ for each hour h from 0 to 23. The distribution of users for a hour h, D_h, is given by the Eq. 14.

$$D_h = x_1, x_2, ..., x_{n_h} \quad \text{where} \quad x_i \sim U(h, h+1) \tag{14}$$

The final distribution D is the union of all D_h and is given by the Eq. 15. Thus, the distribution D contains all the sub-distributions $[D_0, D_1, ..., D_{23}]$.

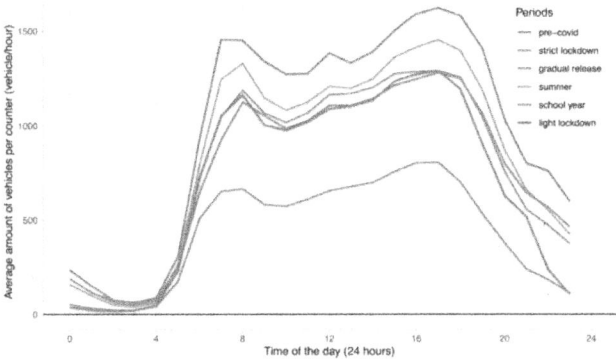

Fig. 2. Real world data from [18].

$$D = D_0 \cup D_1 \cup ... \cup D_{23} \qquad (15)$$

Then, for each user u joining the simulation at time τ_a, a request r_u is established. A query contains information on the point of origin o and destination d in the graph G, the shortest path sp to join the od points, the departure time τ_a in the simulation and the type of user m being the mode, either driver or passenger. We denote a request as the set $r_u = (o, d, sp, \tau_a, m)$

7 Results

In this section, we present the results obtained by the five methods presented in Sect. 5 as well as variants of these methods using one or other of the similarity functions presented in Sect. 4. In total, we compare eight algorithms designed to build pools of users for car-pooling.

The parameters of the tested methods are detailed in Table 2. For the OD Time Alignment method, the $\tau_{\text{threshold}_o}$ and $\tau_{\text{threshold}_d}$ values are equal and are counted in seconds in the *Threshold* column. Similarly, the maximal detour in term of distance and time for the two variants of the Trip Buffering methods are noted in the *Threshold* column. For the OD Similarity method, the value of γ_{pool_o} is equal to the value of γ_o and the value of γ_{pool_d} is equal to the value of γ_d. Note that each of these methods is scalable to any vehicle capacity and is therefore adaptable to different scenarios. In the following, for each of the violin-shaped plots, we carried out 50 experiments, for each experiment we selected a random number, from 10 to 200, of simultaneous requests. The other results were performed on a proportion of simultaneous user requests based on the model presented in Sect. 6.2 with data points every 15 min throughout the day.

Table 2. Methods parameters

Method	Criteria	$v_{capacity}$	Threshold	α	γ_o	γ_d	N	$buff$	w_1	w_2
OD Similarity	δ	4	0.0005	0.0001	5000	5000	–	–	–	–
OD Clustering	δ	4	0.0005	0.0001	5000	5000	–	–	–	–
OD Time Alignment	τ	4	40	–	–	–	–	–	–	–
Trip Similarity	δ, τ	4	0.00055	–	–	–	2	–	0.5	0.5
Trip Similarity Dist	δ, τ	4	0.00055	–	–	–	2	–	0.9	0.1
Trip Similarity Time	δ, τ	4	0.00055	–	–	–	2	–	0.1	0.9
Trip Buffering Dist	δ	4	5000	–	–	–	–	8000	–	–
Trip Buffering Time	τ	4	10	–	–	–	–	30	–	–

7.1 Runtime

In this section we compare the running times of the different methods presented in the Sect. 5.

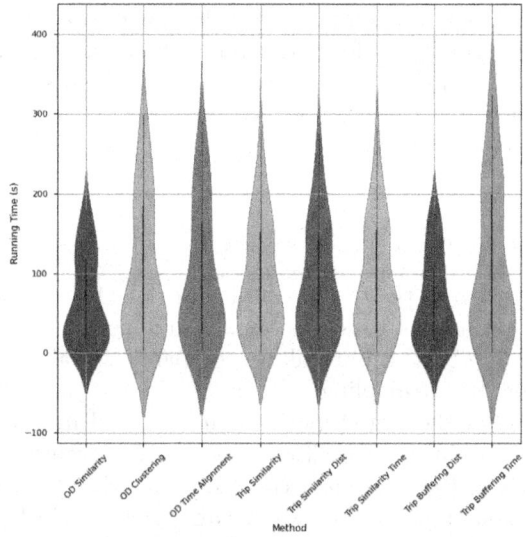

Fig. 3. Average running time of the pooling methods.

On average, the methods OD Similarity and Trip Buffering using the distance similarity function perform best as shown in the Fig. 3. On the other hand, the OD Clustering method and the Trip Buffering method using the time similarity are the two slowest methods on average. The reason why the Trip Buffering method using time similarity is the slowest is due to the extra computation needed to retrieve the driver's travel time. And the reason why the OD Clustering method is slow may be explained by the extra time needed to build the user network and to cluster this network using the Louvain method.

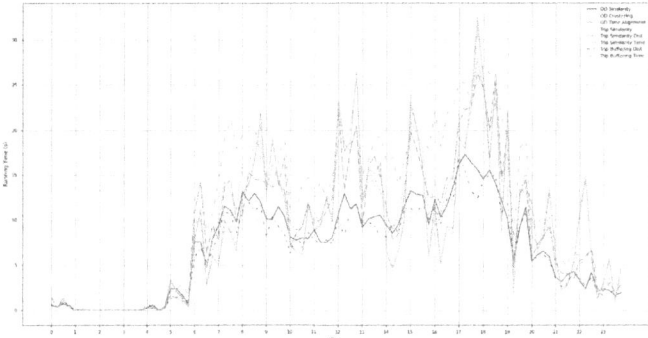

Fig. 4. Running time of the pooling methods according to the number of user requests.

As can be seen in Fig. 4, the execution time for each of the different methods tested depends directly on the number of simultaneous requests received. We also note that the three variants of the Trip Similarity method are much more sensitive to the number of queries than the other methods. Also, the two methods that seem most robust to variations in the number of queries are OD Similarity and Trip Buffering using the distance similarity function. During our simulation, the best method, the Trip Buffering using the detour distance similarity, certifies an average running time of between 3 and 17 s at peak times. It is also worth to note that the execution time of the methods also depends on the size of the graph being processed since all the methods require shortest path calculations to obtain similarity values.

7.2 Requests Satisfaction

In this section, we compare the average size of pools created as well as the average satisfaction of the user requests of the different methods presented in Sect. 5.

Figure 5 shows the average number of users, drivers and pedestrians, in each of the pools created. Based on our user request simulation model, we can see that the average vehicle occupancy across all methods exceeds 3 users, indicating that the cars are nearly at maximum capacity. It can be seen that the OD Similarity method is the ones for which the number of users is the most constant. The smaller average pool size in the OD Time Alignment method can be attributed to the random selection of requests. This randomness can result in significant time differences and consequently lower similarities.

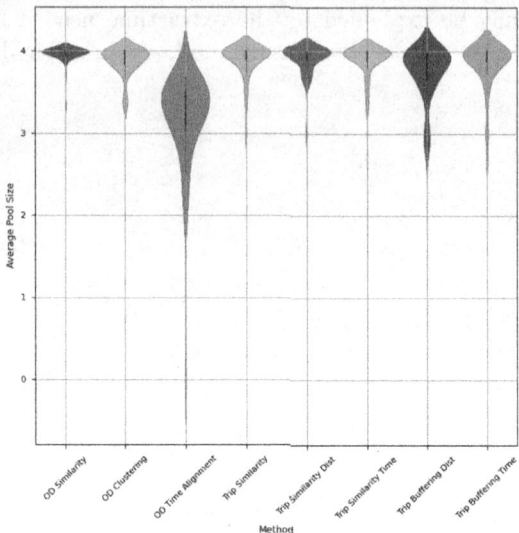

Fig. 5. Average pool size of the pooling methods.

The satisfaction percentage is defined as the ratio between the number of passengers in a pool and the total number of passengers multiplied by 100, overall average results are given by the Fig. 6 and the average results for each hour of the day are given by the Fig. 7. Thanks to these two figures, we can see that, whether in randomized experiments or in the simulation, the OD Similarity method performs most successfully, followed by the two variations of Trip Buffering and Trip Similarity. The very high user demand satisfaction results of the OD Similarity method can be attributed to the fact that the average pool size is highly constant for this method and almost always equals the maximum vehicle capacity $v_{capacity} = 4$. One of the reasons for the poor performance of the OD Clustering method is that it depends on the Louvain algorithm to form the pools, which is not deterministic. Also, on the basis of our user request simulation model, we can see that on average all the methods manage to satisfy at least 30% of requests during rush hours.

Furthermore, it is important to note that the percentage of satisfied users also depends directly on the number of randomly selected drivers and the capacity of their vehicles. Indeed, if there are not enough drivers and/or seats available at time t, then unavoidably some passengers cannot be satisfied. So when there are many requests, the probability of having more drivers increases, as does the percentage of requests satisfied, as shown in Fig. 7. Similarly, when there are fewer requests, only a few drivers are needed to satisfy the majority of requests. As with the execution time criterion, we can see that all methods follow a similar behavior, but this time it's not correlated with the number of requests. In our simulation, the best method, OD Similarity, certifies an average user satisfaction between 35% and 100% during rush hours.

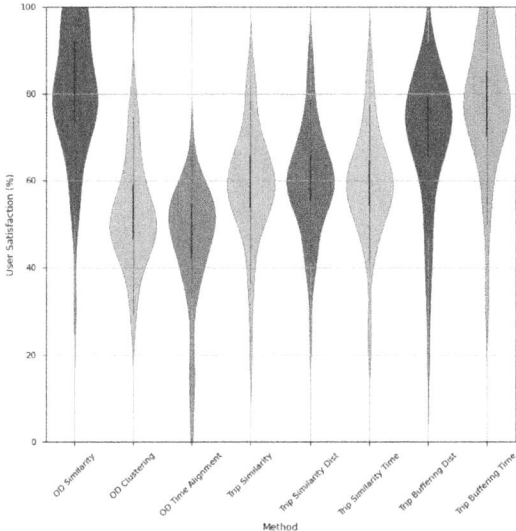

Fig. 6. Average satisfaction percentage of the pooling methods.

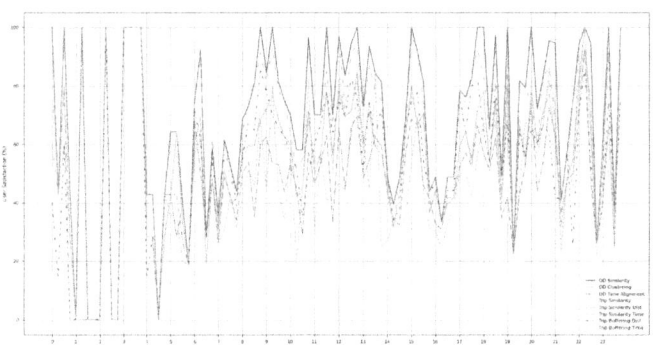

Fig. 7. Average satisfaction percentage of the pooling methods according to the number of user requests.

7.3 Driver Detour Distance

In this section, we compare the different methods based on the criterion of the average detour made by the driver to pick up passengers. The detour is calculated by comparing the length of the direct path with the length of the path where the driver picks up the passenger. The result is the average of these detour distances for all pools of a method, in kilometers.

In Fig. 8, we see that the method with the smallest average driver detour is the variant of the Trip Buffering method using the detour distance as criteria. This result is expected since it is the only method with a constraint on the maximum detour accepted by the driver. Also, We observe that the version of

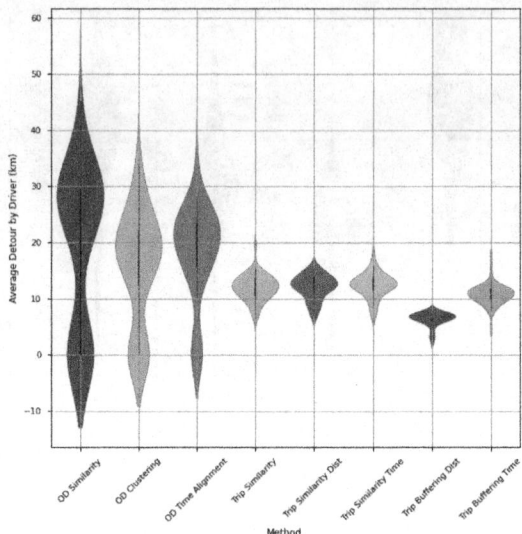

Fig. 8. Average detour distance for the driver by pooling methods.

Trip Buffering based on the detour distance perform better in term of detour than the one based on the detour time. We also note that the three variants of Trip Similarity method perform equally well. This can be explained by the fact that these methods take into account multiple nodes on the path, unlike the OD Similarity and OD Clustering methods, which have the greatest deviation and only take into account the origin and destination.

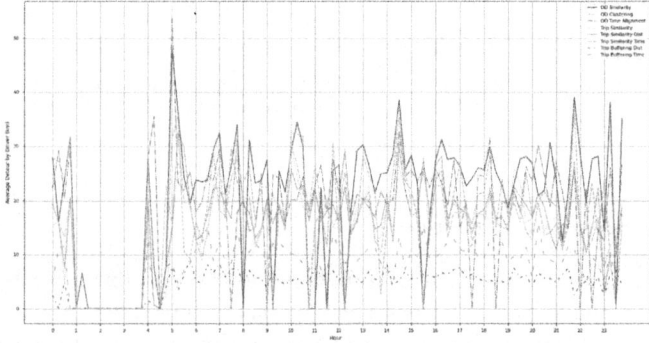

Fig. 9. Average detour distance for the driver by pooling methods according to the number of user requests.

In Fig. 9, we note that for certain instances, the detour distance of several methods is 0. This is due to the fact that the methods use the shortest path calculation to determine similarity, which in some cases does not exist, only the Trip Buffering method is not sensitive to this condition. For instance, for the hours in [2, 4], all the detour distances are 0 because no user request can be satisfied. The best method, the Trip Buffering using the detour distance similarity, certifies an average driver detour between 4 and 10 km during rush hours.

7.4 Passenger Waiting Time

In this section, we compare the different methods on the basis of the average waiting time of passengers before the driver comes to pick them up.

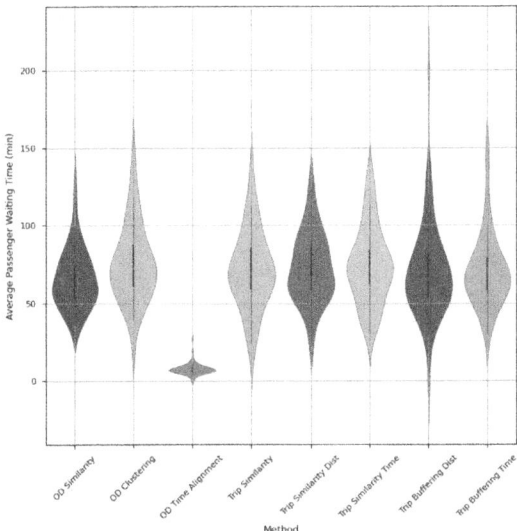

Fig. 10. Average waiting time for the passenger by pooling methods.

Figure 10 shows that all the methods performed similarly in our randomized experiment except for the OD Time Alignment method. Indeed, this method achieves significantly better results because the user pools are composed of passengers and a driver with similar departure times, consequently reducing the average waiting time.

Figure 11 illustrates that passenger waiting time depends on the number of user requests only when this number is below a certain threshold. Indeed, when the number of requests is low, not enough drivers are among the users, and passengers are therefore forced to wait longer. This suggests that increasing the number of participants will lead to better performance in terms of waiting time. This observation has also been made by [20] concerning the matching rate, which also seems to be our case for the percentage of satisfied requests. Once this

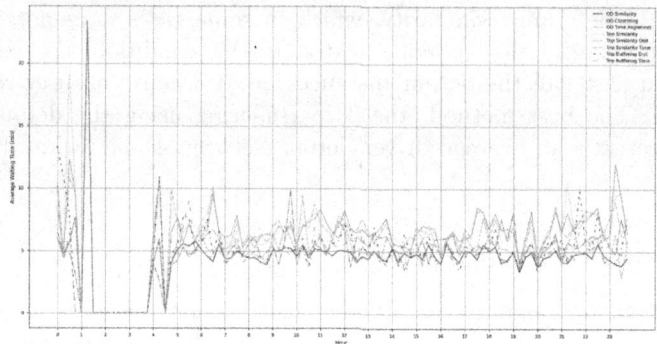

Fig. 11. Average waiting time for the passenger by pooling methods according to the number of user requests.

threshold is reached, the waiting time for each method is stable. Although the OD Time Alignment method is the best on average over 50 random experiments, we can see from Fig. 11 that during our simulation the most constant method is the OD Similarity, certifying an average passenger waiting time between 4 and 6 min during rush hours.

7.5 Overall Comparison

To summarize the results discussed in this section, based on the different parameters presented earlier, we compared the eight methods by establishing a score that is weighted sum of the criteria for the randomized experiment and for the simulation. The weights assigned to the respective criteria are given in Table 3.

Table 3. Criteria weight

Criterion	Weight
Running Time (s)	−1.5
User Satisfaction (%)	1
Average Pool Size	1
Number of Pools	0
Average Detour by Driver (km)	−1
Average Passenger Waiting Time (min)	−1

We are looking for the method that minimizes the execution time, the driver's detour and the passenger's waiting time. This method must also maximize the percentage of satisfied users and the number of users in each car. Thus, the method with the highest score should be the one that best corresponds to the

Table 4. Method scores

Method	Randomized experiment score	Simulation score
Trip Buffering Dist	−99.706951	36.447783
OD Similarity	−100.205970	35.640391
OD Time Alignment	−122.408349	16.648278
Trip Similarity Dist	−158.499280	22.717071
Trip Similarity	−159.199678	22.693013
Trip Similarity Time	−161.330231	22.379183
Trip Buffering Time	−169.952633	33.921981
OD Clustering	−186.394580	18.173433

objectives and preferences pursued in this research. The scores obtained are given in Table 4.

According to the weights chosen, the two methods that best satisfy the criteria are, for both experiments, the Trip Buffering method using distance similarity and the OD Similarity method.

8 Future Works

For future improvements, we would like to modify the current proposed model. First, it would be interesting to take a more extensive period of time to build a more realistic model. Taking, for example, a full year, this would enable us to study in detail the adaptability of the different methods to changes in demand within the road network. Then, the model accuracy could be improved by taking intervals in minutes rather than hours, and by defining uniformly random functions for 5 min intervals for example. In addition, it would be interesting to consider other types of distribution than Uniform, such as a model using a Gaussian mixture, as was done in [3]. Finally, concerning the model, we made the hypothesis that the number of requests in the city of Brussels followed the same distribution as for the entire country of Belgium, therefore it would be more advisable to define a model for a specific city rather than for an entire country, using to do so the population of the city comparatively to the population of whole country, for example.

Still on the subject of the data used for the experiments, in the future we plan to test the different methods and even combinations of methods on a real sample of students from one university. Indeed, in the particular context of universities, solving the ride-sharing challenge becomes important not only for the reasons listed above but also because according to [17] around 78% of students travel alone by car. In addition, according to [13], the propensity to practice peer-to-peer car-pooling is higher among younger people. Thus, we will be able to compare the theoretical results presented here with the results obtained under real world conditions.

Concerning the methods themselves, as a first step, it would be interesting to analyze the evolution of execution times for each method in relation to different road network sizes, as well as to explore scalability issues and execution times across larger datasets or varied urban environments other than the Brussels road network, as some may scale very well and others not. In a second step, we would like to test combinations of methods to try and get the best out of each of them. For example, we could imagine using one method during rush hour and another during calm periods. It would be valuable to analyse if this type of hybrid strategy could have a positive impact on the percentage of satisfied requests in the system.

Finally, regarding the preferences and objectives studied here, in the future we would like to take into account the criterion of destination arrival time guarantees for passengers and drivers. Indeed, even if the waiting time is important, such a system can only be viable in practice if it offers guarantees on the arrival time. This new parameter should therefore be taken into account in the scoring of the different methods.

9 Conclusion

We have proposed a comprehensive comparison of five methods for forming user pools for ride-sharing. In addition, we presented two different similarity functions, one based on time and the other on distance. We have proposed variants of the five initial methods using these two similarity functions, and finally compared the eight methods obtained. These methods were evaluated according to five main criteria: running time, percentage of request satisfaction, pool size, passenger waiting time and driver detour distance. In our experiments, whether randomized or during simulation, the Trip Buffering method using the distance similarity function always outperformed the other methods. Our results indicate that the Trip Buffering method using the distance similarity function is the most effective for optimizing the key metrics in ride-sharing. The presented results were based on a user request simulation model that we have proposed for Belgium, further supporting the robustness and applicability of our findings.

Acknowledgments. This project was supported by the FARI - AI for the Common Good Institute (ULB-VUB), financed by the European Union, with the support of the Brussels Capital Region (Innoviris and Paradigm).

References

1. Aydin, O.F., Gokasar, I., Kalan, O.: Matching algorithm for improving ride-sharing by incorporating route splits and social factors. PLoS ONE **15**(3), 1–23 (2020). https://doi.org/10.1371/journal.pone.0229674
2. Baudru, J., Bersini, H.: Heuristic optimal meeting point algorithm for car-sharing in large multimodal road networks. In: Proceedings of the 10th International Conference on Vehicle Technology and Intelligent Transport Systems. VEHITS (2024). https://doi.org/10.5220/0000186800003702

3. Bei, X., Zhang, S.: Algorithms for trip-vehicle assignment in ride-sharing. In: Proceedings of the AAAI Conference on Artificial Intelligence, vol. 32, no. 1 (2018). https://doi.org/10.1609/aaai.v32i1.11298
4. Biying, Y., Ye, M., Meimei, X., Baojun, T., Bin, W., Jinyue, Y., Yi-Ming, W.: Environmental benefits from ridesharing: a case of Beijing. Appl. Energy **191**, 141–152 (2017). https://doi.org/10.1016/j.apenergy.2017.01.052, https://www.sciencedirect.com/science/article/pii/S0306261917300600
5. Blondel, V.D., Guillaume, J.L., Lambiotte, R., Lefebvre, E.: Fast unfolding of communities in large networks. J. Stat. Mech.: Theory Exper. (10), P10008 (2008)
6. Boeing, G.: OSMnx: new methods for acquiring, constructing, analyzing, and visualizing complex street networks. Comput. Environ. Urban Syst. **65**, 126–139 (2017). https://doi.org/10.1016/j.compenvurbsys.2017.05.004
7. Mattia, G.C., Delmastro, F., Bruno, R.: A machine-learned ranking algorithm for dynamic and personalised car pooling services. In: 2016 IEEE 19th International Conference on Intelligent Transportation Systems (ITSC), pp. 1856–1862 (2016). https://doi.org/10.1109/ITSC.2016.7795857
8. Chi-Chung, T., Chun-Ying, C.: Heuristic algorithms for the dynamic taxipooling problem based on intelligent transportation system technologies. In: Fourth International Conference on Fuzzy Systems and Knowledge Discovery (FSKD 2007), vol. 3, pp. 590–595 (2007). https://api.semanticscholar.org/CorpusID:589680
9. De Carvalho, V.R., Golpayegani, F.: Satisfying user preferences in optimised ridesharing services:. Appl. Intell. **52**, 11257–11272 (2022). https://api.semanticscholar.org/CorpusID:246221061
10. Dijkstra, E.: A note on two problems in connexion with graphs. Numerische Mathematik **1**, 269–271 (1959). http://eudml.org/doc/131436
11. Fagundez, G., Massobrio, R., Nesmachnown, S.: Online taxi sharing optimization using evolutionary algorithms. 2014 XL Latin American Computing Conference (CLEI), pp. 1–12 (2014). https://api.semanticscholar.org/CorpusID:35643693
12. Furuhata, M., Dessouky, M.M., Ordóñez, F., Brunet, M.E., Wang, X., Koenig, S.: Ridesharing : the state-of-the-art and future directions (2013). https://api.semanticscholar.org/CorpusID:17974805
13. Gärling, T., Gärling, A., Johansson, A.: Household choices of car-use reduction measures. Transp. Res. Part A: Policy Pract. **34**(5), 309–320 (2000). https://doi.org/10.1016/S0965-8564(99)00039-7, https://www.sciencedirect.com/science/article/pii/S0965856499000397
14. Hagberg, A., Swart, P., Schult, D.: Exploring network structure, dynamics, and function using networkx. Technical report, Los Alamos National Lab.(LANL), Los Alamos, NM (United States) (2008)
15. Ivanova, et al.: Sustainable urban mobility in the EU: no substantial improvement is possible without member states' commitment. European Court of Auditor (2020)
16. Ketabi, R., Alipour, B., Helmy, A.: Playing with matches: Vehicular mobility through analysis of trip similarity and matching (2018)
17. Luè, A., Colorni, A.: A software tool for commute carpooling: a case study on university students in Milan. Int. J. Serv. Sci. **2** (2009). https://doi.org/10.1504/IJSSCI.2009.026540
18. Macharis, C., Tori, S., Séjournet, A., Keseru, I., Vanhaverbeke, L.: Can the COVID-19 crisis be a catalyst for transition to sustainable urban mobility? Assessment of the medium- and longer-term impact of the COVID-19 crisis on mobility in brussels. Front. Sustain. **2** (2021). https://doi.org/10.3389/frsus.2021.725689
19. Mandl, N., et al.: Annual European union greenhouse gas inventory 1990-2017 and inventory report 2019. European Environment Agency (2019)

20. Neda, M., Jayakrishnan, R.: A real-time algorithm to solve the peer-to-peer ridematching problem in a flexible ridesharing system. Transp. Res. Part B: Methodological **106**, 218–236 (2017). https://doi.org/10.1016/j.trb.2017.10.006
21. Psaraftis, H.N.: A dynamic programming solution to the single vehicle many-to-many immediate request dial-a-ride problem. Transp. Sci. **14**(2), 130–154 (1980). http://www.jstor.org/stable/25767975
22. Rodenbach, J., Matthijs, J., Seeuws, B., Ryvers, S., Decombel, S.: Impact report, car-sharing in Belgium in 2022 (2022). https://www.autodelen.net/wp-content/uploads/2023/03/Impact-report-Car-sharing-in-Belgium-in-2022.pdf
23. Schrank, D., Eisele, B., Lomax, T.: 2012 urban mobility report. Texas Transportation Institute; Southwest Region University Transportation Center (U.S.) (2012). https://rosap.ntl.bts.gov/view/dot/61387
24. Shirley, C., et al.: Emissions of carbon dioxide in the transportation sector. Congressional Budget Office, Nonpartisan Analysis for U.S. Congress (2022)
25. Silwal, S., Gani, M.O., Raychoudhury, V.: A survey of taxi ride sharing system architectures, pp. 144–149 (2019). https://doi.org/10.1109/SMARTCOMP.2019.00044
26. Thaithatkul, P., Seo, T., Kusakabe, T., ASAKURA, Y.: A passengers matching problem in ridesharing systems by considering user preference. J. Eastern Asia Soc. Transp. Stud. **11**, 1416–1432 (2015). https://doi.org/10.11175/easts.11.1416
27. Toth, P., Vigo, D., Toth, P., Vigo, D.: Vehicle Routing: Problems, Methods, and Applications, second edition (2014)
28. Wojciechowski, J., et al.: Air pollution: Our health still insufficiently protected. European Court of Auditor (2018)
29. Yatnalkar, P., Narman, H.: A matching model for vehicle sharing based on user characteristics and tolerated-time (2019). https://doi.org/10.1109/HONET.2019.8908058

Acceptance of Automated Vehicles in Logistics and Beyond: A Survey-Based Investigation

Guglielmo Papagni[1], Setareh Zafari[1(✉)], Johann Schrammel[1], and Manfred Tscheligi[1,2]

[1] Austrian Institute of Technology, Center for Technology Experience, Giefinggasse 2, 1210 Vienna, Austria
setareh.zafari@ait.ac.at
[2] University of Salzburg, Center for Human-Computer Interaction, Jakob-Haringer-Straße 8/Techno 5, 5020 Salzburg, Austria
https://www.ait.ac.at/

Abstract. Automated vehicles (AVs) are becoming an integral part of several industries, including logistics, which is already undergoing substantial transformations. As logistics operations extend beyond logistics hubs and into city streets, AVs are expected to soon interact not only with operators, but also with lay people on a daily basis. Introducing AVs into society poses many technical and social challenges. This paper addresses user acceptance of AVs in the context of logistics operations, particularly focusing on the shared use of open roads. To conduct the investigation, a user survey was designed, and 178 observations were collected. Our results show that participants' familiarity about AVs had a significant positive impact on all acceptance factors except usefulness and security. Furthermore, their age also had a positive effect on perceived security. Our qualitative results help to refine these findings, showing nuances that connect acceptance to transparent communication, environmental issues, and use in mixed traffic.

Keywords: Automated Vehicles · Acceptance · Logistics · Public Perception

1 Introduction

As reflected by the increasing number of related publications and the growing interest, both academic and among the public, automated vehicles (AVs) are expected to revolutionize most sectors of the transportation industry, as well as private transportation [1,14]. The logistics sector is one of the contexts of application that is spearheading the transition towards adopting AVs and that is playing a central role in realizing the "Industry 4.0" vision, intended as the profound transformation processes that, by means of automation, digitalization and so forth are taking over the industrial world [3,10,38].

At least two main reasons can be identified to explain why logistics and AVs are so deeply intertwined. First, the logistics sector is one of the pillars not only of international trade, but also of modern society and its economy [30]. Automation promises to increase efficiency by optimizing processes, which in turn would be of great benefit to logistics operations. Therefore, logistics stakeholders have all the interests to become key players in the development and implementation of the relevant technologies. Second, while AVs for private use are more often discussed by the media because of their resonance with the public [19,29], logistics hubs offer far better conditions for testing and initial adoption. This is because logistics operations are already highly standardized, repetitive, and, to a certain extent, predictable, and these features suit well AVs' current capabilities [12]. Furthermore, the relatively limited human presence in logistics hubs (compared to, for instance, city streets), the fact that most people in the hubs are trained employees, and the easier implementation of intelligent infrastructure reduce the chances of unexpected events and accidents making such environments better and safer testing grounds than open roads. At the same time, however, logistics operations are hardly confined to private grounds and hubs. Once AVs are put into action, many of them (e.g., when used for last-mile delivery) will access public roads and interact with other road users. Therefore, studying how AVs are integrated into the network of logistics operations is not only useful to understand how the industry will move forward, but also to peek into people's reaction to working alongside and coexisting with AVs.

For this radical transformation to happen successfully, several different challenges must be faced. While being often deeply intertwined, these challenges differ in nature and include technical (concerning e.g., safety, reliability, and efficiency) and societal ones (related to e.g., public acceptance, environmental impact, and narratives). To address some of the most pressing of these latter challenges and to answer the questions of how people perceive AVs in logistics operations and the implications of sharing roads with them, in this paper we report on the results of a study conducted to investigate people's perception of AVs' impact on the logistics industry and acceptance in everyday life.

The study consisted of a short survey targeting the public and potential AVs' users. A total of 178 complete observations were collected and analyzed. We found that participants' familiarity with AVs yielded positive results in terms of reliability, safety, trust, and overall acceptance. Positive correlations between participants' age and perception of security also emerged. Additionally, the qualitative analysis of the answers to the only open question shows cautious optimism, at least regarding the use of AVs in logistics contexts. Use in mixed traffic, the evolution of the job market, and environmental concerns among others emerge as key challenges that require further efforts and discussion. In the remainder of the paper, Sect. 2 discusses existing work related to acceptance of AVs. Section 3 describes the experimental methodology, while results are presented in Sect. 4 and discussed in Sect. 5. Final considerations, limitations, and directions for future work are then presented in Sect. 6.

2 Related Work

Acceptance of AVs is explained primarily through the technology acceptance model (TAM) [9] and the advanced unified theory of acceptance and use of technology (UTAUT) [36]. According to these theoretical frameworks, studies found perceived ease of use, trust, and social influence are effective predictors of behavioral intentions with perceived usefulness having the greatest effect.

Several studies have adapted and extended these models with regards to AVs. For instance, Osswald and colleagues (2012) took additional aspects of safety and anxiety into consideration and proposed a theoretical car technology acceptance model (CTAM) [26]. Hutchin and colleagues (2017) proposed a model for safety critical technology acceptance (SCTAM) including elements, derived from a survey, such as benefits, automation schema, and confidence [16]. Kapser and Abdelrahman (2020) extended on the UTAUT2 framework to include risk perception and adapted the model for use in the context of last-mile delivery. They found pricing to be the most significant indicator. Performance expectancy, facilitating conditions, perceived risk, social influence, and hedonic motivation were also significant indicators [18]. Seuwou and colleagues (2020) proposed a theoretical AV technology acceptance model and found performance and safety of AVs as well as consumer's trust in the manufacturers of AVs to be the main drivers for AVs adoption [34].

Research also tackles the public perspective towards AVs. Du and colleagues (2022) studied public misconceptions about AVs and found that those holding more misconceptions about AVs were more receptive to AVs, while fewer misconceptions about AVs were associated with more skepticism [11]. Lee and Hess (2022) showed the association of different dimensions of public concerns namely, safety, privacy, and data security with specific demographic variables [20]. Hilgarter and Granig (2020) studied public response after riding an autonomous shuttle and found experience with AVs and speed as main factors influencing the perception of safety [15].

Regarding logistics, some studies investigate acceptance of automation technologies in general. For instance, Capkin and Azarko (2023) report positive findings (particularly among younger participants) regarding the potential to increase reliability and efficiency through automation, but also concerns related to safety, security, and job losses [5]. Concerning AVs in logistics operations, Neubauer and Schauer (2018) identified acceptance factors specific to logistics contexts like "job relevancy" (focusing on employees' training and the availability of new jobs) and "social dimension" (i.e., how different stakeholder groups differ in terms of technology acceptance) [25]. Building on these insights, Rosic and colleagues (2021) developed an automated road transport logistics acceptance model (ARTLAM) by extending standard dimensions of existing models for technology acceptance with broader concepts of trustworthiness and facilitating conditions to provide deeper understanding of logistics requirements [32].

Among experimental studies, expected increases in overall process efficiency are reported as a key indicators of acceptance of AVs also in the context of logistics [6]. Dallasega and colleagues (2022) conducted a large-scale survey among supply chain stakeholders and identified organizational capabilities (e.g., retraining of employees), interconnection and material flow (e.g., customer and provider supply chain connections), and processes automation (e.g., automated warehousing) [8] as main trajectories for improvement. Rose and colleagues (2022) conducted a stakeholder study and found that participants were quite positive towards AVs in the context of port logistics, particularly to deal with the shortage of drivers, and identify legislation improvements, smart and connected infrastructure, and restructuring of organization processes as key requirements for acceptance [31]. The issue of drivers' scarcity also emerges in other studies, in which AVs were identified as the solution [24].

At the same time, other researchers describe more negative reactions to introducing AVs in logistics. For instance, Fröhlich and colleagues (2018) report negative attitudes among truck drivers towards carrying out non-driving related tasks in the cabin of autonomous trucks. As possible explanations, they suggest uncertainty regarding the trucks' actual performance, as well as a persisting fear that AVs might eventually jeopardize drivers' jobs. However, drivers also saw the potential in terms of improvements of the workflow [13]. In a study investigating 5G-enables autonomous logistics vehicles, Li and colleagues (2023) found that, while positive towards the technology, drivers were reticent to relinquish all types of control (i.e., remote or monitoring) over the vehicles [21].

3 Methodology

Stemming from the acknowledgement that people's perception of new technologies plays a key role in determining future adoption, this study aimed to capture public perception of AVs, in terms of factors determining acceptance and primarily (but not exclusively) in the context of logistics operations.

3.1 Survey Instrument Design

The survey was designed with the "LimeSurvey"[1] software and presented in English and French. The survey link was distributed through multiple channels (such as newsletters, posts on LinkedIn, and other social media) by several stakeholders in multiple rounds, to recruit as wide a variety of participants as possible and avoid sampling biases. As they clicked the link, participants were briefed about the purpose of the study. If they agreed to participate, they were informed that the collected data was fully anonymized and had to give their explicit and informed consent. Then, the experimental procedure, which in total took roughly 10 min, started. Participants had to first watch a short video of one of the vehicles employed in the AWARD EU Project[2] driving autonomously on an open road. The video also had background narration and subtitles briefly explaining the challenges and potential benefits of autonomous technologies (see Fig. 1).

---

[1] https://www.limesurvey.org/.
[2] https://award-h2020.eu/.

Fig. 1. One of the automated vehicles employed in the research project is shown to participants while in operation on an open road.

After the video, participants were asked a total of 13 5-point Likert scale questions divided in two groups. While each question targeted one of the facets of technology acceptance as envisioned in the ARTLAM model, the first group of questions primarily addressed AVs in the context of logistics operations, while the second focused on a broader assessment of participants' perception. Specifically, the first part (seven questions) aimed to collect participants' expectations in terms of usefulness (for operators and logistics in general), ease of use, impact on operations efficiency, reliability, and safety (for operators and other road users). The second group (six questions) addressed AVs' vulnerability (to cyber-attacks), predictability, trustworthiness, impact on the job market, and overall expected acceptance on both the individual and societal levels. Then, participants were presented with one last non-mandatory open question asking for any further considerations, concerns, or expectations regarding the future of AVs. A set of standard demographic questions (gender, age, nationality), a 5-point Likert scale question to self-assess their familiarity with AVs (where "1" was "not familiar at all" and "5" was "very familiar"), and one last question asking whether they were involved in the AWARD Project closed the experimental procedure.

3.2 Participants

Survey data were collected in multiple rounds using LimeSurvey between late January and mid-May 2024. Overall, 254 responses were collected. Out of these responses, 178 observations were complete and were used for the following analysis of results. Participants were mostly of French (53%) and Austrian (20%) nationality. The remaining respondents selected a variety of countries (primarily European, i.e., Belgium n = 6, Netherlands n = 5, Finland n = 4, Germany n =

4, and a few others). About two-thirds of the participants were male (n=111), one-third female (n=62), and only a small number selected the non-binary option or preferred to not disclose their gender. As illustrated in Fig. 2, participants' age spread across all age groups, although the younger age groups were noticeably larger. Furthermore, nearly half of the participants (49.4%) stated that they were either familiar (26.4%) or very familiar (23%) with AVs (i.e., scores of "4" and "5" according to the 5-point Likert scale described above). Finally, as the study targeted both the project stakeholders and potential users, approximately one-fifth of the participants had a connection to the AWARD Project.

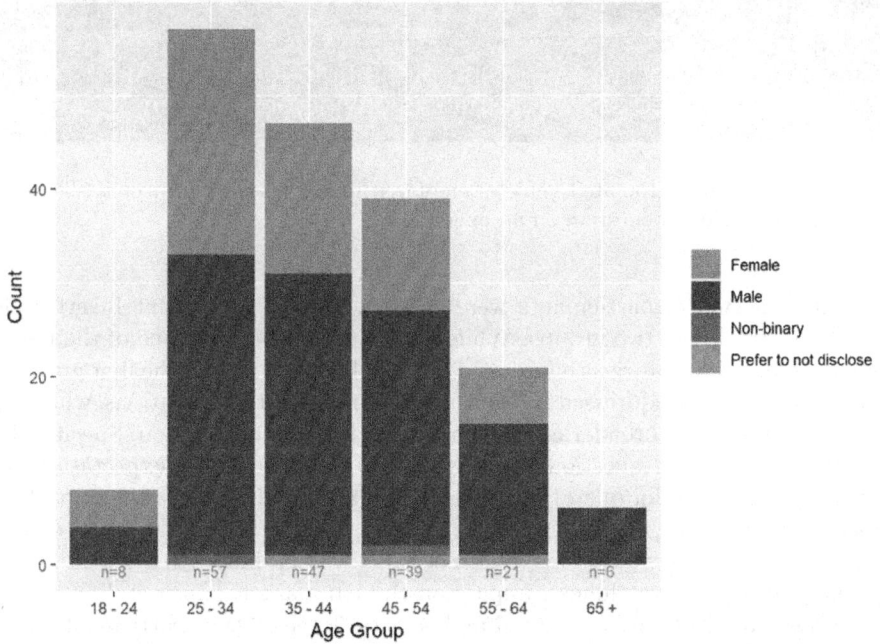

Fig. 2. Age distribution of survey participants (n=178).

4 Results

4.1 Quantitative Analysis

We calculated summary scores for selected related items, derived from the ART-LAM Model. The resulting dimensions are ease of use, perceived usefulness, reliability, safety, security, effects on employment, trust, and acceptance. The summary scores are calculated in a way that higher numbers must always be interpreted as more positive results towards AVs.

In order to further analyze the results, we calculated a multivariate multiple linear regression with all dimensions mentioned above as dependent variables and age, gender and familiarity with AVs as independent variables. Due to the very low number of responses in the gender category non-binary the regarding cases were removed from the analysis. Similarly, cases where respondents did not disclose their gender information were not included in the analysis. The overall MANOVA using Pillai test statistic and Type II sums of squares shows a strong significant impact of familiarity with AVs ($p=0.007$), but no significant overall impact of age ($p=0.223$) and gender ($p=0.539$). Further analysis of the influence of the independent variables on the individual dimensions using multiple regression models (see Table 1 below) shows that familiarity with AVs is significant for the reliability, safety, trust, and acceptance dimensions. These indicate that the more familiar the respondents are with AVs, the higher is their perception about the reliability, safety, trust, and acceptance of AVs. Besides familiarity with AV, the only other possibly relevant influence is of age on security, however for this dimension the overall model is only marginally significant ($p=0.093$). This might imply that older respondents are more concerned regarding the security of AVs. Overall, we want to also note that the expectations regarding the effects of AVs leaned mostly towards the positive side on most dimensions.

Table 1. : Summary of the linear regression models for the different dimensions.

Dimension	P-Value for factor			Model Metrics	
	Gender	Age	Familiarity	P-value	Adj.R^2
Ease of Use	0.953	0.623	0.008	0.061	0.025
Perceived usefulness	0.827	0.659	0.062	0.282	0.004
Reliability	0.328	0.126	0.004	0.008	0.051
Safety	0.115	0.161	0.001	0.001	0.110
Security	0.970	0.032	0.161	0.093	0.020
Effects on Employment	0.419	0.150	0.043	0.062	0.025
Trust	0.401	0.060	0.003	0.004	0.059
Acceptance	0.681	0.062	0.036	0.041	0.031

4.2 Qualitative Analysis

Some relevant insights emerged also from the analysis of participants' answers to the optional open question. A total of 79 answers were collected and coded. The analysis was conducted using the software "ATLAS.ti 23"[3], with an inductive thematic approach. 18 codes emerged, largely reflecting the structure of the survey.

---

[3] https://atlasti.com/.

Participants expressed a wide range of feelings regarding AVs in logistics and interactions on open roads. The most relevant envisioned benefits are the expected increases in process efficiency, particularly for the logistics industry. To a certain extent, safety is also expected to be higher due to AVs' predictability, especially if increasing the number of AVs implies reducing the number of human drivers. Another expected benefit is the improvement of working conditions resulting from reduced working hours (particularly at night) and the possibility of operating and controlling the vehicles remotely. AVs were also seen as an answer to the current shortage of drivers. However, some participants noted how AVs' full potential may not be expressed under constant human supervision, as this would result in the same number of hours as driving manually.

Participants also brought up several open challenges and questions. One key topic concerns the interaction between AVs and manually driven vehicles in mixed traffic. Indeed, while AVs were mostly seen as predictable (and safe), human unpredictability was perceived as an issue. Improving AVs' ability to interpret human behavior or, alternatively, creating separate road lanes were suggested as possible solutions. The considerable investments required to ensure AVs' usefulness and reliability (in mixed traffic and harsh weather), as well as to implement "smart" infrastructures were seen as an obstacle, but interest from investors was expected to grow with time. Better regulations, clear paths for liability allocation, stronger international cooperation, and standardization were seen as crucial enablers needing further efforts.

Regarding acceptance, few participants were skeptical regarding society's actual need for AVs, but the majority found them potentially useful. Several participants stressed the importance of clear and transparent communication to the public to avoid misconceptions concerning developments, potential benefits and risks, and privacy. Likewise, how AVs will affect the job market should also be clearly communicated, as losing jobs to automation is a common fear. In general, acceptance was seen as dependent on time, as several participants believed that there will be a transition phase, during which AVs shall be further tested and proven reliable, safe, and efficient. Some also noted that acceptance will vary depending on people's age, with older generations struggling more to adapt. Finally, AVs' environmental impact, which according to some participants is not clear yet (i.e., whether their introduction will have a significant carbon footprint and abuse of precious resources), is expected to steer public opinion.

Since the most interesting results from the quantitative analysis emerged in relation to participants' familiarity with AVs, we further investigated the qualitative answers with "familiarity level" as a discriminant. The analysis of the answers (n=36) from participants who stated to be familiar and very familiar with AVs (i.e., "4" and "5" on the scale) reveals similar results to the analysis of the full set of answers, with no specific themes emerging.

5 Discussion

The most important finding of this study shows that the more familiar the respondents are with AVs, the higher is their perception about reliability, safety, effects on employment, trust, and acceptance of AVs. This finding extends on the existing work on the role of familiarity with technology and acceptance. For instance, Wicki (2021) found that familiarity by trial with AVs did not increase acceptance significantly [37]. In our case, participants were considered familiar if they had preexisting experience with AVs, rather than after a single or short-term exposure. Therefore, in line with other studies (see [7]), our finding suggests that more in depth knowledge of AVs does indeed yield more positive attitudes towards AVs. This understanding shall help to improve predictions of the impact of awareness and familiarity on acceptance of new technologies.

Our finding has further implications also for another topic that is often related to the rise of AVs, that is, people's fear of losing jobs. In fact, more familiar participants were also more positive towards AVs' impact on the job market. Recent evidence suggests that AVs can revolutionize logistics job profiles, by helping to cope with drivers' shortage [24,31] and providing better working conditions for operators via retraining and remote working [2,4]. Some occupations will likely be replaced by automation, but in many cases, this might translate to better conditions. At the same time, the fear of losing jobs to AVs shall be considered carefully. In their open answers, several participants suggested clear and transparent communication to handle and reduce this fear, that is, showing people how working conditions will evolve and likely improve. AVs can positively contribute to job profiles by making logistics operations safer, and reducing the physical and mental strain by limiting the repetitive and dangerous tasks, hence making these jobs more interesting and attractive for current and future workers. In general, communication should also be tailored to help the public build a more accurate understanding of AVs' potential benefits and limitations as well as, importantly, understand AVs' environmental impact.

Gender was found to have no significant effect on our studied dimensions. However, we found a significant effect of age on security, as older respondents were more concerned regarding the security of AVs. While in our case the difference emerged only in relation to security, our finding support previous studies which found age to be a discriminant factor for technology acceptance in logistics contexts [5,33]. This finding may partly be explained as one of the facets of an age-related "digital divide", which implies gaps between older and younger generations in access to, acceptance, and adoption of new technologies [35]. Furthermore, the fact that AVs are a kind of technology that, unlike computers or cellphones, are not yet fully materialized in people's everyday lives might explain the persistence of this gap [23].

Understanding these underlying mechanisms is crucial for addressing public concerns and promoting the acceptance and safe integration of AVs into society. People with knowledge about AVs expressed more positive attitudes towards AVs on a variety of acceptance factors. In turn, as several participants mentioned the need for better communication around AVs, one key message from this study concerns the fact that people who are currently not in contact with these technologies should be provided clear and transparent information so that they can develop informed opinions on AVs. This is aligned with existing work suggesting that initial trust and acceptance are determined by personal dispositions, rather than actual knowledge of how technologies work [22,27,28]. As transparency and communicating the benefits of emerging technologies could mitigate negative effect of perceived risk [17], showing AVs potential and limitation to lay persons by means of training, transparent educational campaigns, or hands-on demonstrations could improve acceptance.

Another insight that emerges from our study is that time is a crucial factor on many levels. First, this means giving people time to get used to the presence of AVs and learn how to interact with them. Second, time is required for the further necessary technological improvements that will enable AVs to merge safely with normal traffic and to respond to unpredictable situations. Finally, time is a key factor for the development of better regulations (both for the deployment of AVs and the evolution of job conditions), for attracting more investors, and for the implementation of smart and connected infrastructures. As these conditions progressively materialize, 'transitory' solutions (e.g., communication, training of employees and so on) must be put in place to help with this epochal transformation.

6 Conclusions and Limitations

This paper reports on the results of a study conducted to investigate people's acceptance of AVs in the context of logistics operations and shared use of public roads. Our approach positions itself at the intersection between acceptance of AVs in broad terms and in the specific context of logistics operations. The reason is that, while logistics hubs offer a suitable testing ground for AVs, due to being enclosed environments with more controllable conditions, logistics operations will hardly be confined to these hubs. In turn, this means that AVs employed by logistics operators will interact with other road users. Therefore, it is crucial to capture people's attitudes towards these vehicles across different domains.

To this end, the results of our study help to shed light on the dynamics of AVs acceptance. Our main finding regards the effects that participants' familiarity with AVs had on perceived reliability, safety, trust, and acceptance. The fact that high familiarity yielded high scores suggests that knowledge does make a difference. As it emerged repeatedly in the answers to the open questions, the public narrative around AVs should focus on conveying a clear and transparent message, highlighting what needs to be improved, but also the potential benefits in terms of safety, efficiency and working conditions. Additionally, participants stated that AVs' environmental impact is not clear. With the increasing public awareness for climate change, how this topic is addressed may steer acceptance of AVs significantly and should therefore receive special attention.

Our results also show how technology acceptance may indeed be age dependent. While the only significant correlation that emerged in our case concerned age and perceived security, this can be interpreted as resulting from a persisting "digital divide". Future work shall investigate whether such a difference in perception emerges also in contexts of long-term direct exposure with AVs.

Our study also has a few limitations. First, while the investigation here reported was informed by the ARTLAM model, it did not provide validation to the model itself. While the reason for this choice was to have the necessary flexibility to investigate participants' perception of AVs both within and outside the logistics context, future work shall accurately test all the model items to compare its validity with other technology acceptance models. Furthermore, our participants' sample was not very balanced, as slightly more than half of the people who took part in the study were of French nationality. Future work shall assess acceptance on a more balanced sample.

Finally, our findings also raise several questions in need of further investigation. For instance, are AVs capable of avoiding human errors or are we introducing new types of risks and errors that might have worse impact? Shall legal and ethical consideration follow the technical development of AVs or rather lead? This would be a fruitful area for further work.

Acknowledgments. The author(s) declare financial support was received for the research, authorship, and/or publication of this article. This work is part of the project AWARD, which has received funding from the European Union's Horizon 2020 research and innovation programme under grant agreement No 101006817. The content of this paper reflects only the author's view. Neither the European Commission nor CINEA is responsible for any use that may be made of the information it contains.

Disclosure of Interests. The authors have no competing interests to declare that are relevant to the content of this article.

References

1. Ahmed, H.U., Huang, Y., Lu, P., Bridgelall, R.: Technology developments and impacts of connected and autonomous vehicles: an overview. Smart Cities **5**(1), 382–404 (2022)
2. Autor, D.H.: Why are there still so many jobs? the history and future of workplace automation. J. Econ. Perspect. **29**(3), 3–30 (2015)
3. Barreto, L., Amaral, A., Pereira, T.: Industry 4.0 implications in logistics: an overview. Procedia Manuf. **13**, 1245–1252 (2017)
4. Bessen, J.E.: How computer automation affects occupations: technology, jobs, and skills. Boston Univ. school of law, law and economics research paper, pp. 15–49 (2016)
5. Capkin, S.O.K., Azarko, A.: Acceptance evaluation of automated logistics services: case study in Rome. Transp. Res. Procedia **72**, 4436–4443 (2023)
6. Carlan, V., Naudts, D., Audenaert, P., Lannoo, B., Vanelslander, T.: Toward implementing a fully automated truck guidance system at a seaport: identifying the roles, costs and benefits of logistics stakeholders. J. Shipp. Trade **4**(1), 12 (2019)
7. Chikaraishi, M., Khan, D., Yasuda, B., Fujiwara, A.: Risk perception and social acceptability of autonomous vehicles: a case study in Hiroshima, Japan. Transp. Policy **98**, 105–115 (2020)
8. Dallasega, P., Woschank, M., Sarkis, J., Tippayawong, K.Y.: Logistics 4.0 measurement model: empirical validation based on an international survey. Ind. Manag. Syst. **122**(5), 1384–1409 (2022)

9. Davis, F.D.: Perceived usefulness, perceived ease of use, and user acceptance of information technology. MIS Quart., 319–340 (1989)
10. Dekhne, A., Hastings, G., Murnane, J., Neuhaus, F.: Automation in logistics: big opportunity, bigger uncertainty. McKinsey Q **24** (2019)
11. Du, M., Zhang, T., Liu, J., Xu, Z., Liu, P.: Rumors in the air? Exploring public misconceptions about automated vehicles. Transp. Res. Part A: Policy Prac. **156**, 237–252 (2022)
12. Fröhlich, et al.: Towards a comprehensive understanding of stakeholder requirements for automated road transport logistics. In: AutomationXP@ CHI (2021)
13. Fröhlich, P., Sackl, A., Trösterer, S., Meschtscherjakov, A., Diamond, L., Tscheligi, M.: Acceptance factors for future workplaces in highly automated trucks. In: Proceedings of the 10th International Conference on Automotive User Interfaces and Interactive Vehicular Applications, pp. 129–136 (2018)
14. Gandia, R.M., et al.: Autonomous vehicles: scientometric and bibliometric review. Transp. Rev. **39**(1), 9–28 (2019)
15. Hilgarter, K., Granig, P.: Public perception of autonomous vehicles: a qualitative study based on interviews after riding an autonomous shuttle. Transport. Res. F: Traffic Psychol. Behav. **72**, 226–243 (2020)
16. Hutchins, N., Hook, L.: Technology acceptance model for safety critical autonomous transportation systems. In: 2017 IEEE/AIAA 36th Digital Avionics Systems Conference (DASC), pp. 1–5. IEEE (2017)
17. Jandl, C., Zafari, S., Taurer, F., Hartner-Tiefenthaler, M., Schlund, S.: Location-based monitoring in production environments: does transparency help to increase the acceptance of monitoring? Prod. Manuf. Res. **11**(1), 2160387 (2023)
18. Kapser, S., Abdelrahman, M.: Acceptance of autonomous delivery vehicles for last-mile delivery in Germany-extending utaut2 with risk perceptions. Transp. Res. Part C: Emerg. Technol. **111**, 210–225 (2020)
19. Lanzer, M., Baumann, M.: How media reports influence drivers' perception of safety and trust in automated vehicles in urban traffic. In: Proceedings of the Human Factors and Ergonomics Society Europe, pp. 63–74 (2022)
20. Lee, D., Hess, D.J.: Public concerns and connected and automated vehicles: safety, privacy, and data security. Humanit. Soc. Sci. Commun. **9**(1), 1–13 (2022)
21. Li, S., Zhang, Y., Edwards, S., Blythe, P.T.: Exploration into the needs and requirements of the remote driver when teleoperating the 5G-enabled level 4 automated vehicle in the real world-a case study of 5g connected and automated logistics. Sensors **23**(2), 820 (2023)
22. Li, X., Hess, T.J., Valacich, J.S.: Why do we trust new technology? A study of initial trust formation with organizational information systems. J. Strateg. Inf. Syst. **17**(1), 39–71 (2008)
23. Morris, M.G., Venkatesh, V.: Age differences in technology adoption decisions: implications for a changing work force. Pers. Psychol. **53**(2), 375–403 (2000)
24. Müller, S., Voigtländer, F.: Automated trucks in road freight logistics: the user perspective. In: Advances in Production, Logistics and Traffic: Proceedings of the 4th Interdisciplinary Conference on Production Logistics and Traffic 2019 4, pp. 102–115. Springer (2019)
25. Neubauer, M., Schauer, O.: Human factors in the design of automated transport logistics. In: Advances in Human Aspects of Transportation: Proceedings of the AHFE 2017 International Conference on Human Factors in Transportation, 17–21 July 2017, The Westin Bonaventure Hotel, Los Angeles, California, USA 8, pp. 1145–1156. Springer (2018)

26. Osswald, S., Wurhofer, D., Trösterer, S., Beck, E., Tscheligi, M.: Predicting information technology usage in the car: towards a car technology acceptance model. In: Proceedings of the 4th International Conference on Automotive User Interfaces and Interactive Vehicular Applications, pp. 51–58 (2012)
27. Papagni, G., Koeszegi, S.: Understandable and trustworthy explainable robots: a sensemaking perspective. Paladyn, J. Beh. Robot. **12**(1), 13–30 (2020)
28. Papagni, G., de Pagter, J., Zafari, S., Filzmoser, M., Koeszegi, S.T.: Artificial agents' explainability to support trust: considerations on timing and context. AI Soc. **38**(2), 947–960 (2023)
29. Penmetsa, P., et al.: How is automated and self-driving vehicle technology presented in the news media? Technol. Soc. **74**, 102290 (2023)
30. Placek, M.: Topic: Logistics industry worldwide — statista.com. https://www.statista.com/topics/5691/logistics-industry-worldwide/#topicOverview (2023). Accessed 25 June 2024
31. Rose, H., Lange, A.K., Hinckeldeyn, J., Jahn, C., Kreutzfeldt, J.: Investigating the requirements of automated vehicles for port-internal logistics of containers. In: International Conference on Dynamics in Logistics, pp. 179–190. Springer (2022)
32. Rosic, J., Hammer, F., Gafert, M., Fröhlich, P.: Acceptance is in the eye of the stakeholder: gathering the needs for automated road transport logistics. In: 13th International Conference on Automotive User Interfaces and Interactive Vehicular Applications, pp. 207–209 (2021)
33. Rześny-Cieplińska, J., Szmelter-Jarosz, A., Moslem, S.: Priority-based stakeholders analysis in the view of sustainable city logistics: evidence for Tricity, Poland. Sustain. Urban Areas **67**, 102751 (2021)
34. Seuwou, P., Chrysoulas, C., Banissi, E., Ubakanma, G.: Measuring consumer behavioural intention to accept technology: Towards autonomous vehicles technology acceptance model (AVTAM). In: Trends and Innovations in Information Systems and Technologies, vol. 18, pp. 507–516. Springer (2020)
35. Van Dijk, J.: The digital divide. John Wiley & Sons (2020)
36. Venkatesh, V., Morris, M.G., Davis, G.B., Davis, F.D.: User acceptance of information technology: toward a unified view. MIS Q., 425–478 (2003)
37. Wicki, M.: How do familiarity and fatal accidents affect acceptance of self-driving vehicles? Transport. Res. F: Traffic Psychol. Behav. **83**, 401–423 (2021)
38. Willems, L.: Understanding the impacts of autonomous vehicles in logistics. Dig. Transf. Log.: Demystify. Impacts Fourth Ind. Revolut., 113–127 (2021)

Modeling Autonomous Delivery Robots Under Framework of Automated Driving System Using Dynamic Transport Assignment: A State-of-the-Art

Yousuf Dinar[1(✉)], Carsten Gertz[1], Jacqueline Bianca Maaß[1], and Elena Queck[2]

[1] Institute for Transportation Planning and Logistics, Hamburg University of Technology, 21073 Hamburg, Germany
{Yousuf.dinar,gertz,jacqueline.maass}@tuhh.de
[2] Zwickau University of Applied Sciences, Institute for Energy and Transport, 08056 Zwickau, Germany
elena.queck@fh-zwickau.de

Abstract. The integration of Autonomous Delivery Robot (ADR) into the current logistics system, widely viewed as a promising last-mile solution, represents a significant advancement in modern delivery methods. Before deployment in practical applications, Autonomous Delivery Robots (ADR) undergo comprehensive testing phases, including Modeling and Simulation (M&S), Closed-Track Testing, and Open-Road Testing. The microscopic transportation simulation is a reliable test among different M&S methods. The ADR lacks a dedicated framework aimed at the Operation Design Domain (ODD) of Automated Driving Systems (ADS), unlike standard Automated Vehicle (AV). To achieve improved and predictable road safety, effective management of traffic conditions, and ultimately secure broader acceptance, ADR modeling should align with the appropriate Operational Design Domain (ODD), particularly for microscopic transportation simulation. The Operational Design Domain (ODD) ensures safe and reliable operation by considering factors such as physical infrastructure, operational limits, objects, connectivity, environmental conditions, and zones, though not all are applicable in microscopic transportation simulations. Implying the intermodal transportation for the ADR in commercially available Graphical User Interface (GUI) based simulators is not yet in practice. Therefore, there is a notable absence of implementation method. Dynamic Public Transport Assignment (formerly known as Dynamic Transit Assignment) is one of the few methods that addresses overlapping traffic agents and road users in the widely used microscopic simulator, Aimsun. This study proposed a set of ODD when utilizing Dynamic Public Transport Assignment to establish a logistics system that facilitates intermodal transportation of ADR. The results show ADR based network coverage require longer travel time for delivery than intermodal transport added scenario. This paper improves how ADR is modeled in suggested ODD and looks at how Dynamic Public Transport Assignment affects intermodal logistics, giving a foundation for modelers and researchers, making it state-of-the-art.

Keywords: Autonomous delivery robot · Automated driving system · Microscopic transportation simulation · Dynamic Public Transport Assignment

1 Introduction

In response to the shifting nature, requirements, and demands of modern industrial operations, cutting-edge technologies are being seamlessly integrated into the logistics system. The rising dependence on e-commerce and business-to-consumer (B2C) transactions demands a delivery method that is sustainable, safe, innovative, and economical [1]. The "missing link" between wholesale logistics and consumers appears to be Autonomous Delivery Robot (ADR), which are expected to significantly address the last-mile problem in the near future [2].

The critical importance of executing three specific pre-implementation tests: Modeling and Simulation (M&S), Closed-Track Testing, and Open-Road Testing, which together constitute indispensable evaluations prior to deploying the vehicle for real-world use [3]. For Modeling and Simulation (M&S), unlike regular Autonomous Vehicle (AV), there are no comprehensive guidelines serving as a framework for Automated Driving Systems (ADS) focused on ADR. This highlights the absence of essential components for ADS in the ADR operational domain. However, the vital components of the ADS for the autonomous vehicles such as Tactical Maneuver Behavior, Operational Design Domain (ODD) Elements, Object and Event Detection and Response (OEDR) Behavior and Failure Mode Behaviors [4]. The components in Table 1 are identified as the key parts that together make up the core features of a typical ADS test scenario. The ODD significantly choose the components of simulation scenarios in microscopic transport simulation environment. Others transport simulation such as macroscopic or vehicle centric analysis can cover other vital components exclusively.

Table 1. Summary of vital components of ADS [4]

Vital components	Description
Tactical Maneuver Behavior	Tactical maneuver behaviors relate to the immediate control related tasks the ADS is executing as part of the test (e.g., lane following, lane change, turning)
Operational Design Domain (ODD)	The relevant ODD elements generally define the operating environment in which the ADS is navigating during the test (e.g., roadway type, traffic conditions, or environmental conditions)
Object and Event Detection and Response (OEDR)	OEDR capabilities relate directly to the objects and events the ADS encounters during the test (e.g., vehicles, pedestrians, traffic signals)
Failure Mode Behaviors	Some tests may include injection or simulation of errors or faults that induce failures at various stages within the ADS's functional architecture

Implementing ADR in the simulation interface necessitates an appropriate provision within the ADS components for evaluation, ensuring the testing is widely accepted and reliable. In this study, the ODD of the ADS for ADR is standardized and employed for modeling.

The implementation of an intermodal transport system for the ADR logistics network remains a complex task in commercial simulation packages, such as Aimsun and PTV VISSIM. In the context of intermodal scenarios, there are several simulation methods capable of replicating real-world situations, yet specific use cases are currently absent. Approaches such as External Agent Interface (EAI), Dynamic Public Transport Assignment, and Aimsun Ride represent robust methodologies for enhancing modeling capabilities in intermodal user case. This study covered the Dynamic Public Transport Assignment in the Aimsun interface. The results show that, with the right rules and technical knowledge, setting up the components for the ADS is possible, ensuring safe intermodal transport. This study offers a framework for simulating ADR in GUI-based microscopic transport simulations, while also tackling two gaps in previous ADR research: the inclusion of ODD and the integration of ADR within intermodal transport systems.

2 Literature Review

No major study has focused on the modeling and simulation (M&S) of ADR in pre-implementation tests, particularly those involving GUI-based microscopic transport simulations. Some studies have explored simulations centered on surrogate testing, laboratory testing, data-driven and agent-based analysis [5]. The integration of intermodal transport for ADR remains unaddressed, and earlier studies have similarly neglected to incorporate the Operational Design Domain (ODD) into simulations. This omission results in simulations that are both unrealistic and incomplete in representing real-world ADR conditions.

2.1 Dynamic Public Transport Assignment

Dynamic Public Transport Assignment is a sophisticated feature within the Aimsun transportation simulator. It allows pedestrians to choose from various public transport options, optimizing their travel time by considering both cost and proximity to transport stops [6]. Dynamic Public Transport Assignment tackles the intermodal transport challenge in ADR simulations by ensuring that delivery robots function in a way that closely resembles human behavior, or at least approximates it. In Aimsun, pedestrians exhibit behavior based on the Social Force model [6]. The components of Social Force model are given in the Table 2. This model plays a crucial role in various actions, such as following others, changing lanes, boarding and alighting from buses, crossing the road, and performing other maneuvers [1]. To create a new behavior, extensive study of ADR is essential before generating the model.

Table 2. Components of Social Force model [6, 7]

Behavior	Description	Aimsun Interface
Agents wants to reach destination comfortably	Agents do not take detour, choose shortest path and no intentional delay	Driving Force Relaxation: models the acceleration and aggressiveness. The shorter the relaxation time, the more pushy pedestrians will act Interaction Strength: models how quickly the exerted force decreases with the distance Interaction Range: models the influence of distance on the repulsive force
The motion of an agent is effected by other agents	Agent is influenced from other agents depending on the speed and density	Anisotropy of Interaction: represents the effective field of view, which takes the form of a dimpled limaçon
Agents are sometimes attracted by other agents (friends, street artists, etc.) or objects (e.g. window displays)	Agents become closure the other agents	Implemented in the form of existing behavior
Agents are sometimes disattracted by other agents or Objects	Agents avoid being closer to other agents	Implemented in the form of existing behavior

2.2 Operational Design Domain

The Operational Design Domain (ODD) defines specific scenario parameters for conducting ADS testing, providing a realistic and practical approach to evaluate performance. Figure 1 shows the a few components in ODD classification framework. The features of ODD were classified into seven general categories for testing any new automated transport system [4].

1. L3 Conditional Automated Traffic Jam Drive.
2. L3 Conditional Automated Highway Drive.
3. L4 Highly Automated Low Speed Shuttle.
4. L4 Highly Automated Valet Parking.
5. L4 Highly Automated Emergency Take-Over.
6. L4 Highly Automated Highway Drive.
7. L4 Highly Automated Vehicle/TNC.

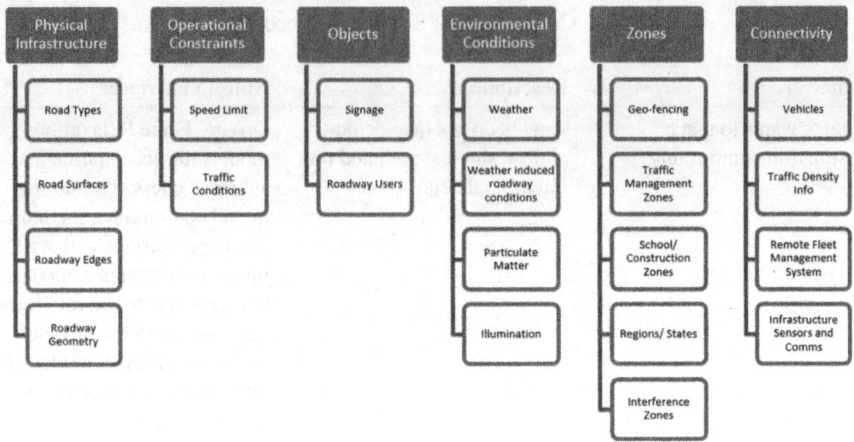

Fig. 1. A portion of the ODD Classification framework [4]

3 Experimental Setup

3.1 Study Area

The study area focuses on the inner network of Lauenburg, comprising several key bus routes, including lines 138, 238, 338, 438, and a dedicated shuttle service. Figure 2 depicts the transport network of the Lauenburg. This network spans an area of 9.54 km^2. The journey for the Autonomous Delivery Robots (ADR) begins at PEZ, from where they walk to the Lauenburg ZOB bus stop. From there, they take the bus to the Museum stop and continue on to Rathaus. Upon reaching the Museum and Rathaus stops, the ADRs proceed to their designated drop-off points. The operation of Autonomous Delivery Robots (ADR) follows an intermodal transport-based logistics approach. They cover considerable distances on foot and utilize buses from selected stops, effectively reducing

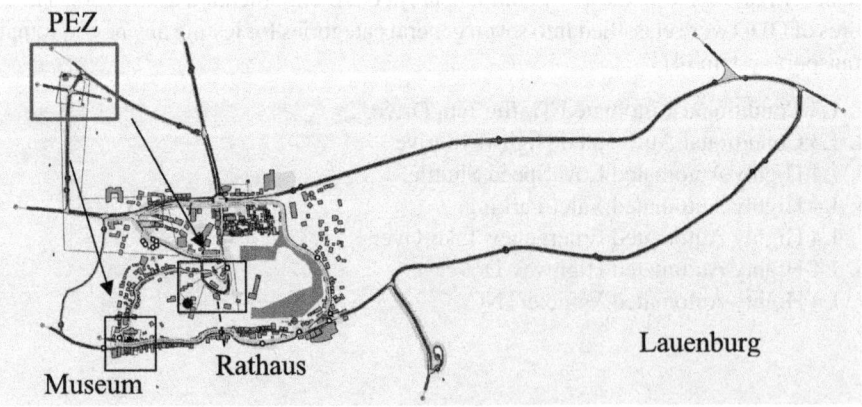

Fig. 2. Network of the study area

delivery travel time. Additionally, integrating ADRs into intermodal transport optimizes the use of available space on buses, offering further efficiency benefits.

3.2 Simulation Input

For this study a few data were collected as primary source and some as secondary source. Primary data were motor and pedestrian traffic counting, number of commuters, vehicle composition, routing in intersection and data related to the ADR. Secondary data are traffic signal data, public transport schedule, number of accident etc. The Fig. 3 shows the targeted ADR for this study which is prepared by the Institute of Logistics Engineering of Hamburg University of Technology. Table 3 shows the input used for the modeling of ADR for this study.

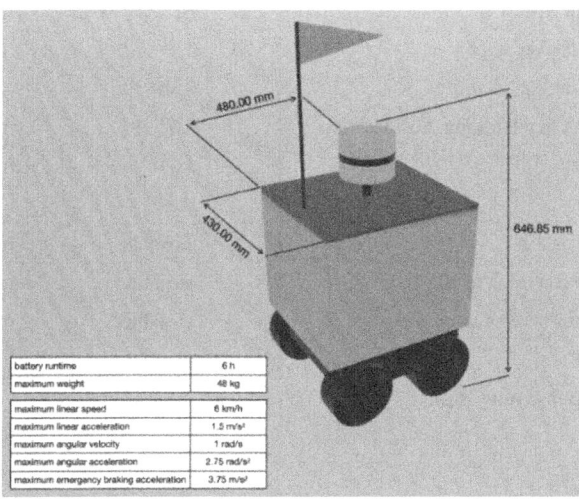

Fig. 3. Study model of ADR [8]

3.3 Scenario

To understand the impact of incorporating ADRs into the current transport system, two scenarios were thoroughly examined: delivery solely by walking robots and delivery by intermodal-enhanced delivery robots. The latter scenario, which combines walking with bus usage during the simulation, required the implementation of Dynamic Public Transport Assignment. In this study, the shuttle bus line was utilized as the public transport mode. Figure 4 depicts the shuttle bus connections within the study area. The total number of trips was divided into four for intermodal-enhanced delivery. In the delivery robot-only scenario, these four trips were consolidated into a single trip. The ADR model used in this study operates at Autonomous Level 5, enabling it to navigate obstacles such as path selection and sudden breakdowns. For the purposes of this study, failures or fallback scenarios were not considered.

Table 3. Input for modeling the ADR

Input Data - Autonomous Delivery Robot (ADR)		
	Value	Unit
Physical Properties		
Length	480	mm
Width	430	mm
Height	646.85	mm
Max Weight (including goods)	48	kg
Maximum Linear Speed	6	km/h
Maximum Linear Acceleration	1.5	m/s2
Maximum Lateral Speed	3	km/h
Maximum Angular Velocity	1	rad/s
Maximum Angular Acceleration	2.75	rad/s2
Maximum Emergency Braking Acceleration	3.75	m/s2
Max Deceleration	1	m/s2
Energy Data		
Charging time	variable	mAh/hr
Energy consumption rate (while standing)	variable	mAh/hr
Energy consumption rate (while moving)	variable	mAh/hr
Run time	6	hr
Car Following Behaviour Input		
Clearance (Front)	0.5	m
Clearance (Lateral)	0.5	m
Maximum give away time	30	sec
Speed Limit Acceptance	1	
Reaction Time	10	sec
Safety Margin Factor	1	
Headway aggressiveness	1	
PCU	0.5	
Gap acceptance	1	
Safety distance (Stand/dynamic)	400	mm
Operational Parameter		
Delivery pick up	60	sec
Delivery drop off	60	sec

(*continued*)

Table 3. (*continued*)

Input Data - Autonomous Delivery Robot (ADR)		
	Value	Unit
Bus in time (empty/full bus)	30	sec
Bus out time (empty/full bus)	30	sec
Time from waiting to initial boarding position	10	sec
Parking time	10	sec
Breakdown time	30	sec

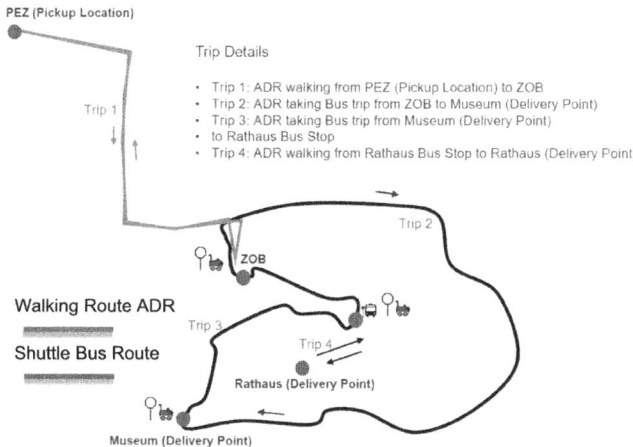

Fig. 4. Total intermodal transport plan for the ADR

3.4 ODD in Simulation

Since ADRs differ from other automated vehicles, the standard ODD framework had to be adapted. During the simulation, only a limited number of inputs from the ODD framework were found to be applicable. Table 4 presents the inputs utilized under the ODD framework which are adopted for this study. It addresses the model and simulation components of the ODD framework.

Table 4. Inputs utilized under the ODD framework checklist

ODD Framework Checklist	
Physical Infrastructure	
Road Types	
Divide Highway	No
Urban	Yes
Parking	No
Managed Lanes	Yes
On-Off Ramp	No
Emergency Evacuation Route	No
Intersection	Yes
Roadway Surfaces	
Asphalt	Yes
Concrete	No
Roadway Edges and Markings	
Lane Markers	No
Temporary Lane Markers	No
Shoulder	No
Lane Barriers	No
Rails	No
Operational Constraint	
Speed Limits	
Minimum Speed Limit	0 km/h
Maximum Speed Limit	= < 6 km/h
Traffic Conditions	
Traffic Density	Normal Traffic Volume

4 Result

The impact of integrating ADR into the transportation system was analyzed under two scenarios: delivery exclusively by walking robots and delivery through intermodal-enhanced robots. To achieve this, the scenarios were conducted both without and with the use of public transport. The travel time for each trip in both scenarios was calculated, as shown in Table 5.

Table 5. Travel Time for two scenarios: delivery solely by walking robots and delivery by intermodal-enhanced delivery robots

Scenario	Trip Segment	Travel Time
Delivery solely by walking robots	Trip 1: The ADR walks from PEZ (Pickup Location) to Rathaus (Delivery Point), with a stop at the Museum along the way	42 min
Delivery by intermodal-enhanced delivery robots	Trip 1: ADR walking from PEZ (Pickup Location) to ZOB	9 min 22 s
	Trip 2: ADR taking Bus trip from ZOB to Museum (Delivery Point)	7 min 12 s
	Trip 3: ADR taking Bus trip from Museum (Delivery Point) to Rathaus Bus Stop	2 min 12 s
	Trip 4: ADR walking from Rathaus Bus Stop to Rathaus (Delivery Point)	3 min
	Total Travel Time	21 min 46 s

To simulate real-world conditions, the average speeds of the ADR and shuttle bus were set at 5 km/h and 15 km/h, respectively. However, actual speeds fluctuated due to narrow roads, uneven surfaces, and pedestrian traffic. Figure 5 highlights two additional performance indicators: potential conflicts and the number of stops for both scenarios. The interaction between ADR and pedestrians was considered to analyze these indicators. The results show that when ADR travels on foot throughout the entire network, it encounters more conflicts and stops more frequently due to its lower speed and limited sidewalk space. In contrast, when ADR is integrated with the bus system, it experiences smoother travel with fewer stops and conflicts.

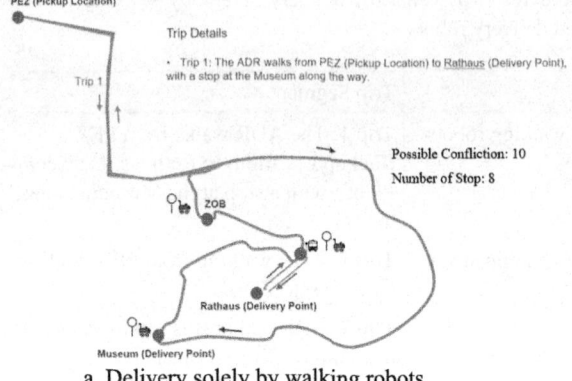

a. Delivery solely by walking robots

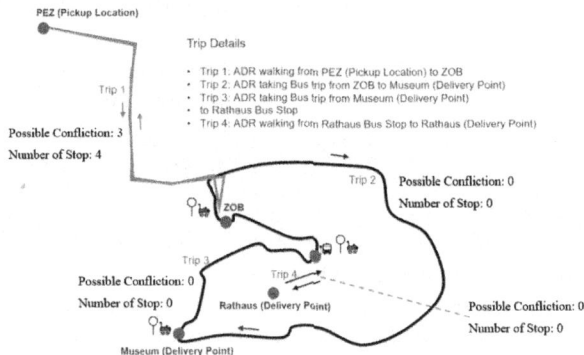

b. Delivery by intermodal-enhanced delivery robots

Fig. 5. Possible Confliction and number of stop between ADR and pedestrian for two scenarios: a. delivery solely by walking robots and b. delivery by intermodal-enhanced delivery robots

5 Conclusion and Future Study

Integrating ADR with the public transport system offers substantial benefits compared to using ADR independently. From the perspective of delivery robot operations, this integration reduces travel time, minimizes the number of stops, and decreases potential conflicts, as ADRs are effectively removed from the network while onboard the bus, thus having no impact on road traffic. However, delays in ADR boarding or disembarking can disrupt the bus schedule, which is particularly important when ADR units need to board and exit at multiple stops. Furthermore, prolonged ADR walking times in traffic may result in extended waits at intersections with pedestrian crossings.

This study employed a fixed route for both ADR walking and the shuttle bus. However, dynamic route selection for ADR walking could lead to significant variations. Furthermore, dynamic bus route choices may have a considerable impact on traffic conditions, either improving or worsening them. This is a key area for further investigation in future research. Queue lengths at intersections with pedestrian crossings tend to increase after the introduction of ADRs. To fully understand this effect, scenarios without ADRs should be analyzed, though it is essential to first explore the logistics demand. However, due to the lack of real-world data, researchers currently estimate logistics demand based on arbitrary assumptions. This area requires more thorough investigation.

References

1. Jafari, A., Liu, Y-C.: A heterogeneous social force model for personal mobility vehicles on futuristic sidewalks. Simul. Modelling Practice Theory., **131**, 102879 (1 February 2024). Available at: https://doi.org/10.1016/j.simpat.2023.102879
2. Garus, A., Christidis, P., Mourtzouchou, A., Duboz, L., Ciuffo, B.: Unravelling the Last-Mile Conundrum: A Comparative Study of Autonomous Delivery Robots, Delivery Bicycles, and Light Commercial Vehicles in 14 Varied European Landscapes. Sustainable Cities and Society, **105490**, (1 May 2024). Available at: https://doi.org/10.1016/j.scs.2024.105490
3. Hoffmann, T., Prause, G.: On the regulatory framework for last-mile delivery robots. Machines., **6**(3), 33 (2018). Available at: https://doi.org/10.3390/machines6030033
4. Thorn, E., Kimmel, S.C., Chaka, M.: A framework for automated driving system testable cases and scenarios. Proceedings of the 26th International Technical Conference on the Enhanced Safety of Vehicles (ESV). (1 September 2018). Available at: https://rosap.ntl.bts.gov/view/dot/38824/dot_38824_DS1.pdf
5. Poeting, M., Schaudt, S., Clausen, U.: Simulation of an optimized last-mile parcel delivery network involving delivery robots. Lecture notes in logistics. pp. 1–19 (2019). Available at: https://doi.org/10.1007/978-3-030-13535-5_1
6. Aimsun. Simulating Pedestrians—Aimsun Next Users Manual. Available at: https://docs.aimsun.com/next/22.0.2/UsersManual/PedestrianSimulator.html#Dyn_PT_Assignment
7. Helbing, D., Molnár, P.: Social force model for pedestrian dynamics. Phys. Rev. E Stat. Phys. Plasmas Fluids Related Interdisc. Topics., **51**(5), 4282–4286 (1 May 1995). Available at: https://doi.org/10.1103/physreve.51.4282
8. Thiel, M., Ziegenbein, J., Blunder, N., Schrick, M., Kreutzfeldt, J.: Institute for Technical Logistics, Hamburg University of Technology (TUHH) From Concept to Reality: Developing Sidewalk Robots for Real-World Research and Operation in Public Space [journal-article]. Logistics Journal: Proceedings. pp. 1–2 (2023). Available at: https://doi.org/10.2195/lj_proc_thiel_en_202310_01

Virtual Variable Baseline Stereo Vision Algorithm

Chia-Chiun Kuo and Liang-Kuang Chen(✉)

National Taiwan University of Science and Technology, Taipei, Taiwan
lkchen@mail.ntust.edu.tw

Abstract. This research presents a novel variable baseline stereo vision system to overcome the limitations of traditional fixed baseline technology in complex environments. Traditional stereo vision systems struggle with large depth variations and visual dead zones due to their fixed baselines. In response to the scenario where UAVs are equipped with a single camera and take multiple images from different locations to simulate stereo vision, a modified algorithm is proposed. This algorithm adjusts the baseline according to the target's distance and angle, improving depth resolution. Through improved calibration techniques and sensing systems, the algorithm accurately identifies the target's 3D position. Experimental results show that, compared to traditional methods, the variable baseline system significantly enhances distance measurement accuracy in complex environments, demonstrating potential in applications such as mountain rescue and wildlife conservation. Future work will focus on enhancing the system's intelligence and expanding its application scope and practical value.

Keywords: Stereo vision · variable baseline · UAV

1 Introduction

1.1 Research Motivation

Stereo vision technology, which provides depth and spatial position information of objects, has been widely applied in various fields such as UAV (unmanned aerial vehicle) navigation, autonomous vehicle driving, and robotic vision. However, traditional stereo vision systems typically use a fixed baseline design, which often leads to decreased measurement accuracy and lack of flexibility when used in a wide variety of scenarios. Consider using a single camera on a UAV to perform the stereo vision function, two images taken at different position can be used to emulate the two images required, assuming that the baseline information can be accurately measured by other sensing devices. Therefore, a variable "virtual baseline" concept can be formed. Researching into variable baseline stereo vision technology aims to address these issues by dynamically adjusting the distance and angle between the cameras, to adapt for different environmental requirements, thus providing a key solution to these problems.

1.2 Literature Review
Three-Dimensional Positioning Technology
Three-dimensional (3D) positioning technology holds significant importance in fields such as navigation, autonomous driving, robotics, and geographic information systems (GIS), e.g., [1]. Traditional technologies like GPS, LiDAR, and Time-of-Flight (ToF) have been widely used in 3D positioning but each has its limitations, e.g., [2]. Stereo vision technology simulates human binocular disparity to perceive depth, characterized by low cost, ease of implementation, and high performance. It is particularly suitable for applications in dynamic environments. Next, the specific applications and development of two stereo vision localization techniques in 3D localization will be discussed.

Fixed Baseline Stereo Vision Positioning
Fixed baseline stereo vision systems are widely used in many fields due to their stability and accuracy. Ishibashi [3] utilized fixed baseline technology to operate robotic arms on underwater vehicles, demonstrating its feasibility in extreme and visually constrained underwater environments, although the operational range was limited to the arm's working area. Mustafah et al. [4] employed fixed baseline systems to precisely locate indoor UAVs, showcasing their potential in GPS-denied environments; however, their operational range remained confined to indoor settings, e.g., [5–7]. Therefore, it is evident that fixed baseline systems lack the flexibility and adaptability needed for environments with varying conditions.

Variable Baseline Stereo Vision Positioning
Variable baseline stereo vision systems dynamically adjust the distance between cameras, allowing for more precise depth measurement and enhancing the system's flexibility and efficiency in various applications. Nakabo et al. [8] demonstrated how a movable camera system on a linear slider could effectively track high-speed moving targets. Zhu et al. [9] focused on fire scene applications, experimenting with different baseline lengths to study their impact on fire source localization. Sanket et al. [10] developed a novel variable baseline stereo vision system that improved UAV navigation and obstacle avoidance efficiency in complex environments, e.g., [11–14].

In the literature, three-dimensional positioning technology is widely used in fields including navigation, autonomous driving, robotics, and geographic information systems. Traditional methods such as GPS, LiDAR, and Time-of-Flight have different limitations, while stereo vision technology offers low cost, ease of implementation, and high efficiency, making it ideal for applications in dynamic environments. Fixed baseline stereo vision systems are stable and accurate but lack the flexibility in varying conditions. Variable baseline systems improve depth measurement adaptively by changing camera baseline, which will be critical for tasks like target tracking and UAV navigation. However, most previous research focuses on baseline distance adjustments, neglecting the impact of camera angles, leading to positioning blind spots in complex scenarios.

1.3 Research Objective

The objective of this research is to investigate a 3D positioning algorithm based on variable baseline stereo vision. Building upon existing technology, this algorithm incorporates dynamically adjustable and extendable baseline distances and angles, enabling it to adapt for various distance and environmental requirements. The algorithm is originally intended for use in a variety of UAV inspection tasks. Emphasizing the advantages of long-distance positioning, it is particularly suitable for complex and hazardous environments.

2 Principles of Variable Baseline Stereo Vision Algorithm

2.1 Camera Calibration

Camera calibration utilizes multiple known world coordinates and their corresponding image coordinate points to determine the camera's intrinsic parameters, extrinsic parameters, and distortion coefficients. The matrix relationship is described by Eq. (1).

$$Z_c \begin{bmatrix} u \\ v \\ 1 \end{bmatrix} = \begin{bmatrix} \frac{f}{d_x} & 0 & C_x & 0 \\ 0 & \frac{f}{d_y} & C_y & 0 \\ 0 & 0 & 1 & 0 \end{bmatrix} \begin{bmatrix} R & T \\ 0^T & 1 \end{bmatrix} \begin{bmatrix} X_w \\ Y_w \\ Z_w \\ 1 \end{bmatrix} \quad (1)$$

The specific steps of the camera calibration process are as follows: First, a checkerboard pattern is used as the calibration object, and multiple images of the checkerboard are taken from different angles and distances to ensure coverage of all directions in the camera's field of view. Next, these images are loaded into MATLAB's Camera Calibrator, and the actual size and dimensions of each square on the checkerboard are inputted. The system then analyzes the images, detects the checkerboard's corner points, and calculates the camera's intrinsic parameters (such as focal length and principal point) as well as the distortion coefficients. Finally, the calibration results are reviewed and validated by checking the corner re-projection errors to ensure calibration accuracy.

The calibration process in this study was conducted using MATLAB's Camera Calibrator before the actual measurement applications, and its accuracy was verified through the following two methods.

Verification of Internal Parameters Using Pixel Coordinate Error
To verify whether the internal parameters obtained from the Camera Calibrator are sufficiently accurate, the results of Eq. (1) were compared with the data from standard calibration templates, the chessboard image. Given the obtained extrinsic parameters, the known world coordinates are substituted to obtain the corresponding pixel coordinates, and the error can be evaluated.

The extrinsic parameters are composed of a 3D rotation matrix (R) and a translation matrix (T). In this study, the experimental setup places the camera on a three-axis motion platform. Assuming the image plane (camera) and the world coordinate plane (chessboard) are parallel to each other, the rotation matrix (R) can be set as the identity matrix. The translation matrix is the sum of the displacement of the motion platform $\begin{bmatrix} x_d & y_d & z_d \end{bmatrix}^T$, the initial distance between the motion platform and the image plane $\begin{bmatrix} 0 & 0 & z_1 \end{bmatrix}^T$, and the offset $\begin{bmatrix} x_i & y_i & 0 \end{bmatrix}^T$ between the projection of the image plane's origin onto the chessboard and the world coordinate origin. Under these conditions, the ideal state is described by Eq. (2) (Fig. 1).

$$T = \begin{bmatrix} x_d & y_d & z_d \end{bmatrix}^T + \begin{bmatrix} 0 & 0 & z_1 \end{bmatrix}^T + \begin{bmatrix} x_i & y_i & 0 \end{bmatrix}^T \quad (2)$$

Fig. 1. Experimental Setup for Verification of Internal Parameters Using Known External Parameters

After obtaining the extrinsic and intrinsic parameters from the Camera Calibrator, both are substituted into Eq. (1). By selecting three points in the image and substituting their world coordinates, the pixel coordinate error can be calculated to evaluate the accuracy of the intrinsic parameters. The results are shown in Table 1.

Table 1. Pixel Coordinate Error

World Coordinates (mm)	Calculated Pixel Coordinates (pixels)	True Pixel Coordinates	Error (pixels)
(0,0,0,1)	(464.56,259.41)	(470,247)	(−5.44,12.41)
(24,24,0,1)	(536.27,331.24)	(541,317)	(−4.73,14.24)
(−24,−24,0,1)	(392.86,187.58)	(399,176)	(−6.14,11.58)

The above method can obtain the pixel coordinate error based on the known world coordinates after calculating the intrinsic parameters. However, the reverse calculation, that is, obtaining the world coordinate error from the known pixel coordinates, faces additional difficulty because the non-square matrix is not invertible. Using pseudo-inverse provides a convenient work-around solution, however, it will not be possible to distinguish whether the observed error is caused by the intrinsic parameters or the pseudo-inverse matrix. Therefore, in this study a different verification process is proposed.

Verification of Intrinsic Parameters Using World Coordinate Error

After the intrinsic parameters, the distortion correction is performed. Assuming that the actual u-axis and v-axis differ from the ideal u-axis and v-axis by an angle θ, and introducing scaling factors α and β respectively, we finally add the optical offsets x_0 and y_0 to the ideal image coordinates (u, v), resulting in the following Eq. (3).

$$(u, v) = (\alpha \frac{x_c f}{z_c} - \alpha \frac{y_c f}{z_c} \cot\theta + x_0, \beta \frac{y_c f}{z_c \sin\theta} + y_0) \qquad (3)$$

Multiplying both sides by z_c and removing the rows that are all zeros, we obtain an invertible square matrix, as shown in Eqs. (4) and (5).

$$z_c \begin{bmatrix} u \\ v \end{bmatrix} = \begin{bmatrix} \alpha f & -\alpha f \cot\theta & 0 \\ 0 & \frac{\beta f}{\sin\theta} & 0 \end{bmatrix} \begin{bmatrix} x_c \\ y_c \\ z_c \end{bmatrix} + z_c \begin{bmatrix} x_0 \\ y_0 \end{bmatrix} \qquad (4)$$

$$\begin{bmatrix} x_c \\ y_c \end{bmatrix} = \begin{bmatrix} \frac{f_x}{z_c} & \frac{s}{z_c} \\ 0 & \frac{f_y}{z_c} \end{bmatrix}^{-1} \left(\begin{bmatrix} u \\ v \end{bmatrix} - \begin{bmatrix} x_0 \\ y_0 \end{bmatrix} \right) \qquad (5)$$

According to Eq. (5), when the depth z_c is known, the calculated intrinsic parameters can be substituted into the equation. This allows the calculation of the camera coordinates from the pixel coordinates. After translating these coordinates using the known extrinsic parameter matrix, the world coordinates can be obtained. The error can then be subsequently calculated by subtracting the theoretical values from these computed world coordinate values, thus verifying the accuracy of the intrinsic parameters.

This experiment was conducted with two settings, measuring depths of 61 cm and 36 cm. Starting from the chessboard coordinate (0, 0), measurements were taken in units of two squares to the right and one square downward, with a total of 30 feature points measured each time. The results are shown in the Fig. 2.

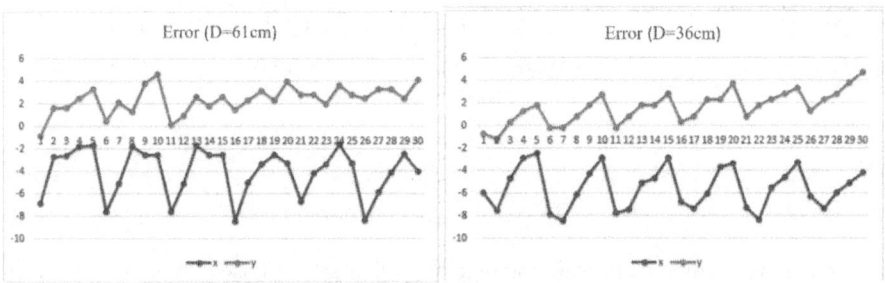

Fig. 2. World Coordinate Error for Two Depths

To improve the intrinsic parameters accuracy, the proportion of the chessboard in the image is enlarged by using a bigger chessboard. The world coordinate error was then recalculated at a depth of 61 cm and showing smaller errors, as shown in Fig. 3. It can be observed that the error significantly decreased. Although the maximum error still reached

6 mm occasionally, the errors in the front rows generally reduced to below 4 mm, and even below 2 mm in several cases. The comparisons confirmed that the intrinsic parameters calculated by MATLAB's Camera Calibrator have reasonable accuracy.

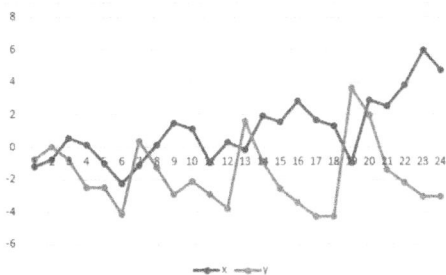

Fig. 3. World Coordinate Error at a Depth of 61 cm After Optimization

2.2 Stereo Vision Algorithm

The following section presents the derivations of traditional stereo vision positioning, to facilitate the illustrations of the proposed modifications of this research. As shown in Fig. 4, assume that the left and right cameras are positioned on the same plane, with parallel optical axes and identical focal lengths f. The origin of the world coordinate system is located at the left camera. Let b represent the baseline between the cameras, and P denote the position of the point to be measured in the world coordinates. Let (x_l, y_l) and (x_r, y_r) be the image coordinates of point P in the left and right cameras, respectively. We have

$$\frac{z}{f} = \frac{x}{x_l} \tag{6}$$

$$\frac{z}{f} = \frac{x-b}{x_r} \tag{7}$$

From Eqs. (6) and (7), we can derive the following Eq. (8).

$$z = \frac{f*b}{(x_l - x_r)} = \frac{f*b}{d} \tag{8}$$

Similarly, the stereo vision algorithm with different depths is also based on the same geometric relationships, as shown in the right-hand side of Fig. 4.

Regarding the stereo vision algorithm with angled cameras, this study refers to the camera skew angle formula presented in the literature [15]. It first requires three test points P_j, P_{j+1} and P_{j+2} that have the same depth and are equally spaced in the x-direction. From the geometric relationships, the following equations can be derived:

$$\begin{cases} \tan(\alpha_{10} + \beta_{1j+1}) - \tan(\alpha_{10} + \beta_{1j}) \\ = \tan(\alpha_{10} + \beta_{1j+2}) - \tan(\alota_{10} + \beta_{1j+1}) = \frac{\Delta x}{z} \\ \tan(\alpha_{20} + \beta_{2j+1}) - \tan(\alpha_{20} + \beta_{2j}) \\ = \tan(\alpha_{20} + \beta_{2j+2}) - \tan(\alpha_{20} + \beta_{2j+1}) = \frac{\Delta x}{z} \end{cases} \tag{9}$$

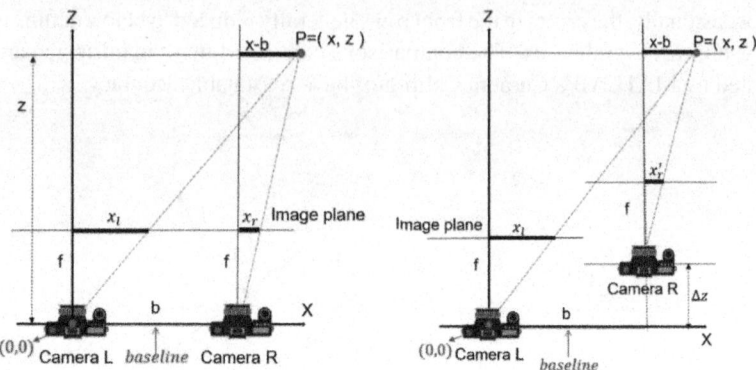

Fig. 4. Traditional (Left) and Stereo Vision Algorithm with Different Depths (Right) Diagrams

α_{10}, α_{20} are the skew angles of the left and right cameras, respectively, while β_{1j} and β_{2j} are the angles between the optical centers of the left and right cameras and the test point.

$$\beta_{1j} = \tan^{-1}(\frac{u_1 - x_{10}}{f_1}) \tag{10}$$

$$\beta_{2j} = \tan^{-1}(\frac{u_2 - x_{20}}{f_2}) \tag{11}$$

where u_1 and u_2 are the pixel coordinates of the test point, x_{10} and x_{20} are the positions of the optical axis centers of the cameras, and f is the focal length (Fig. 5).

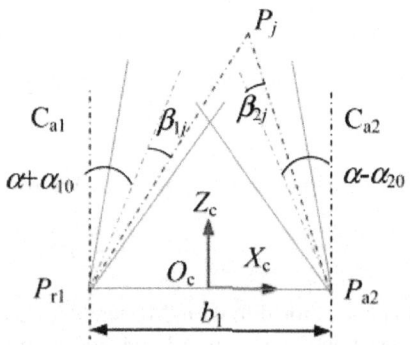

Fig. 5. Camera Skew Angle Calculation [15]

Finally, by rearranging Eq. (9), we obtain the residuals of the feature points to construct the objective functions F_1 and F_2. By minimizing F_1 and F_2, the optimal α_{10} and α_{20} can be determined to reduce the error.

Variable Baseline Stereo Vision Positioning

The above algorithm was modified by assuming that the positions of the two cameras can be on different planes and not parallel to each other, thus allowing emulating the stereo vision using UAVs. Let the left camera be positioned at the origin of the coordinate system, with the x-axis pointing due north and the y-axis pointing due east. This coordinate system is referred to as the North-East-Down (NED) frame. The following is the model and its derivations follow (Fig. 6):

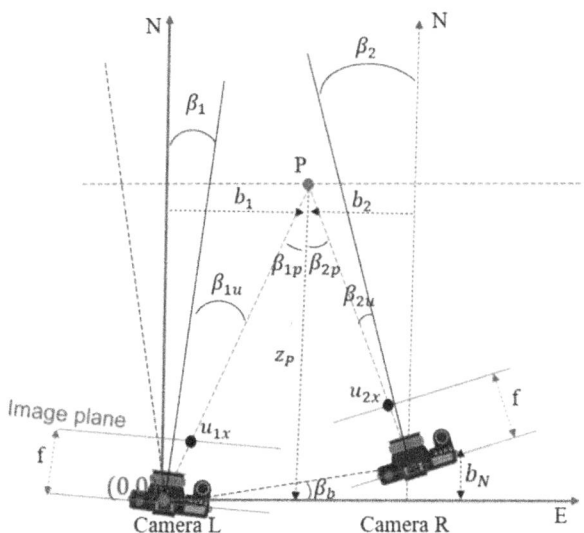

Fig. 6. Variable Baseline Stereo Vision Positioning Algorithm Model (NED Frame)

Let the angles between the perpendiculars from the left and right cameras to the baseline and the test point be β_{1p} and β_{2p}, respectively. The baseline b is divided into its components b_N (north) and b_E (east). From the geometric relationships of the baseline, we can resultingly obtain Eq. (16), and after rearranging, we obtain the depth formula for this model, Eq. (17).

$$\beta_{1p} = \beta_{1u} + \beta_1 \tag{12}$$

$$\beta_{2p} = \beta_{2u} + \beta_2 \tag{13}$$

$$b_N = b \sin \beta_b \tag{14}$$

$$b_E = b \cos \beta_b \tag{15}$$

$$b_1 + b_2 = b_E = z_p \tan \beta_{1p} + z_p \tan \beta_{2p} - b_N \tan \beta_{2p} \tag{16}$$

$$z_p = \frac{b_E + b_N \tan \beta_{2p}}{\tan \beta_{1p} + \tan \beta_{2p}} \tag{17}$$

3 Emulating UAV Scenarios for Algorithm Verification

3.1 Camera β_1 Calibration

Using the camera skew angle calculation formula described in Sect. 2.2, calibrations could be performed by finding three test points in the actual scene that have the same depth and are equally spaced in the x-direction. However, in practical experimental scenarios, feature points do not always exhibit such regular distributions. The known angles α_{10} and α_{20} from this formula correspond to the skew angles β_1 and β_2 of the left and right cameras, respectively. Therefore, it is assumed that the distances between the points are known and unequal, and redefine the equations based on the specific distances of each point. Similarly, suppose there are three feature points P_j, P_{j+1} and P_{j+2} with the same depth, and the distances between the three points are d_{01} and d_{12} respectively. We redefine the corresponding trigonometric relationships to calculate the residuals, as defined below (Fig. 7).

$$\begin{cases} \tan(\alpha_{10} + \beta_{1j+1}) - \tan(\alpha_{10} + \beta_{1j}) \\ = \frac{d_{01}}{d_{12}}\left(\tan(\alpha_{10} + \beta_{1j+2}) - \tan(\alpha_{10} + \beta_{1j+1})\right) \\ \tan(\alpha_{20} + \beta_{2j+1}) - \tan(\alpha_{20} + \beta_{2j}) \\ = \frac{d_{01}}{d_{12}}\left(\tan(\alpha_{20} + \beta_{2j+2}) - \tan(\alpha_{20} + \beta_{2j+1})\right) \end{cases} \tag{18}$$

Fig. 7. Left and Right Camera Real-World Scenario Emulating the Chessboard

Similarly, by rearranging Eq. (18), we could obtain the residuals of the feature points to construct the objective functions F_1 and F_2. By minimizing F_1 and F_2, the optimal α_{10} and α_{20} can be determined to reduce the error.

In this study, the scenes were selected in the campus buildings, with a railing emulating a chessboard as the target object. Six sets of feature points were taken on the railing. For each set of feature points, the distance between the first and second points was one railing segment, and the distance between the second and third points was two railing segments, in order to emulate the un-usability of chessboard in real scenes. The verification results are shown in Fig. 8. By averaging the values from the six sets, we obtained $\alpha_{10} = 13.545$ and $\alpha_{20} = -7.423$.

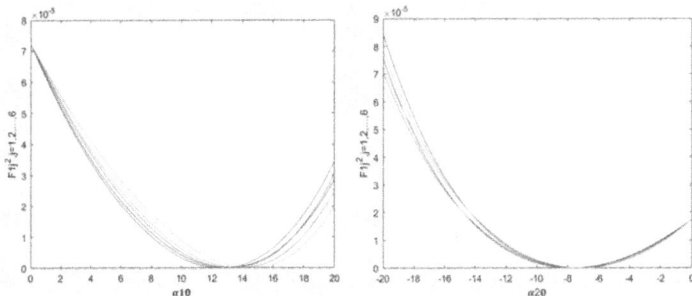

Fig. 8. Curves of Squared Residuals for α_{10} and α_{20} (Verification with Targets at Unequal Distances)

3.2 Sensitivity Analysis

To quantify the impact of each parameter on system performance, we increased each parameter by 1% above their original values, and observed the effect of these changes on the output depth z. However, since 1% increment of the initial values of β_1 and β_2 are too small, that is, an angle change of only fractions of a degree would be unrealistic, in this analysis, the increment for the two angle parameters (β_1 and β_2) are both set to 1 deg. to observe their effects. The results of this analysis are shown in Fig. 9.

This bar chart shows the change in output corresponding to each input variation. It can be observed that the angles between the camera optical center and the north direction (β_1 and β_2) have a significant impact on the output depth z, showing the sensitivity to the orientation relationship between the two cameras.

4 Experimental Results

The experimental testing scenarios were chosen to be a playground and a building environment, as shown in Fig. 10. A total station (THINRAD TTS-102R10X) was used to provide accurate 3D positioning information of the target objects under considerations. The total station measurements are considered as the ground truths since the accuracy is significantly higher than the stereo vision algorithm proposed in this research. The experimental setup parameters were $b_N = 0.0856$ m, $b_E = 2.8643$ m, $\beta_1 = 0.979°$, and $\beta_2 = 0.507°$. Five target object points were selected from the scenes, and the positioning results are shown in Table 2.

By analyzing the experimental results, we obtained the following observations:

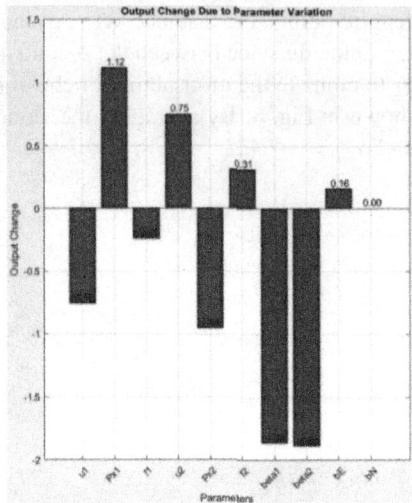

Fig. 9. Sensitivity Analysis Results (β_1 and β_2 Variation Set to 1 Degree)

Fig. 10. Experimental Feature Points of the Left and Right Cameras

Table 2. The Results of Experiment

point	True N Direction Value (m)	True E Direction Value (m)	Calculated N Direction Value (m)	Calculated E Direction Value (m)	N Direction Error (%)	E Direction Error (%)
P1	1.3571	0.6835	1.3736	0.6907	1.21	1.05
P2	1.3484	0.8053	1.3962	0.8002	3.55	0.62
P3	3.0815	−0.159	3.0213	−0.171	1.95	7.56
P4	4.3931	−0.903	4.3418	−0.9583	0.16	7.56
P5	1.3574	0.8968	1.4164	0.9211	4.35	2.70

Accuracy: The proposed 3D algorithm can accurately locate most feature points under different angles and distances, with acceptable errors typically in the N direction (depth). However, the error in the E direction is larger, especially at point P5 in the experiments, where the errors reached 14.57% and 18.26%.

Sources of Error: Errors may arise from various sources, including camera calibration errors, image processing errors, and alignment between the total station and the camera system. In the third experiment, the error at feature point P5 significantly increased due to the longer distance, which may imply the limitations of long-distance measurement with current camera resolutions and baseline settings.

Recommendations for Improvement: To further improve the system's accuracy, it is recommended to increase the baseline length for long-distance measurements to enhance stability. Additionally, using higher precision cameras and stricter calibration methods can help reduce errors.

5 Conclusion and Future Work

This study developed a novel virtual variable baseline stereo vision positioning algorithm to address the limitations of traditional fixed baseline technology in complex environments. We reviewed existing technologies and designed an algorithmic system for potential UAV platforms, potential applications such as bridge and building defect inspection are of interest. After conducting camera calibration and algorithm validation in the lab, we successfully constructed an algorithm that could dynamically adjust the camera baseline to meet various inspection needs. Experimental results demonstrated improved feature point distance measurement accuracy in irregular environments. Experimental tests in real-life scenes showed the system accuracy by comparing with the total station measurements, with N-direction positioning errors generally within acceptable ranges.

Future work will focus on optimizing system performance by enhancing camera calibration and stereo vision algorithms to reduce errors and increase stability and accuracy. We plan to apply the technology to practical scenarios such as traffic monitoring, smart agriculture, and autonomous driving to further test its performance and feasibility. These improvements will allow the technology to provide efficient and precise 3D positioning solutions across a wider range of fields.

References

1. Naik, R.K., Nair, B.B.: Improving GPS based distance measurement accuracy using Machine Learning: an empirical study. In *2019 International Conference on Communication and Electronics Systems (ICCES)*, Coimbatore, India, pp. 1294–1299 (2019)
2. Liu, J., Sun, Q., Fan, Z., Jia, Y.: TOF Lidar development in autonomous vehicle. In *2018 IEEE 3rd Optoelectronics Global Conference (OGC)*, Shenzhen, China, pp. 185–190 (2018)
3. Ishibashi, S.: The stereo vision system for an underwater vehicle. *OCEANS 2009-EUROPE*, Bremen, Germany, pp. 1–6 (2009)
4. Mustafah, Y.M., Azman, A.W., Akbar, F.: Indoor UAV positioning using stereo vision sensor. Proc. Eng. **41**, 575–579 (2012)
5. Baek, H.-S., Choi, J.-M., Lee, B.-S.: Improvement of distance measurement algorithm on stereo vision system(SVS). In *2010 Proceedings of the 5th International Conference on Ubiquitous Information Technologies and Applications*, Sanya, China, pp. 1–3 (2010)

6. Hsu, T.S., Wang, T.C.: An improvement stereo vision images processing for object distance measurement. Int. J. Autom. Smart Technol. **5**(2), 85–90 (2015)
7. Dodge, D., Yilmaz, M.: Convex vision-based negative obstacle detection framework for autonomous vehicles. IEEE Trans. Intell. Vehicles, **8**(1), 778–789 (2023)
8. Nakabo, Y., et al.: Variable baseline stereo tracking vision system using high-speed linear slider. *Proceedings of the 2005 IEEE International Conference on Robotics and Automation*, Barcelona, Spain, pp. 1567–1572 (2005)
9. Zhu, J., Li, W., Da, L.: A variable baseline distance stereo vision system for fire localization based on sub-pixel detection. *2019 9th International Conference on Fire Science and Fire Protection Engineering (ICFSFPE)*, Chengdu, China, pp. 1–9 (2019)
10. Sanket, N.J., Singh, C.D., Asthana, V., Fermüller, C., Aloimonos, Y.: MorphEyes: variable baseline stereo for quadrotor navigation. *2021 IEEE International Conference on Robotics and Automation (ICRA)*, Xi'an, China, pp. 413–419 (2021)
11. Hsia, K.-H., Lienn, S.-F., Su, J.-P.: Height estimation via stereo vision system for unmanned helicopter autonomous landing. *2010 International Symposium on Computer, Communication, Control and Automation (3CA)*, Tainan, Taiwan, pp. 257–260 (2010)
12. Kowalczuk, Z., Merta, T.: Stereo image visualization for VISROBOT system. *2013 18th International Conference on Methods and Models in Automation and Robotics (MMAR)*, Miedzyzdroje, Poland, pp. 794–799 (2013)
13. Kowalczuk, Z., Merta, T.: Three-dimensional mapping for data collected using variable stereo baseline. *2016 21st International Conference on Methods and Models in Automation and Robotics (MMAR)*, Miedzyzdroje, Poland, pp. 1082–1087 (2016)
14. Karrer, M., Chli, M.: Distributed variable-baseline stereo SLAM from two UAVs. *2021 IEEE International Conference on Robotics and Automation (ICRA)*, Xi'an, China, pp. 82–88 (2021)
15. Xu, D., Zhang, D., Liu, X., Ma, L.: A calibration and 3-D measurement method for an active vision system with symmetric yawing cameras. IEEE Trans. Instrument. Measure., **70**, 1–13, Art no. 5012013 (2021)

A Conceptual Framework to Operational Design Domain (ODD)-Based Scenario Generation for Technical Evaluation Obstacle Detection in Automated Shuttle Bus

Kun Gao[1(✉)], Ulrike Weinrich[1], Thomas Riemer[1], and Hans-Christian Reuss[2]

[1] Research Institute of Automotive Engineering and Vehicle Engines Stuttgart, 70569 Stuttgart, Germany
Kun.gao@fkfs.de
[2] Institute of Automotive Engineering (IFS), University of Stuttgart, 70569 Stuttgart, Germany

Abstract. In response to increasing public transportation demands and environmental concerns, the RABus project explores the integration of automated shuttle buses (SAE Level 4) in Friedrichshafen and Mannheim to enhance urban and rural mobility. This research focuses on obstacle detection and avoidance, emphasizing the role of AI in improving automated driving safety. However, limited training data introduces AI uncertainty, affecting system reliability. Traditional safety assessments are insufficient for complex scenarios, leading to the adoption of scenario-based validation methods. This paper presents a scenario description method that combines the Operational Design Domain (ODD) and traffic behavior. By analyzing the ODD and hazard factors, a systematic approach to creating a test scenario catalog is proposed to evaluate the safety and reliability of obstacle detection systems in automated shuttle buses, addressing the need for a robust testing framework.

Keywords: scenario generation · automated shuttle bus · scenario-based testing

1 Introduction

In times when demands on public transport are growing and the environmental impact of traffic is becoming a focus, automated shuttle buses have emerged as a promising solution [1]. Their integration has the potential to improve the quality of life, make traffic more efficient, and reduce environmental impacts. Within the RABus project (Real Laboratory for Automated Bus Operation in Public Transport in Urban and Rural Areas), the deployment of automated shuttle buses (SAE Level 4) in public transport in Friedrichshafen and Mannheim is being researched. Obstacle detection and avoidance are investigated as one of the central, safety-critical driving functions for automated vehicles. The obstacle detection system's task is to precisely and timely detect obstacles that could impair driving safety while avoiding false detections (ghost obstacles) and system false alarms.

Neural networks and AI systems have rapidly evolved, showing strong performance in obstacle detection for automated driving. However, as AI systems rely on limited training data, unpredictability and unreliability may occur in their prediction and decision-making processes due to factors such as data and algorithms [2], leading to AI uncertainty [3]. This uncertainty arises because AI systems are based on limited and imperfect data and cannot fully simulate the complex, dynamic real world. Therefore, validating the safety and reliability of AI systems poses a challenge.

Traditional safety assessment methods, based on miles of safe driving, often fail in rare and extreme scenarios. Both academic and industrial research has increasingly focused on scenario-based validation methods [4–6] to assess the safety of automated driving systems in specific scenarios. However, the scenarios that a vehicle encounters are unlimited, and there is no universally valid method for defining test scenarios. Thus, how to systematically describe scenarios and create a reliable scenario catalog remains unresolved.

Appropriate testing environments are crucial for obtaining reliable safety results for the obstacle detection system. Testing should consider the system's ODD, provide comprehensive coverage of different operating conditions, and include a reasonable mix of normal and critical scenarios.

Currently, the field lacks universally accepted methodologies for defining test scenarios and international standards for ADS validation. To address this gap, this paper explores a test scenario generation methodology based on the ODD, employing a systematic approach for assessing the safety and reliability of obstacle detection systems in Level 4 automated shuttle buses. The paper begins by reviewing existing scenario generation methods and subsequently proposes a novel ODD-based scenario generation methodology. Using the current toolchain, scenario instances are generated within the CARLA Simulator [10].

In the RABus project, from the perspective of end users (automated shuttle bus operators), we use a scenario-based validation method to evaluate the obstacle detection system's safety and reliability. Scenario validation begins with scenario definition and generation. By executing automated driving tasks in these scenarios, key parameters of the vehicle and obstacle detection system can be obtained for an automated driving assessment. Our project results are presented across three papers. In our previous paper [7], we studied the assessment methods for automated driving functions using key performance indicators of the obstacle detection system. This paper focuses on scenario description and generation. Finally, Sect. 3 will discuss data requirements for different test levels (grey-box/black-box testing). It is important to note that the evaluation is conducted from the perspective of end users, rather than vehicle manufacturers or passengers. Figure 1 shows the relationship between the three papers in the RABus project.

A Conceptual Framework to Operational Design   65

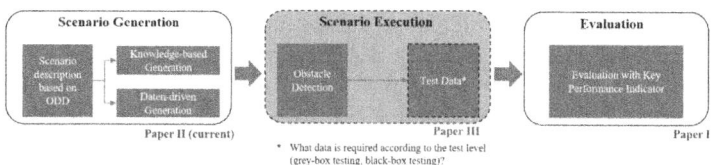

Fig. 1. Relationship between the three papers in the RABus project

2 State of the Art Test Scenario Generation

Before discussing test scenario generation methods, it is important to review the fundamental terminology and current state of development in this field. This paper focuses on evaluating the safety and reliability of obstacle detection systems in automated shuttle buses to support their promotion in Baden-Württemberg. Therefore, this chapter reviews relevant regulations, standards, and state-of-the-art research methods for test scenario generation.

2.1 Scenario and ODD

According to ISO 21448 [8], expected functional safety divides driving scenarios into four quadrants: "known not hazardous", "known hazardous", "unknown hazardous", and "unknown not hazardous". The goal is to minimize the extent of categories 2 (known hazardous) and 3 (unknown hazardous) scenarios, as shown in Fig. 2. Therefore, research focuses on discovering unknown scenarios to reduce the scope of categories 2 and 3.

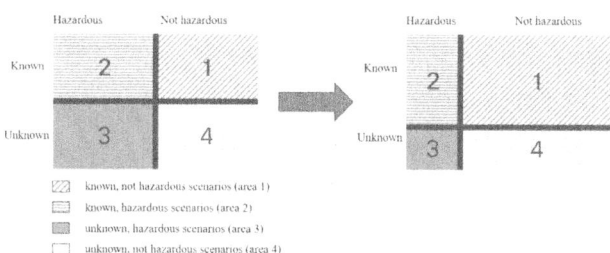

Fig. 2. Evolution of the scenario categories from ISO 21448 [8]

Scenarios simulate real-world or hypothetical driving situations to test and validate the response of automated driving systems under complex conditions. For evaluating perception systems, it is necessary to conduct thorough scenario testing, including both nominal and critical scenarios. A critical scenario involves one or more risk factors that increase the probability or severity of injury [9]. Scenarios can be categorized according to their level of abstraction into Functional Scenario → Abstraction Scenario → Logical Scenario → Concrete Scenario [11]. To facilitate ADS testing within virtual environments, scenarios must be transformed into a machine-readable format. ASAM Open-SCENARIO [12] provides a methodology for describing virtual scenes for structured depiction and sharing.

SAE J3016 [13] defines ODD as the "operating conditions under which a given driving automation system is specifically designed to function, including environmental, geographical, and time-of-day restrictions". A clearly defined ODD ensures that the vehicle operates within its safe operational range, reducing risks and increasing public confidence in automated driving. As automated driving technology advances, the ODD definition should be updated and expanded.

The Target Operational Domain (TOD) can extend beyond the ODD. Therefore, to assess safety and reliability, tests should cover the system's performance within and beyond the ODD, including how it handles operations outside the ODD (e.g., rejecting ADS activation). An instantiation of the ODD, combined with desired behaviors and traffic participant descriptions, provides a scenario definition for ADS. The design of scenarios and ODDs influences each other. The detailed definition of an ODD helps determines the scenarios to be tested, while scenario testing results help adjust and optimize ODD boundaries. Therefore, generating test scenarios involves iterating the ODD description and traffic participant behavior to enhance ADS safety and reliability.

2.2 Scenario-Based Testing

Scenario-based testing methods can reduce driving mileage to a subset of specific scenarios, allowing for targeted evaluation of ADS performance and behavior. Test scenarios may be virtual, which helps reduce development costs and increase efficiency, or real, as real-world tests are essential for reliable safety conclusions. Increasing research focuses on scenario testing. Based on literature studies [14–16], the general process of scenario-based testing is outlined (Fig. 3).

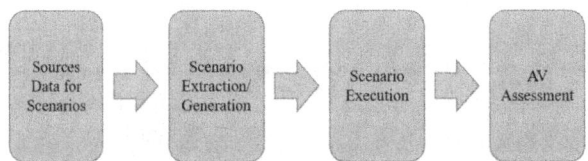

Fig. 3. General process of scenario-based testing [14]

The ISO 34502 [9] proposes a test framework for scenario-based safety assessment for highly automated driving on highways. It details the process of scenario-based safety assessment during ADS development, suitable for highways with limited access but not for automated shuttle buses (SAE Level 4) in urban and rural environments.

Scenario generation provides specific test cases, ensuring coverage within the ODD while considering dangerous and rare scenarios. Zhang et al. [17] summarize scenario generation methods, which can be broadly categorized into knowledge-based and data-driven approaches. Knowledge-based methods use expert knowledge, while data-driven methods derive scenarios from datasets or real scenes [18].

The Safety Pool Scenario Database [19] is the world's largest for automated vehicle scenarios, providing over 250,000 scenarios, including those for UN R157 ALKS [20], using ASAM OpenDRIVE [21] and OpenSCENARIO [12] formats for virtual validation.

However, Safety Pool scenarios focus on whole-vehicle safety, not individual subsystems such as perception. Evaluating obstacle detection for shuttle buses requires unique scenarios from a perception perspective. Moreover, Safety Pool data may not always fit the specific ODD definitions. A method focused on the ODD is needed for reliable safety test evidence.

This paper proposes a framework for scenario generation, based on requirements for obstacle detection systems and ODD definitions, to provide a scenario test catalog for SAE Level 4 shuttle buses.

3 Test Scenario Generation Framework

3.1 Scenario Description Based on ODD

In the RABus project, SAE Level 4 automated shuttle buses operate in Mannheim and Friedrichshafen. Specific urban driving scenarios are considered for both cities, with a separate ODD for each location. Ensuring the integrity of automated vehicle tests involves designing scenarios that cover all aspects of the ODD or ensuring the system can recognize scenarios beyond it.

Scenarios are "test cases" for ADS, illustrating the behavior and motion of dynamic elements within the ODD [22]. For vehicle operation, the behavior of the subject vehicle and traffic participants within the ODD must be considered.

Test scenarios can be described as:

Scenario = ODD + Subject vehicle behavior + Dynamic elements Motion + Dynamic elements Location.

ODDs are divided into:

- Scenery.
- Environmental conditions.
- Dynamic elements.

The subject vehicle's behavior can be moving straight, turning left, or right. Dynamic elements are modeled using a 9-grid system in Fig. 4 and have independent states, such as cut in, cut out, moving towards, away, or stopping, based on Zhang et al. [23].

Fig. 4. Position of dynamic elements

Scenarios are defined across six levels:

1. Scenery.
2. Environmental conditions.
3. Dynamic elements.
4. Behavior of the subject vehicle.
5. Position of dynamic elements.
6. Motion of dynamic elements.

3.2 Scenario Generation Framework

Given the variety of driving scenarios and the dependence of AI-based obstacle detection on training data, assessing safety and reliability, particularly in rare and hazardous scenarios, is crucial. Scenario generation must consider both critical and potentially dangerous cases.

Even within a defined ODD, real-world scenarios are nearly infinite. The scenario catalog must therefore include data from various sources, such as natural driving, regulatory, and critical scenarios.

Two common scenario generation methods are knowledge-based and data-driven. Knowledge-based methods rely on expert knowledge and standards to create intuitive scenarios but may underestimate risk factors. Data-driven methods use real or virtual driving data and algorithms like reinforcement learning to efficiently generate scenarios, uncovering weaknesses that may be missed otherwise, but they may lack interpretability.

Combining both methods addresses their shortcomings and creates a comprehensive scenario catalog.

Figure 5 illustrates the proposed framework for scenario generation, integrating both knowledge-based and data-driven approaches:

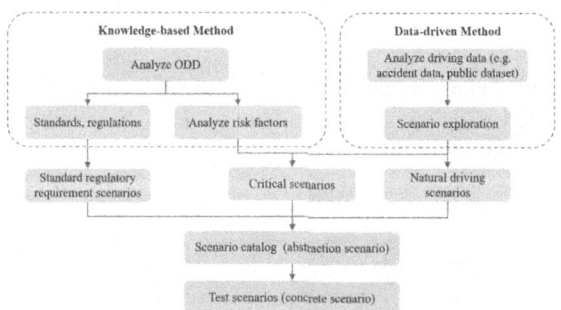

Fig. 5. Framework of scenario generation

Combined Scenario Generation Method. The scenario generation process starts with analyzing the ODD, referencing standards and regulations, and analyzing risk factors. This leads to the creation of regulatory, critical, and natural driving scenarios using both knowledge-based and data-driven approaches.

1. Knowledge-based Method: This approach relies on expert knowledge, standards (e.g., NCAP), and ODD analysis to generate standard regulatory scenarios and identify critical scenarios. The four key risk factors considered during scenario analysis include:
   - Scenarios due to the characteristics of the vehicle/sensors themselves and their installation.
   - Scenarios due to the characteristics of the environment.
   - Scenarios due to the characteristics of the detected targets.
   - Scenarios due to occlusions.
2. Data-driven Method: Driving data, such as accident records, public datasets (e.g., KITTI [24], Safety Pool [19], NDOT Crash Data [25]), are analyzed using algorithms like K-Means clustering, optimization, or reinforcement learning to identify potential hazardous scenarios, explore scenario space, and determine system limitations.

The outcomes of both methods are integrated into a scenario catalog that contains abstraction scenarios. These abstraction scenarios are then refined with specific parameters to generate concrete test scenarios that can be used for evaluation in virtual or real testing environments.

Finally, the abstraction scenarios are refined by specifying all concrete parameters, such as vehicle speed, vehicle position, weather conditions, and lighting conditions. These concrete scenarios can then be directly converted into definitive test scenarios in a virtual environment or a test track environment for actual testing execution.

4 Implementation and Discussion: Scenario Generation in Virtual Environment

4.1 Feasibility Results of the Method

Using the scenario generation methods described, a large number of scenario test catalogs can be generated for evaluating the performance of automated driving obstacle detection systems. Scenarios can be real or virtual.

This chapter introduces examples of generating virtual scenarios through ODD analysis and potential risk analysis, demonstrating the feasibility of using ASAM OpenDRIVE, OpenSCENARIO, and CARLA Simulator for test scenario generation. CARLA Simulator, based on Unreal Engine, can model scenarios according to developers' requirements. Currently, scenarios are generated using built-in CARLA maps, but future developments may include customized scene maps aligned with the ODD of automated shuttle buses.

Case 1: Potential hazardous scenario in residential areas. In the RABus project, the safety of automated buses in residential areas is studied. For this example, hazardous scenarios in residential areas are generated using a knowledge-based approach. Typical roads are low speed with roadside parking. A functional scenario is: A pedestrian suddenly appears from a blind spot on a residential road, potentially leading to a hazardous event.

First, define the scenery, environmental conditions, and dynamic elements in the ODD, where:

- Scenery: a residential area, low-speed road, straight road.
- Environmental conditions: daytime, clear.
- Dynamic elements include the subject vehicle, pedestrians, and cars stopped at the roadside.

Next, define the behavior and states of the dynamic elements within the scenario:

- Behavior of the subject vehicle: maintaining a constant speed and moving straight ahead.
- Position of dynamic elements: all are in front of and to the right of the subject vehicle (position number 3 in Fig. 4).
- Motion of dynamic elements: the pedestrian is cutting in, and the stopped vehicles at the roadside are stationary.

In CARLA Simulator, this scenario can lead to the ego vehicle failing to recognize a pedestrian due to occlusion from parked cars, presenting a challenge to the obstacle detection system (Fig. 6).

Fig. 6. Potential hazardous scene in the virtual environment (occlusions)

Case 2: Potential hazardous scenarios on highway. In Case 2, we used a data-driven approach to generate potential hazardous scenarios. The data originated from the "City of Las Vegas - Traffic Crashes Data" provided by the Nevada Department of Transportation. This dataset includes details such as accident time, type of collision, weather, lighting, road conditions, and vehicle behavior. Although the records are manually collected and may be incomplete, they still provide valuable references for scenario generation.

We analyzed the ODD and filtered the dataset to focus on highway environments. Relevant scenario parameters (weather, lighting, road status, collision type, and vehicle behavior) were subjected to K-Means clustering to identify hazardous scenarios. A typical scenario involves a rear-end collision on a dual-lane highway at night during rainy conditions.

The scenario description is as follows:

First, define the ODD:

- Scenery: highway, with high speed, straight road.

- Environmental conditions: nighttime, raining.
- Dynamic elements include the subject vehicle and another car.

Next, define the behavior and states of the dynamic elements:

- Behavior of subject vehicle: moving straight ahead.
- Position of dynamic elements: directly in front of the subject vehicle (position number 2 in Fig. 4).
- Motion of dynamic elements: moving towards the subject vehicle.

In CARLA Simulator, this scenario can be generated as shown in Fig. 7. On a rainy night, the vehicle in front moves slowly. If the subject vehicle fails to decelerate, a collision occurs.

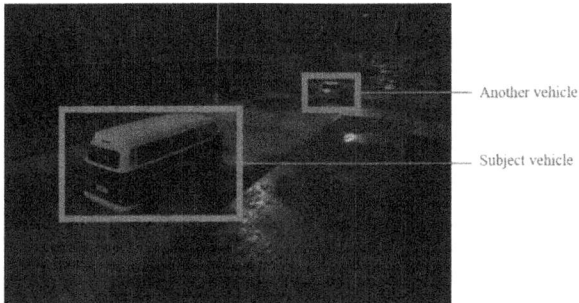

Fig. 7. Potential hazardous Scene in Virtual Environment (Rainy Night)

Here we demonstrate the extraction of typical accident scenarios from accident data using a data-driven approach, and the generation of these scenarios in a virtual environment to validate the reliability and safety of obstacle detection systems.

4.2 Limitation of the Method

Although the two cases demonstrate the feasibility of the proposed method, limitations remain.

A key issue with scenario generation is ensuring the generated scenarios are sufficient for validating safety. ISO 21448 aims to minimize, not eliminate, unknown hazards. By considering the ODD and integrating knowledge-based and data-driven methods, diverse scenarios can be generated. However, fully defining the ODD is still challenging due to the lack of practical guidelines. Therefore, combining other methods (e.g., accident data and road rules) is necessary to enrich the scenario set, and the ODD can be updated continuously based on test results.

If the ODD is fully defined, all scenarios can be represented through parameters, allowing the calculation of scenario coverage within the ODD and supporting the completeness of testing. This is a focus for the next phase of the project.

5 Conclusion and Outlook

This work, part of the RABus project, aims to evaluate the safety and reliability of obstacle detection systems in automated buses. Scenario testing, using analysis of the ODD and hazard factors, effectively generates test scenarios. We proposed a combined knowledge-based and data-driven method, generating virtual scenarios using CARLA Simulator and tools like ASAM OpenDRIVE and OpenSCENARIO.

Future work will refine the ODD to cover more scenarios, such as complex urban environments, diverse weather, and varying traffic conditions. A richer scenario catalog will provide a comprehensive assessment of the obstacle detection system. For data-driven methods, aligning scenarios with the ODD is crucial.

Virtual simulators offer a controlled setting, but integrating real-world data will further enhance reliability. Future research will explore combining simulator and real-world testing for a holistic evaluation.

This paper focuses on scenario generation. The next step involves executing these scenarios and assessing driving functions. Determining the minimum data required for evaluation and integrating it with performance metrics (e.g., detection rates) is essential. We will continue developing an evaluation system from the end user's perspective.

Acknowledgment. This work was performed in the project "RABus—Reallabor für den Automatisierten Busbetrieb im ÖPNV in der Stadt und auf dem Land", which is funded by the Ministry of Transport Baden-Württemberg. We would like to thank the project AI2ISO, which was funded by the Ministry of Economic Affairs, Labor and Tourism Baden-Württemberg, for its support in the scenario-based validation approach.

References

1. Riener, A., Appel, A., Dorner, W., Huber, T., Kolb, J. C. und Wagner, H., Hrsg.: Autonome Shuttlebusse im ÖPNV: Analysen und Bewertungen zum Fallbeispiel Bad Birnbach aus technischer, gesellschaftlicher und planerischer Sicht. Berlin, Heidelberg: Springer Berlin Heidelberg (2020)
2. Abrecht, S., Hirsch, A., Raafatnia, S. und Woehrle, M., "Deep Learning Safety Concerns in Automated Driving Perception," 2023. https://doi.org/10.48550/arXiv.2309.03774
3. Tomsett, R., Preece, A., Braines, D., Cerutti, F., Chakraborty, S., Srivastava, M., Pearson, G. und Kaplan, L.: Rapid trust calibration through interpretable and uncertainty-aware AI. Patterns, Bd. 1, Nr. 4, 100049 (2020)
4. PEGASUS-Gesamtmethode, 2019. [Online]. Available: https://www.pegasusprojekt.de/files/tmpl/Pegasus-Abschlussveranstaltung/PEGASUS-Gesamtmethode.pdf. [visited on April.17.2024]
5. Bagschik, G.: Systematischer Einsatz von Szenarien für die Absicherung automatisierter Fahrzeuge am Beispiel deutscher Autobahnen. Universitätsbibliothek Braunschweig (2022)
6. Sun, J., Zhang, H., Zhou, H., Yu, R., Tian, Y.: Scenario-based test automation for highly automated vehicles: a review and paving the way for systematic safety assurance. IEEE Trans. Intell. Transp. Syst. 23(9), 14088–14103 (2022)
7. Gao, K., Weinrich, U., Riemer, T., Reuss, H.: Technical evaluation of the obstacle detection for automated shuttle buses. SAE Technical Paper 2023-01-1227, 2023, https://doi.org/10.4271/2023-01-1227

8. ISO 21448:2022 Road vehicles—Safety of the intended functionality (2022)
9. ISO 34502:2022 Road vehicles Test scenarios for automated driving systems Scenario based safety evaluation framework (2022)
10. Dosovitskiy, A., Ros, G., Codevilla, F., Lopez, A., Koltun, V.: CARLA: An Open Urban Driving Simulator. PMLR, pp. 1–16 (2017)
11. ISO 34501:2022 Road vehicles Test scenarios for automated driving systems Vocabulary (2022)
12. ASAM OpenSCENARIO2.0.0, 2022
13. SAE J3016 Taxonomy and Definitions for Terms Related to Driving Automation Systems for On-Road Motor Vehicles (2021)
14. Riedmaier, S., Ponn, T., Ludwig, D., Schick, B., Diermeyer, F.: Survey on scenario-based safety assessment of automated vehicles. IEEE Access **8**, 87456–87477 (2020)
15. Schuldt, F.: Ein Beitrag für den methodischen Test von automatisierten Fahrfunktionen mit Hilfe von virtuellen Umgebungen. Universitätsbibliothek Braunschweig (2017). https://doi.org/10.24355/DBBS.084-201704241210
16. Zhang, X., et al.: Finding critical scenarios for automated driving systems: a systematic mapping study. IEEE Trans. Software Eng. **49**(3), 991–1026 (2023)
17. Zhang, X., Khastgir, S., Asgari, H., Jennings, P.: Test framework for automatic test case generation and execution aimed at developing trustworthy AVs from both verifiability and certifiability aspects, pp. 312–319 (2021)
18. Menzel, T.: From functional to logical scenarios: detailing a keyword-based scenario description for execution in a simulation environment. 2019 IEEE Intelligent Vehicles Symposium (IV), pp. 2383–2390 (2019)
19. Safety Pool—Powered by Deepen AI and WMG University of Warwick
20. UN Regulation No. 157—Uniform provisions concerning the approval of vehicles with regard to Automated Lane Keeping Systems, [Online]. Available: https://op.europa.eu/en/publication-detail/-/publication/36fd3041-807a-11eb-9ac9-01aa75ed71a1. [visited on April.17.2024]
21. ASAM OpenDRIVE BS 1.8.0 Specification (2023)
22. Zhang, X., Khastgir, S., Tiele, J.-K., Takenaka, K., Hayakawa, T., Jennings, P.: ODD and behavior based scenario generation for automated driving systems. IEEE Access **12**, 10652–10663 (2024)
23. Zhang, X., Khastgir, S., Jennings, P.: Scenario description language for automated driving systems: a two level abstraction approach. 2020 IEEE International Conference on Systems, Man, and Cybernetics (SMC), Toronto, ON, Canada, pp. 973–980 (2020), https://doi.org/10.1109/SMC42975.2020.9283417
24. Geiger, A., Lenz, P., Stiller, C., Urtasun, R.: Vision meets robotics: the KITTI dataset. Int. J. Robot. Res. **32**(11), 1231–1237 (2013). https://doi.org/10.1177/0278364913491297
25. Nevada Department of Transportation. Traffic Crash Data. [Online]. Available: https://www.dot.nv.gov/safety/traffic-crash-data. [visited on April.16.2024]

Precise Driver's Drowsiness Detection Using a Combination of Proven Methods with a Single Layer Neural Network

Ghazal Abdolbaghi and Alireza Yazdizadeh(✉)

Electrical Engineering Department, Shahid Beheshti University, Tehran, Iran
g.abdolbaghi@mail.sbu.ac.ir, a_yazdizadeh@sbu.ac.ir

Abstract. Drowsiness reduces reaction time, leading to potentially fatal accidents. Most existing research focuses on only one symptom of drowsiness, which often results in false alarms. This paper introduces a novel approach for real-time drowsiness detection. The proposed method employs four deep learning architectures based on convolutional neural networks: AlexNet for environmental feature extraction, ResNet50V2 for hand gesture recognition, VGG-FaceNet for facial feature extraction, and FlowImageNet for behavioral feature analysis. To improve the voting system, the paper suggests using a layer with adjustable learning weights to reduce false results and enhance accuracy. Testing the static method on the NTHUDDD dataset and a custom dataset demonstrates that the proposed approach achieves higher accuracy (97.25% and 96.75% respectively) compared to existing methods.

Keywords: Convolutional Neural Network (CNN) · Drowsiness Detection · Static Method

1 Introduction

Drowsiness is a transitional state between consciousness and wakefulness, leading to slower reaction times and memory loss [1]. According to the US National Highway Traffic Safety Administration, drowsiness is responsible for approximately 100,000 traffic accidents annually worldwide, resulting in over 1,500 fatalities and more than 70,000 injuries [2]. This issue is prevalent globally.

Drowsiness significantly impacts driving safety, and drowsiness detection systems are crucial as they provide warnings before drowsiness reaches a critical and dangerous level.

Machine Learning (ML) has been utilized in many fields, providing significant advantages such as high accuracy, adaptability to diverse datasets, effectiveness with both small and large datasets, and scalability in terms of data volume and computational power [3]. In drowsiness detection systems, similar to other domains, machine learning techniques have been implemented. Various models, including Convolutional Neural Networks (CNNs), Long Short-Term Memory (LSTM) networks, and Recurrent Neural Networks

(RNNs), are employed to train on datasets and learn crucial features associated with drowsiness.

Datasets used in the development of driver drowsiness detection systems can be categorized into three main types: (1) vehicle-based, such as steering wheel angle, lateral and longitudinal acceleration, (2) facial-based, including rapid blinking and yawning, and (3) biological signals. Among these, biological signals are highly accurate but intrusive, as they require sensors to be attached to the driver, which can be inconvenient and distracting while driving.

Despite numerous approaches proposed by researchers for identifying drowsiness, current detection methods still face substantial challenges. Focusing on a single drowsiness symptom often fails to yield reliable and accurate results. Also, as static structures in machine learning do not have any memory, they are unable to detect drowsiness during a period of time. To address these issues, this paper introduces a novel approach for detecting drowsiness.

The primary contributions of this paper are as follows:

1. This paper employs four deep learning structures based on convolutional neural networks: AlexNet for environmental feature extraction, ResNet50V2 for hand gesture recognition, VGG-FaceNet for facial feature extraction, and FlowImageNet for behavioral feature analysis. Each structure is pretrained using transfer learning, with additional layers added to enhance results, thus leveraging the combined advantages of an integrated framework.
2. Instead of using a simple average voting mechanism to combine the four structures, this paper achieves better results by adding a layer to the outputs of these structures and adjusting the corresponding weights. This approach efficiently minimizes error and enhances accuracy.

The remainder of the paper is organized as follows: Sect. 2 reviews related work. Section 3 provides an overview of the proposed method within the entire detection framework. Section 4 presents experimental results for the training and testing processes of the proposed method. Finally, Sect. 5 concludes the paper.

2 Related Works

In facial-based approaches, two common strategies are used to detect drowsiness: facial landmarks and machine learning techniques. Techniques based on facial landmarks usually involve calculating the Eye Aspect Ratio (EAR) in the eye region and the Mouth Aspect Ratio (MAR) in the mouth region to detect yawning and predict drowsiness.

In [4], a real-time drowsiness detection system was developed using facial features, specifically the Eye Aspect Ratio (EAR) and Mouth Aspect Ratio (MAR). Videos were recorded in a real driving environment using a camera, then converted into frames. Facial landmarks were extracted from these images using a Haar Cascade classifier and the Dlib library, along with a logistic regression classifier algorithm. The EAR and MAR were calculated from these facial landmarks. By combining Dlib with the logistic regression model, they achieved an accuracy of 92%. However, using the Haar Cascade classifier with the logistic regression model resulted in a lower accuracy of approximately 86%, primarily due to the Haar Cascade classifier's weakness in detecting eyes. A drawback of this method is that using MAR and EAR metrics for drowsiness detection may not achieve high accuracy, as the threshold values can vary from person to person.

In [5], the focus for drowsiness detection is on the eyes. The researchers utilized the MediaPipe library [6] to extract the eye region and trained three deep learning architectures—ResNet50v2, InceptionV3, and VGG-16—to detect drowsiness from the eye region in real-time. The selection of the optimal region in this work is highly precise. The implementation process involves using MediaPipe's face mesh method to identify 468 three-dimensional facial landmarks in real-time to detect the face in the image. From these landmarks, four points are selected to form the optimal region. ResNet50v2 achieved the highest accuracy of approximately 99.71%. However, this work, like many others, predominantly emphasized the eye area and overlooked other potential indicators of drowsiness.

In [7], four deep learning architectures were used to develop a drowsiness detection system, employing a simple averaging method to combine their outputs. The Flow-ImageNet model was used to learn facial expressions and behavioral features, such as nodding and head movements, by processing deep optical flow images. AlexNet was utilized to extract environmental features, like the presence of glasses and variations in lighting conditions. The VGG-FaceNet model focused on extracting facial features, including the lips, cheeks, eyebrows, and pupil diameter. Lastly, ResNet was used to capture hand gestures related to yawning. A threshold value of 0.24 was set for averaging the outputs, with values above this threshold indicating drowsiness. The ensemble model achieved an accuracy of about 85%, which is relatively low and indicates that the model's performance might not be very precise, as it lacked sophisticated judgment.

Most studies concentrate on the eyes and mouth areas while overlooking other indicators of drowsiness, such as raised eyebrows, head movements, and drooping cheeks, which can affect prediction accuracy.

To evaluate driver drowsiness, we adopt the levels outlined in [8], which are based on [9] with some difference's scales. The authors defined drowsiness on a scale with five levels, as shown in Table 1. The levels of drowsiness are shown (see Fig. 1).

Table 1. Drowsiness levels [8].

Drowsiness Levels	Features
1-Not Drowsy	Line of sight moves fast and frequently. Facial movements are active, accompanied by body movements.
2-Slightly Drowsy	Line of sight moves slowly. Lips are open.
3-Moderately Drowsy	Blinks are slow and frequent. There are mouth movements.
4-Significantly Drowsy	There are blinks that seem conscious. Frequent yawning.
5-Extremely Drowsy	Eyelids close. Head tilts forward or falls backward.

3 Framework

This paper centers on Convolutional Neural Networks (CNNs), which are renowned for their capability to extract features directly from raw data and have shown fantastic performance. These networks are effective at building models that remain robust to transformations in the input data. The framework is shown in (see Fig. 2).

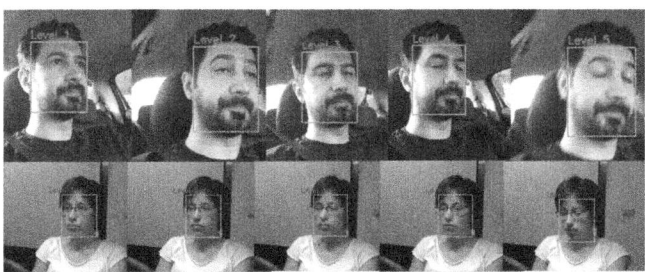

Fig. 1. Five levels of drowsiness in two datasets.

In this study, rather than relying on a simple voting method, we implement a smart weighting approach by adding a layer at the output of the four models. Initially, transfer learning is used to enhance the models' accuracy by adding additional layers. Instead of using a basic voting mechanism, a single-neuron layer is introduced, where weights for each model are adjusted according to static rules.

Next, we examine the layers added to each structure through the transfer learning method and present the weight adjustment rules for each branch in the static mode of the ensemble model. We choose Exponential Linear Unit (ELU) as the activation function as we get better result with it.

It is important to note that for all models, we utilize the sparse categorical entropy loss function, which is well-suited for multi-class classification and yields superior results compared to other loss functions. The formula for this loss function is provided in Eq. (1). The formula of this loss function is shown in Eq. (1) in which y_i is the true label and $\hat{y}_i$ is the predicted value.

$$H(\hat{y}, y) = -\frac{1}{N(num\_class)} \sum_{i=1}^{N} \log H(\hat{y}_i, y_i) \tag{1}$$

The layers illustrated, are incorporated into the pre-trained ResNet50V2, VGG-FaceNet, AlexNet, and FlowImageNet using the transfer learning method to achieve improved results (see Figs. 3, 4, 5, and 6). We calculate the weighting rules in static mode to create the smart weighting model. The static weighting rule for structures are presented in Eqs. (2), (3), (4), and (5) respectively. The static weighting rule is shown in Eq. (4) in which η is the learning rate, ReLU' is the derivation of activation function, and the X parameters are the inputs. The weights Δw_{kj}^l laws in the following equations

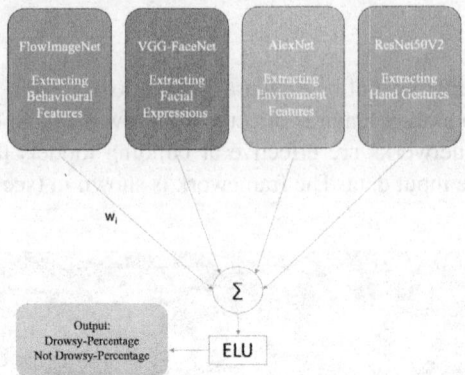

Fig. 2. Framework of the proposed method.

in this structure and others are based on these parameters.

$$\Delta w = \eta.[\sum_{j=1}^{7}(\frac{1}{7}\sum_{i=1}^{7}(I - \frac{e^{O_j}}{\sum_{k=1}^{c} e^{O_k}})...).ReLU'(X_{Conv,ijk}).X_{input,l} \quad (2)$$

$$\Delta w = \eta.(\sum_{l=1}^{5}(\sum_{j=1}^{5}(\frac{1}{5}\sum_{i=1}^{5}(I - \frac{e^{O_j}}{\sum_{k=1}^{c} e^{O_k}})).\frac{\partial O_j}{\partial X_{dense1,l}})...\frac{\partial X_{Conv12,ijk}}{\partial X_{Conv13,ijk}}).ReLU'(X_{Conv13,ijk}).X_{input,l} \quad (3)$$

$$\Delta w = \eta.(\sum_{j=1}^{18}(\frac{1}{18}\sum_{i=1}^{18}(I - \frac{e^{O_j}}{\sum_{k=1}^{c} e^{O_k}})).\frac{\partial O_j}{\partial X_{dense1,l}})...\frac{\partial X_{Maxpool,l}}{\partial X_{Conv5,ijk}}).ReLU'(X_{Conv5,ijk}).X_{Input,l} \quad (4)$$

$$\Delta w = \eta.(\sum_{l=1}^{3}(\sum_{j=1}^{3}(\frac{1}{3}\sum_{i=1}^{3}(I - \frac{e^{O_j}}{\sum_{k=1}^{c} e^{O_k}})).\frac{\partial O_j}{\partial X_{dropout,l}})...\frac{\partial X_{Maxpool,l}}{\partial X_{Conv5,ijk}}).$$

$$ReLU'(X_{Conv5,ijk}).X_{Input,l} \quad (5)$$

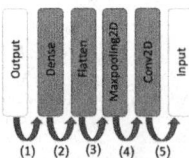

Fig. 3. The layers added to a pre-trained ResNet50V2 structure.

4 Experimental Results

We have evaluated our proposed method using various datasets, first NTHUDDD which is a well-known dataset in drowsiness and a custom dataset that we compiled (see Fig. 7).

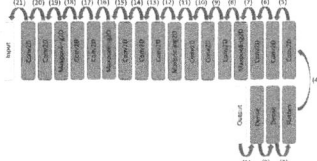

Fig. 4. The layers added to a pre-trained VGG-FaceNet structure.

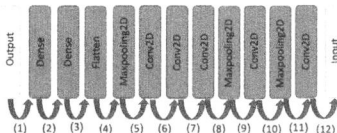

Fig. 5. The layers added to a pre-trained AlexNet structure.

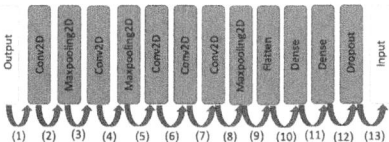

Fig. 6. The layers added to a pre-trained FlowImageNet structure.

Fig. 7. Frames of our dataset.

4.1 Evaluation Metrics

Given that our task involves classification, we utilized a confusion matrix to evaluate the performance of the proposed model. Specifically, we considered four metrics: accuracy, precision, recall, and F1-score.

Based on the calculations using the specified formulas, the results are summarized in Table 2. It is presented the confusion matrix, which includes four key parameters: true positive (TP), true negative (TN), false positive (FP), and false negative (FN), illustrating two different scenarios (see Fig. 8). It is displayed the accuracy and loss metrics throughout the training process for these scenarios (see Figs. 9 and 10). The primary goal during network training is to minimize the loss (whether in terms of error or cost) observed in the output when the training data is processed. As evidenced by the training process in both scenarios, the accuracy for both training and validation steadily increases, while the loss for training and validation consistently decreases, indicating that the proposed method is functioning effectively. As we see, the values of TP and TN in scenario 1 for predicting drowsy and not_drowsy states are 96% and 98%, respectively (see Fig. 8).

Table 2. The results of evaluation parameters on the proposed method with two datasets.

Num	Scenario	Label	Precision	Recall	F1-Score	Support	Accuracy
1	Static Structure on NTHUDDD	Drowsy	0.964	0.9803	0.971	1010	0.9725
		Not-Drowsy	0.981	0.9649	0.974	1153	
2	Static Structure on our Dataset	Drowsy	0.976	0.9595	0.9655	1010	0.9675
		Not-Drowsy	0.959	0.976	0.968	1153	

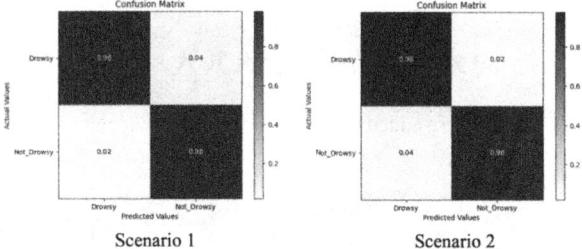

Scenario 1　　　　　　　　Scenario 2

Fig. 8. Confusion matrix for two scenarios.

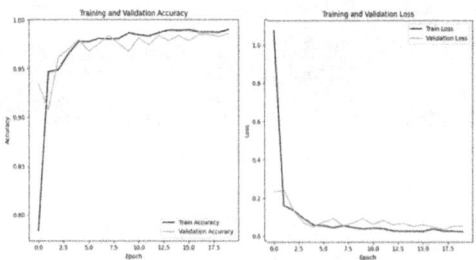

Fig. 9. Metrics for scenario1: Accuracy and Loss.

Tables 3 and 4 highlight the differences in accuracy between two approaches: simple voting and the static model. The results clearly show that the static model yields significantly better accuracy across both datasets. As we see, the implementation results of the proposed structure on these datasets (see Fig. 11). A notable advantage of this structure is its ability to accurately detect drowsiness even when yawning occurs with hands covering the face, as reflected by the impressive accuracy (see Fig. 12).

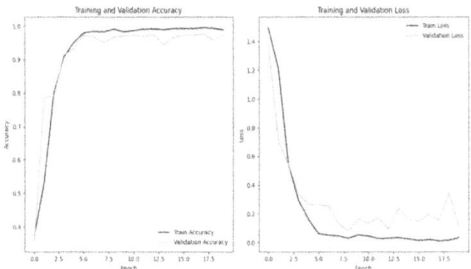

Fig. 10. Metrics for scenario2: Accuracy and Loss.

Table 3. Comparison of accuracy between two states on NTHUDDD dataset.

Structure	Accuracy	Voting	Static Structure
AlexNet	92.73	87.47	**97.25**
VGG-FaceNet	83.03		
ResNet50v2	84.12		
FlowImageNet	90		

Table 4. Comparison of accuracy between two states on our dataset.

Structure	Accuracy	Voting	Static Structure
AlexNet	92	86.49	**96.75**
VGG-FaceNet	81.93		
ResNet50v2	82.68		
FlowImageNet	89.35		

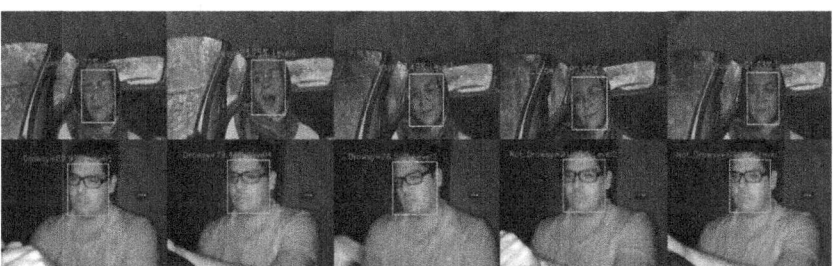

Fig. 11. Evaluation of proposed structure on two datasets.

5 Conclusion

This study utilized four convolutional neural network architectures—AlexNet, ResNet50V2, FlowImageNet, and VGG-FaceNet. The results from these networks were combined using static neural network approaches, leading to the final outcome. These frameworks were initially tested on the NTHUDDD dataset, which is widely recognized as a leading resource for sleepiness detection research. The results demonstrated that

Fig. 12. Detecting drowsiness even yawning with hand.

the proposed approach outperforms the individual methods. Following this, the method was evaluated on a dataset created specifically for this study, confirming its robustness across different datasets. Notably, the static technique delivered significantly better results compared to earlier methods, such as averaging. Finally, the static architecture achieved accuracies of 97.25% and 96.75% on the two datasets, respectively. In future work, the goal is to adapt this system for signal data, which differs from image data.

References

1. Awais, M., Badruddin, N., Drieberg, M.: A hybrid approach to detect driver drowsiness utilizing physiological signals to improve system performance and wearability. Sensors, **17**(9), 1991 (2017); Author, F., Author, S.: Title of a proceedings paper. In: Editor, F., Editor, S. (eds.) CONFERENCE 2016, LNCS, vol. 9999, pp. 1–13. Springer, Heidelberg (2016)
2. "NHTSA." Accessed: Apr. 16, 2024. [Online]. Available: https://www.nhtsa.gov/risky-driving/drowsy-driving
3. Sharma, N., Sharma, R., Jindal, N.: Machine learning and deep learning applications-a vision. Global Trans. Proc. **2**(1), 24–28 (2021)
4. You, F., Gong, Y., Tu, H., Liang, J., Wang, H.: A fatigue driving detection algorithm based on facial motion information entropy. J. Adv. Transp. **2020**, 1–17 (2020)
5. Florez, R., Palomino-Quispe, F., Coaquira-Castillo, R.J., Herrera-Levano, J.C., Paixão, T., Alvarez, A.B.: A CNN-based approach for driver drowsiness detection by real-time eye state identification. Appl. Sci. **13**(13), 7849 (2023)
6. Grishchenko, I., Ablavatski, A., Kartynnik, Y., Raveendran, K., Grundmann, M.: Attention mesh: high-fidelity face mesh prediction in real-time. arXiv preprint arXiv:2006.10962, 2020
7. Dua, M., Shakhshi, Singla, R., Raj, S., Jangra, A.: Deep CNN models-based ensemble approach to driver drowsiness detection. Neural Comput. Appl., **33**, 3155–3168 (2021)
8. Phan, A.-C., Trieu, T.-N., Phan, T.-C.: Driver drowsiness detection and smart alerting using deep learning and IoT. Internet of Things **22**, 100705 (2023)
9. Kitajima, H., Numata, N., Yamamoto, K., Goi, Y.: Prediction of automobile driver sleepiness. 1st report. Rating of sleepiness based on facial expression and examination of effective predictor indexes of sleepiness., Trans. Jpn. Soc. Mech. Eng. Ser. C **63**, 3059–3066 (1997)

FusionSis: Analysis and Evaluation Framework for Fusion Safety State of Connected and Automated Vehicles Under Cyber Attacks

Bowen Zheng[1], Shichun Yang[1], Weifeng Gong[1], Haoran Guang[1], Yi Shi[2], Mingjie Chen[1], and Yaoguang Cao[2](✉)

[1] School of Transportation Science and Engineering, Beihang University, Beijing 100191, China
[2] State Key Lab of Intelligent Transportation System, Beihang University, Beijing 100191, China
caoyaoguang@buaa.edu.cn

Abstract. As connected and autonomous vehicles (CAVs) become more integrated into complex networks and systems, which introduce both safety and security concerns. To address these challenges, it is crucial to develop a fusion safety approach that combines cybersecurity, functional safety, and the safety of the intended functionality. One of the key difficulties in this approach is the diversity of cyber attacks and their potential impact on CAVs. However, evaluating and analyzing the fusion safety in practical environments remains challenging. To address this, we present a simulation framework designed to assess CAV resilience against cyber attacks while maintaining safety operation. The framework can simulates vehicle's status both in physical and cyber field. A diverse cyber attacks with various driving scenarios are supported. Based on framework's current capability, we propose eight fusion safety evaluation metrics to measure CAV performance under attack conditions. These metrics encompass aspects of functional safety, data security and driving comfortableness. Three demonstration scenarios show the effectiveness of the proposed framework in assessing the fusion safety state of CAVs, contributing to the development of more secure and resilient fusion safety applications.

Keywords: Fusion Safety · Cybersecurity · Simulation Framework · Safety Evaluation · Connected and Autonomous Vehicle Security

1 Introduction

The development of Connected and Automated Vehicles (CAVs) has brought significant advancements in automotive technology. As CAVs become increasingly reliant on complex networks and systems, it has also introduced new safety and security challenges, such as function safety, safety of intended functionality (SOTIF), cybersecurity, and the combined problems in between [18]. These concerns have become the most significant challenge to the mass application of automotive driving functions.

To mitigate the risks and threats, various methods have been developed for vehicle engineers during product design and early development stage. Historically, automotive cybersecurity efforts have focused primarily on static and expert knowledge-based analysis, such as Threat Analysis and Risk Assessment (TARA) methodology from ISO/SAE 21434 and refined versions based on TARA [2]. A significant challenge in implementing these methodologies is the precise description and quantification of threats and risks. This is particularly difficult during the early stages of vehicle development and design, when there are no prototypes available for actual testing and no controllers for detailed analysis. However, the integration of advanced connectivity and automation technologies demands a more comprehensive approach that considers the interplay between functional safety, SOTIF, and cybersecurity. This need has given rise to the concept of fusion safety [18] (or operational safety from PAS 1880), which addresses the combined safety and security challenges in a holistic manner. In fusion safety concept, researchers claim that security and safety are indivisible. Cyber attacks can not only result in the loss of vehicle information data in the cyber field but may also degrade or compromise the functional safety performance of vehicles, preventing them from meeting their Operational Design Condition (ODC). This can cause harm to occupants or other road users in the physical field. Therefore, researchers recommend designing novel development processes and toolchains to introduce more analytical dimensions into TARA, aiming to fully understand these threats in the early stages of vehicle design.

Some researchers have noted the complexity of cyber attacks and their potential impact on autonomous vehicles, so they started to use simulators to test the impact of vehicle operation under in-vehicle bus network attacks [10] or sensor attacks [5,7,11]. With the advent of Vehicle-to-Everything (V2X) technology, the challenges mentioned in fusion safety have become more apparent. Researchers are increasingly concerned about dynamic cybersecurity risks to vehicle attacks, such as replay attacks [8] or spoofing attacks [19] at specific scenarios. These works aim to replicate real-world attack scenarios and evaluate the system's response. Because of the different focus of the research or the risks to be analyzed, these works often use different tools for detailed, targeted modeling. Despite these advancements, the diversity of cyber attacks and their potential consequences on CAVs are often underestimated by the scope of evaluation.

Recognizing the limitations of existing methods, this paper introduces FusionSis, a simulation framework designed to assess and analyze the fusion safety state of CAVs under cyber attacks. This framework provides multiple interfaces for initiating cyber attacks and also supports the deployment of cybersecurity protection algorithms similar to those used in real vehicles, all within a specified simulation scenario. The goal of FusionSis is to offer quantification and

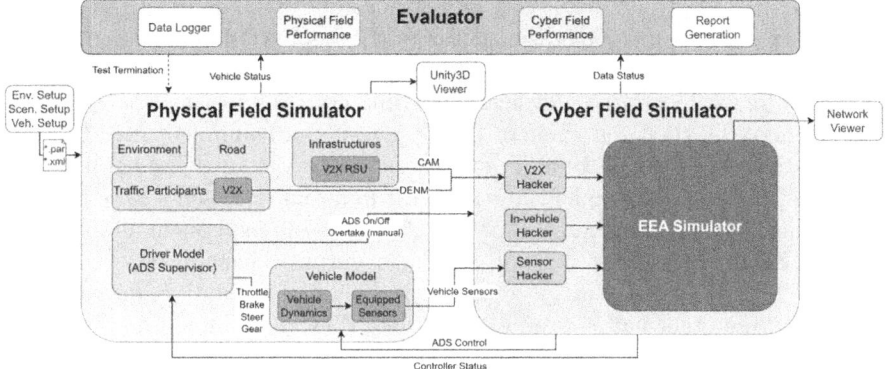

Fig. 1. Overall system architecture of FusionSis with key data flow.

visualization for fusion safety assessment and evaluation during the early stage of vehicle development. Specifically, our contributions include:

- We develop a comprehensive simulation framework to analyze and evaluate the fusion safety state of CAVs under cyber attacks. (Sect. 2)
- We propose eight evaluation criteria to assess the fusion safety state of CAVs, including a data theft evaluation metric. (Sect. 3)
- Three demo scenarios are provided, to show the responsiveness and effectiveness of the framework and evaluation metrics. (Sect. 4)

By addressing these challenges and contributions, we hope to provide a solid foundation for the vision of fusion safety development of CAVs.

2 Architecture of FusionSis

The overall architecture of FusionSis is shown in Fig. 1. Given that fusion safety analysis included risks and threats from both the physical and cyber fields, the FusionSis must be capable of simulating CAV behavior in both realms. Consequently, FusionSis is built around three primary components: the Physical Field Simulator, the Cyber Field Simulator, and the Evaluator, complemented by additional minor peripheral components.

2.1 Physical Field Simulator

The Physical Field Simulator uses a high-resolution ego vehicle dynamics model to generate realistic operational scenarios for CAVs, quantifying safety impacts and providing precise sensor values.

The Physical Field Simulator is responsible for simulating the CAV's intended operational scenario, encompassing its environment, the roads traveled and to be traveled, as well as other traffic participants and intelligent infrastructure with which the CAV interacts. Scenario setup can be specified through the modification or expansion of Operational Design Domain (ODD) definition to take advantage from the deliverable [14] by current SOTIF development process in enterprise. In related works, Physical Field Simulator is sometimes referred to as a "data source" [11] or "data generator" [5], and it can be implemented by modifying autonomous driving simulators such as CARLA [4] or SUMO [13]. In this study, we use a similar implementation with Panosim, a closed-source business simulation software. However, we specifically name this component as the Physical Field Simulator because it serves a broader purpose aligned with the concept of Fusion Safety. Additionally, it generates only half of the data required for the simulation; the other half is produced by the Cyber Field Simulator. Therefore, referring to it simply as a data source or generator would not accurately reflect its role and contribution.

Another key component of the Physical Field Simulator is the ego vehicle simulation. A high-resolution vehicle dynamics model is essential to accurately simulate the CAV's movement in the physical field, enabling the quantification of safety impacts and providing precise sensor values. Incorporating a driver model [9], rather than relying solely on autonomous driving functions, is highly recommended for several reasons. Firstly, the driver model can be use as an "supervisor" of autonomous driving function that is immune from cyber field influences. This model can enable or disable the autonomous driving function at specific times, allowing engineers to compare the outcomes of fully automated driving versus human-machine co-driving. Secondly, the driver model can serve as a manual control interface, such as in a driving simulator, enabling actual human drivers to conduct Driver-In-the-Loop (DIL) simulations.

2.2 Cyber Field Simulator

The Cyber Field Simulator is designed to model the communication processes around and networks within the CAV. The detailed modeling of the CAV's Electronic/Electrical Architecture (EEA) is essential for achieving realistic simulations of cyber attacks. Only by accurately replicating the intricate details of the EEA can we ensure that the enough types of cyber attacks can be supported and effects of cyber attacks are realistically portrayed. Thus, the Cyber Field Simulator's main role is to model the EEA of the ego vehicle to generate the data flow that will be sent through various communication links of the CAV in this scenario.

The core module of the Cyber Field Simulator is the EEA Simulator, which models the data flows in the vehicle network, including vehicle buses, wired and wireless networks, and signals from various sensors. Because the EEA of the target research CAV is different, the final modeling of the EEA Simulator will show a very big difference. Aside from the difference, the communication methods, communication protocols, and equipped sensors used by various CAVs are always similar. For vehicle bus networks such as Controller Area Network (CAN) and CAN with Flexible Data-Rate (CANFD), a man-in-the-middle (MITM) module

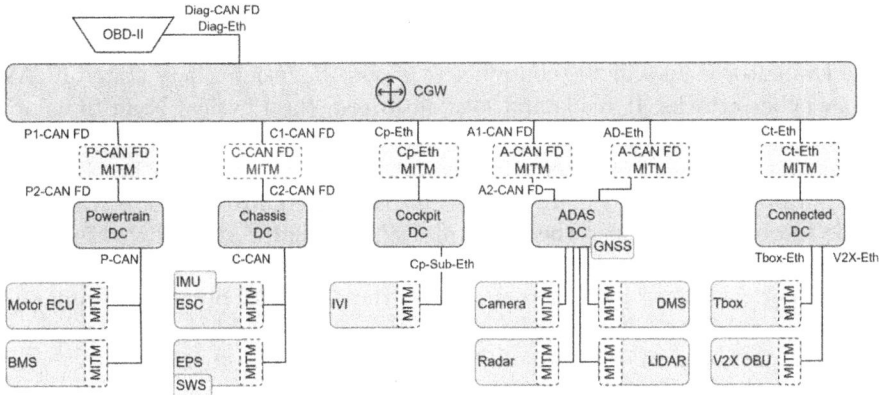

Fig. 2. An implementation instance of the Electrical/Electronic Architecture Simulator based on domain controller architecture.

is the most recommended backdoor device of implementation. During simulation, MITM module can conduct various types of commonly used cyber attacks to the vehicle network, including data injection, replay, delay, and denial of service (DoS) attack. As for the network modeling, although the team claims the finer the better, implementing the full OSI level [6] of the network during the actual simulation execution, modeling the protocols at each layer, is very time consuming and difficult, and may be overengineering. Therefore, we recommend that all networks present in the EEA Simulator should at least implement application layer protocols to meet their basic functions, and ready to implement common cyber attacks. For most focused networks, all levels of the network protocol should be implemented, but can be disabled to improve simulation performance when the specific test case is not care. Full layer implementation can also be achieved using a real controller, enabling Hardware-In-the-Loop (HIL) simulations. Sensors and controllers are also suggested of integrating a MITM module for ease of initiating noise and error injection attacks. However, sensor-targeted attacks may also need the customization of test scenario [7] from Physical Field Simulator. In this study, based on our target modeling vehicle, which is a domain controller based CAV, the EEA Simulator implementation is shown in Fig. 2.

As for the modeling of controllers, we suggest using hierarchical modeling methodology to model the CAV's controllers, following the architecture form of common operating system such as AUTOSAR [15]. This can make the modeling of the controller more consistent with the state of the real car, and if there are already developed Software components (SWCs in AUTOSAR definition), they can be reused directly. Backdoor software can also benefit from this type of implementation, as it isolates the underlying signal implementation from the upper-layer variables, allowing the reusability of the backdoor itself to be quickly migrated to other controllers throughout the EEA Simulator.

2.3 Evaluator

The Evaluator is an scalable container to assess the fusion safety state of CAVs under cyber attacks. It reads and logs data from the Physical Field Simulator and the Cyber Field Simulator and applies a set of predefined evaluation metrics to quantify the impact of cyber attacks on the CAV's overall safety and functionality. In order to expand more evaluation parameters in the future, Evaluator needs to obtain all data of the Physical Field Simulator and Cyber Field Simulator as much as possible, even though most of the data will not be used by the current evaluation indicators. Based on the capability of current framework status, the team have integrated eight easily understandable evaluation criteria to assess the fusion safety state of CAVs, as detailed in Sect. 3.

The Evaluator does not need to operate in real time, as some metrics in the Evaluator require comparing and analyzing current data with historical data to derive meaningful indicator values. Implementing real-time calculations for these metrics offers limited benefits, particularly when the data originate from multiple sources, as frequent reads and writes can impose unnecessary performance overhead on the entire system.

3 Fusion Safety Evaluation Metrics

3.1 Physical Field Metrics

The main impact of the cyber attacks on the physical field is that the driving behavior of the CAV becomes abnormal. Thus, the team has proposed four fusion safety metrics to measure CAV performance under attack conditions.

Longitudinal Accleration Deviation. Acceleration is a critical factor in determining passenger comfort during autonomous driving [3]. Typically, autonomous driving functions limit maximum acceleration and plan trajectories accordingly. Cyber attacks may cause the vehicle to exceed these acceleration limits, potentially leading to passenger discomfort or panic. To isolate the impact of cyber attacks, we measure the deviation, which excludes the effects of normal acceleration within the scenario. The longitudinal acceleration deviation is calculated as follows:

$$\Delta a_{lon,t_i} = a_{lon\_cur,t_i} - a_{lon\_std,t_i} \tag{1}$$

where a_{lon_cur,t_i} represents the current longitudinal acceleration of the CAV at time t_i. a_{lon_std,t_i} is the standard longitudinal acceleration at the same time, obtained from a baseline test case with the same scenario but without any cyber attacks.

Lateral Accleration Deviation. [3] also highlights that lateral motion is the most uncomfortable factor during autonomous driving, surpassing longitudinal acceleration. Thus, we measure the lateral acceleration deviation in the same way as the (1).

$$\Delta a_{lat,t_i} = a_{lat\_cur,t_i} - a_{lat\_std,t_i} \qquad (2)$$

The team suggest that a_{lat_std,t_i} should be obtained along with a_{lon_cur,t_i} at the same time to avoid potential consistency issues.

Collision Risk Score. This score mainly focuses on the collision risk that a vehicle may have under an cyber attack. The Time to Collision (TTC) concept is used here to characterize the risk of impact because it uses relative velocity as a calculation factor, which is often directly related to the severity of the collision. For intuitive understanding, the inverse of TTC is used for the collision risk score, defined as:

$$CRS = max(\frac{v_{ego} - v_1}{x_{ego} - x_1}, \frac{v_{ego} - v_2}{x_{ego} - x_2}, ..., \frac{v_{ego} - v_n}{x_{ego} - x_n}) \qquad (3)$$

where, v_{ego} and x_{ego} is the speed and position of the ego vehicle, and $v_1, v_2, ..., v_n$ and $x_1, x_2, ..., x_n$ are the speed and position of all other traffic participants in the scenario.

Vehicle Field Recovery Time. For a scenario-specific cyber attack, we want to determine how long it takes for the physical state of the vehicle to return to normal after the attack ceases. The Physical recovery time (T_{PhyRec}) is a flexible metric, calculated as follows:

$$T_{PhyRec} = T_{egoPhyConf} - T_{attEnd} - T_C \qquad (4)$$

where T_{attEnd} is the time when the attack ends, and $T_{egoPhyConf}$ is the time when the ego vehicle has been confirmed to have recovered and stable for T_C. However, the definition of what constitutes "recovery" can vary depending on the test case. For example, in a cruise control test, $T_{egoPhyConf}$ could be the time when CAV's velocity returns to target velocity.

3.2 Cyber Field Metrics

For Cyber field, the focus of evaluation is mainly on its influences on vehicle EEA status and data loss.

Channel Load Deviation. The Channel Load Deviation represents the additional load imposed on the vehicle's communication channels due to network attacks. This extra load can be caused by attack packets or by the activation of cybersecurity protection policies. Regardless of the cause, it increases the load rate on the channel, potentially leading to congestion. Since there are multiple communication channels in a vehicle, and their importance varies, the Channel Load Deviation $\Delta \eta_{ego}$ can be calculated using a weighted sum, as follows:

$$\Delta \eta_{ego} = \sum_{i=1}^{n} \alpha_i \times (\eta_{i,cur} - \eta_{i,std}) \tag{5}$$

where $\eta_{i,cur}$ represents network load of channel i at time t_i, and $\eta_{i,std}$ is the standard load for this channel. Each channel has its own weight α_i, with the sum of all weights equal to 1.

Cyber Field Recovery Time. Similar to the Physical recovery time (4), the Cyber recovery time (T_{CyRec}) is also a flexible metric, calculated as follows:

$$T_{CyRec} = T_{egoCyConf} - T_{attEnd} - T_C \tag{6}$$

In most cases, $T_{egoCyConf}$ often represents the time when a specific error in functional diagnostics are no longer displayed.

Data Loss Value. Report shows that cyber attacks aimed at data breaches are expanding from targeting car manufacturers to personal vehicles [12] Therefore, it is necessary to set a measurement on vehicle data and privacy loss. The total loss of value due to cyber attack is calculated as follows:

$$V_{DL} = \sum_{i=1}^{n} f(\beta_i, D_i) \tag{7}$$

where each data should be considered separately depending on the size of the data D_i, and its value function β_i.

Function Loss Ratio. [12] also suggest that cyber attacks can manipulate the vehicle's functionality and expose CAV into the next cyber attack. To better understand the impact of cyber attacks on the CAV's functionality, normal functions and safety-critical functions should be evaluated seperately, as follows:

$$P_{FL} = \langle P_{crit}, P_{norm} \rangle = \langle \frac{L_{crit}}{N_{crit}}, \frac{L_{norm}}{N_{norm}} \rangle \tag{8}$$

where L_{crit} and L_{norm} are the amount of malfunction critical and normal functions respectively; N_{crit} and N_{norm} are the total number of critical and normal functions.

4 Experiment Preparation

In order to validate the effectiveness of the framework and metrics, we have conducted 3 experiments including variety of attacks against CAV.

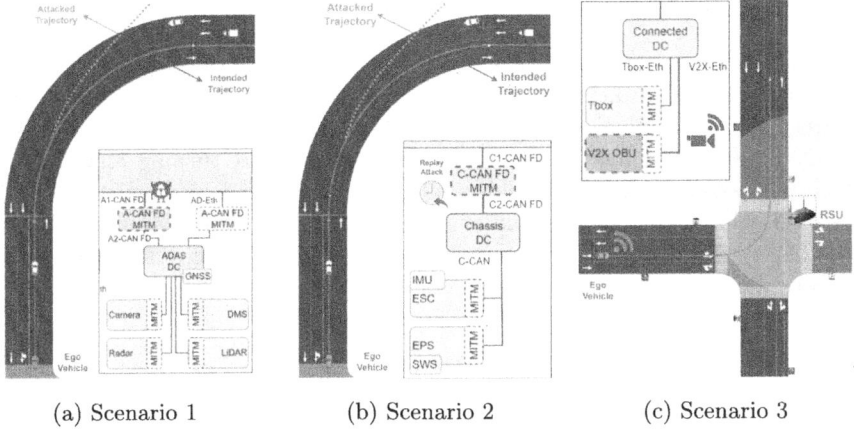

(a) Scenario 1   (b) Scenario 2   (c) Scenario 3

Fig. 3. Sketch of Scenario 1, 2, and 3.

4.1 Experiment Setup

The prototype of FusionSis framework is implemented using Panosim, Simulink and TSMaster [1]. The CAV under test is the default vehicle with default "Smart Driver" agent provided in Panosim. Sensors were equipped to the CAV as shown in Fig. 2 accordingly. Data exchange between software is realized through memory management based on Python. Functionally, the EEA simulator incorporates a total of 35 normal functions ($N_{norm} = 35$) and 75 safety-critical functions ($N_{crit} = 75$). Although this number is considerably smaller than the actual number of functions present in real-world vehicles, it is sufficient for reflecting the intended scope of the test project and the evaluation metrics.

4.2 Scenario 1: DoS Attack

In this test case, a DoS attack is simulated to target the CAV's network while the vehicle is cruising on a curved road. The attacker is the A-CAN FD MITM module pre-configured between Advanced Driver Assistance System (ADAS) Domain Controller and Gateway, as shown in Fig. 3a. The DoS attack aims to block the Lane Keep Assist (LKA) and Adaptive Cruise Control (ACC) messages sent by the ADAS Domain Controller to the Gateway. This could potentially cause the CAV to drive into the opposite lane, leading to a hazardous situation. The Gateway forwards packets based on predefined rules and CAN IDs. Therefore, these DoS packets are not forwarded. The Electric Power Steering (EPS) controller and Motor controller are equipped with self-diagnosis functions. If the LKA or ACC message is absent for one second, the corresponding function will be deactivated, and control will revert to the driver. Until then, the controllers will continue to execute the last received control message.

4.3 Scenario 2: Replay Attack

In scenario 2, the attacker replays the old LKS control message to confuse the EPS controller. The attack is initiating from the C-CAN FD MITM module between Chassis Domain Controller and Gateway, as shown in Fig. 3b. While turning, the MITM module replaces the current target turning angle with an older value-specifically, 1.5 s before, which was recorded at when CAV begins the turn. This may lead to a similar hazardous situation as in scenario 1, where the CAV drives into the opposite lane. However, due to the difference of attack type, evaluation metrics for these two scenarios may be different.

4.4 Scenario 3: Data Breach Attack

Scenario 3 recreates a situation where a Road Side Unit (RSU) at an intersection has been compromised and attempts to steal sensitive data from CAVs passing through the intersection. During simulation, the RSU continuously sends a backdoor activation message throughout its coverage area, which is a circular region with a radius of 23 m. When a CAV with a backdoor installed in its On-Board Unit (OBU) controller enters this coverage area, it silently establishes a connection with the RSU. The OBU then sends the driver's facial image captured by the Driver Monitoring System (DMS) camera to the RSU. Once the CAV exit the RSU coverage area, the data transmission channel is closed. The stolen data primarily consists of the driver's facial images, which are considered core privacy data. However, the captured images have limited head movement and variations. From a data richness perspective, longer video durations do not significantly increase the sensitivity or value of the data. Thus, the data value function $f(\beta_{face}, D_{face})$ for DMS video is set as follows:

$$f(\beta_{face}, D_{face}) = \int_0^{D_{face}} \frac{100}{e^t} dt \qquad (9)$$

where as the duration of video D_{face} increases, data value of video increases, but the rate of increase diminishes after the initial seconds.

5 Simulation Results

The test results for the three scenarios are shown in Figs. 4, 5, and 6. The 4 evaluation metrics that change with time are represented in the line plot on the left, and the remaining 4 numerical metrics are located in the bar plot on the right. To facilitate a deeper understanding of the experimental outcomes, we have also included video recordings of the simulations on Github (https://github.com/kevinuniapp/FusionSis). These videos offer a dynamic perspective on the CAV's behavior under cyber attack scenarios.

In scenario 1, the DoS attack is initiated at 3.5 s, marked by a dashed line in the line plot. The attack immediately blocks all the messages sent between the ADAS Domain Controller and the Gateway, led to the loss 25 safety-critical

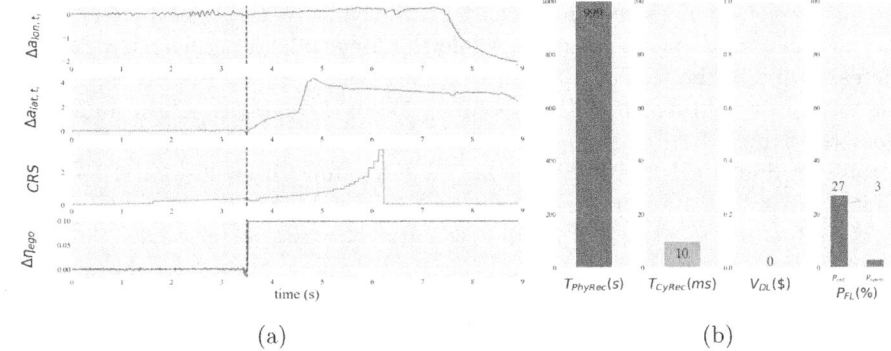

Fig. 4. Scenario 1 results

functions ($P_{crit} = 27\%$) and 1 normal function ($P_{norm} = 3\%$). Given the significant factor of ADAS CAN FD $\alpha_{ADASFD} = 0.15$, the attack increased the busload deviation $\Delta\eta_{ego}$ by about 0.1. The attack prevented the CAV from successfully executing the right turn and maintaining its speed, leading to continuous deviations in longitudinal acceleration $\Delta a_{lon,t_i}$ and lateral acceleration $\Delta a_{lat,t_i}$. Although EPS and Motor automatically deactivated the LKS and ACC functions after 1 s, the lack of driver takeover allowed the vehicle drive into the opposite lane, resulting in a collision and a CRS value exceeding 3. Therefore, the physical field (T_{PhyRec}) of CAV did not recover from the attack. After the DoS attack ceased, the ADAS Domain Controller completed its CAN transceiver self-check and resumed the communication at $T_{CyRec} = 10ms$, quickly recovering from the error state. Fortunately, there is no data breach during the attack ($V_{DL} = 0\$$).

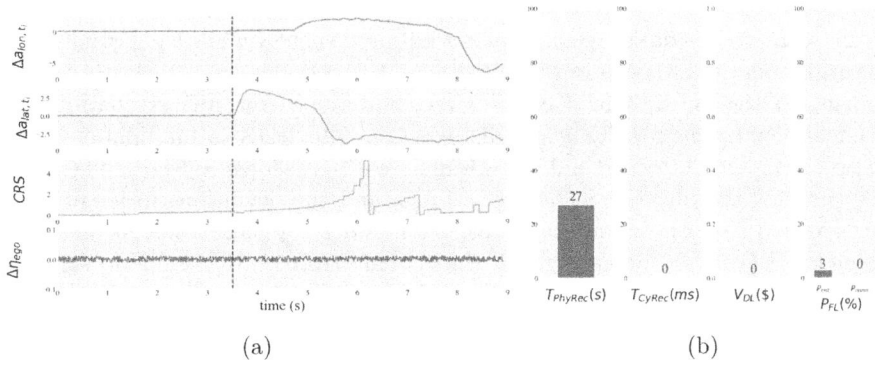

Fig. 5. Scenario 2 results

Scenario 2 is similar to Scenario 1 in terms of attack timing, with the attack initiated at 3.5 s into the simulation. The replay attack leads to similar impacts on the accelerations as seen in Scenario 1. Specifically, the lateral acceleration $\Delta a_{lat,t_i}$ remains same as intended, while the longitudinal acceleration $\Delta a_{lon,t_i}$ increases due to the lack of proper turning. However, in contrast to Scenario 1, the virtual driver takes over the steering 1.5 s after the attack and attempts to steer the vehicle back to its original lane. This takeover maneuver resulted in a collision and a CRS value exceeding 4. Excluding the collision, it took the virtual driver 27 s to return the vehicle to its original lane ($T_{PhyRec} = 27s$). Due to the difference in attack type, the evaluation metrics for the cyber field are significantly different from those in Scenario 1, expect no data has been stolen ($V_{DL} = 0\$$). The replay attack did not add additional busload to the vehicle network $\Delta \eta_{ego}$, nor did it require controllers to recover ($T_{CyRec} = 0ms$) Since the replay attack only manipulated the signals of the steering, it caused the loss of only 2 safety-critical functions($P_{crit} = 3\%$).

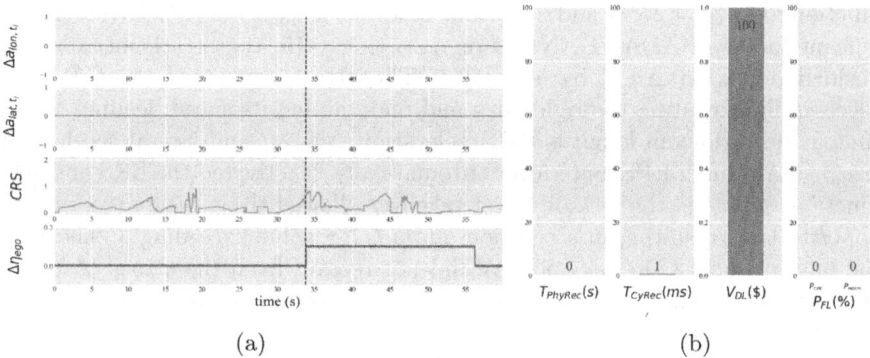

Fig. 6. Scenario 3 results

In scenario 3, a data breach attack is initiated silently while the virtual driver is turning at the intersection. Due to the stealthiness of the attack, there is no impact on the physical field of the CAV, and the virtual driver continues its original driving task without any disturbance. This leads to minimal deviation between the attack record and the normal record, as shown in Fig. 6. The recovery time for the physical field T_{PhyRec} is 0, and the deviations in longitudinal acceleration $\Delta a_{lon,t_i}$ and lateral acceleration $\Delta a_{lat,t_i}$ remain almost 0. The CRS value keeps below 1, indicating that the virtual driver is driving under "safe condition" with causion and no collisions. On the cyber field side, however, the RSU establishes a secret transmission channel between the CAV and itself during the attack for face image transmission. This transmission increases the busload on 3 vehicle ethernet channels: AD-Eth($\alpha_{ADEth} = 0.1$), Ct-Eth($\alpha_{CtEth} = 0.1$) and V2X-Eth($\alpha_{CtEth} = 0.03$), contribute to a total of $\Delta \eta_{ego} = 0.1$ channel deviation. Fortunately, given the large busload redundancy of these channels, the transmission does not influence any functions' normal execution, result in 0 loss ratio for both P_{crit} and P_{norm}. As the CAV leaves the RSU coverage area, the transmission channel is closed, and data transmission timeout and stops, result in an immediately recovery at $T_{CyRec} = 1ms$. The CAV has stayed in the RSU coverage area for $D_{face} = 56.27s - 33.82s = 22.45s$. Substituting Equ. (9) and (7), fields a data loss value of $V_{DL} \approx 100\$$.

6 Conclusion and Future Work

In this work, we introduced FusionSis, a comprehensive framework for assessing the safety of CAVs under cyber attacks. The framework consists of three primary components: Physical Field Simulator, Cyber Field Simulator, and Evaluator. Detailed modeling methodologies and implementation guidance were provided for each component. Eight fusion safety evaluation metrics, including functional safety, data security and driving comfortableness, have been introduced to assess the fusion safety state of CAVs under cyber attacks. These metrics enable a multidimensional analysis of the impact of cyber attacks on CAVs, providing a comprehensive view of the vehicle's safety state. Three demo scenarios validated the framework's functionality and the effectiveness of the metrics. Each scenario demonstrated the distinct impacts of different cyber attacks, from DoS and replay attacks affecting the vehicle's physical control to a data breach attack compromising sensitive private data. The evaluation metrics clearly illustrated the effects of these attacks on the vehicle from multiple dimensions, and can help researchers in distinguishing between various types of attacks and understanding their implications for safety. This work contributes to the development of fusion safety assessment methods for CAVs, paving the way for safer and more safe and secure autonomous vehicles.

For future work, two key areas may be addressed. Firstly, the current limitations of Simulink regarding the modeling of communication channels need to be solved. These limitations include restrictions on the number of channels and inaccuracies in representing communication delays in wired and wireless networks. Alternative approaches to simulating communication shall be explored, including the integration of new software tools or the adoption of novel modeling techniques. Secondly, the evaluation metrics may not yet fully cover the scope required for a comprehensive assessment of fusion safety, including the quantitative evaluation of potential risks and threats. To expand the coverage of the evaluation metrics, the team plans to conduct further research and study related works, such as SCOUT project [17] from europe and Framework V3.0 from JAMA [16]. By incorporating insights from these sources, the team aims to refine and expand the existing metrics to provide a robust and comprehensive evaluation framework for the fusioin safety of CAVs.

Acknowledgement. This work was supported in part by the National Key Research and Development Program of China under Grant 2023YFB3107400 and the National Natural Science Foundation of China (No. U22A202101).

Disclosure of Interests. The authors declare that they have no conflict of interest.

References

1. TOSUN-Shanghai/TSMaster (2024). https://github.com/TOSUN-Shanghai/TSMaster. original-date: 2021-02-09T01:23:59Z
2. Benyahya, M., Lenard, T., Collen, A., Nijdam, N.A.: A Systematic Review of Threat Analysis and Risk Assessment Methodologies for Connected and Automated Vehicles. In: Proceedings of the 18th International Conference on Availability, Reliability and Security, pp. 1–10. ACM, Benevento Italy (2023). https://doi.org/10.1145/3600160.3605084
3. de Winkel, K.N., Irmak, T., Happee, R., Shyrokau, B.: Standards for passenger comfort in automated vehicles: acceleration and jerk. Appl. Ergon. **106**, 103881 (2023)
4. Dosovitskiy, A., Ros, G., Codevilla, F., Lopez, A., Koltun, V.: CARLA: an open urban driving simulator. In: Levine, S., Vanhoucke, V., Goldberg, K. (eds.) Proceedings of the 1st Annual Conference on Robot Learning. Proceedings of Machine Learning Research, vol. 78, pp. 1–16. PMLR (2017). https://proceedings.mlr.press/v78/dosovitskiy17a.html
5. Finkenzeller, A., Mathur, A., Lauinger, J., Hamad, M., Steinhorst, S.: Simutack - an attack simulation framework for connected and autonomous vehicles. In: 2023 IEEE 97th Vehicular Technology Conference (VTC2023-Spring), pp. 1–7. IEEE, Florence (2023). https://doi.org/10.1109/VTC2023-Spring57618.2023.10200555
6. Froehlich, A., Rosencrance, L., Gattine, K.: What is the OSI model? The 7 layers of OSI explained (2021). https://www.techtarget.com/searchnetworking/definition/OSI
7. Hu, Z., et al.: PASS: a system-driven evaluation platform for autonomous driving safety and security. In: Proceedings Fourth International Workshop on Automotive and Autonomous Vehicle Security. Internet Society, San Diego (2022). https://doi.org/10.14722/autosec.2022.23018
8. Iqbal, S., Ball, P., Kamarudin, M.H., Bradley, A.: Simulating malicious attacks on VANETs for connected and autonomous vehicle cybersecurity: a machine learning dataset. In: 2022 13th International Symposium on Communication Systems, Networks and Digital Signal Processing (CSNDSP), pp. 332–337. IEEE, Porto (2022). https://doi.org/10.1109/CSNDSP54353.2022.9908023
9. Klimke, J., Themann, P., Klas, C., Eckstein, L.: Definition of an embedded driver model for driving behavior prediction within the deserve platform. In: 2014 International Conference on Embedded Computer Systems: Architectures, Modeling, and Simulation (SAMOS XIV), pp. 343–350 (2014). https://doi.org/10.1109/SAMOS.2014.6893231
10. Laufenberg, J., Kropf, T., Bringmann, O.: A Framework for CAN communication and attack simulation. In: 2022 IEEE 95th Vehicular Technology Conference: (VTC2022-Spring), pp. 1–7. IEEE, Helsinki (2022). https://doi.org/10.1109/VTC2022-Spring54318.2022.9860568
11. Lauinger, J., Finkenzeller, A., Lautebach, H., Hamad, M., Steinhorst, S.: Attack data generation framework for autonomous vehicle sensors. In: 2022 Design, Automation & Test in Europe Conference & Exhibition (DATE), pp. 128–131. IEEE, Antwerp (2022). https://doi.org/10.23919/DATE54114.2022.9774542
12. Levy, Y.: Upstream 2023 global automotive cybersecurity report. Executive summary, Upstream Security Ltd., Herzliya, Tel Aviv, Israel (2023). https://www.upstream.auto/

13. Lopez, P.A., et al.: Microscopic traffic simulation using sumo. In: The 21st IEEE International Conference on Intelligent Transportation Systems. IEEE (2018). https://elib.dlr.de/124092/
14. Mendiboure, L., Benzagouta, M.L., Gruyer, D., Sylla, T., Adedjouma, M., Hedhli, A.: Operational design domain for automated driving systems: taxonomy definition and application. In: 2023 IEEE Intelligent Vehicles Symposium (IV), pp. 1–6. IEEE, Anchorage (2023). https://doi.org/10.1109/IV55152.2023.10186765
15. Niklas-Höret, M., Rüping, T.: AUTOSAR classic platform (2023). https://www.autosar.org/standards/classic-platform
16. Sato, H., Ozawa, K., Kitahara, E., Kubota, Y.: Automated_driving_safety_evaluation_framework_ver3.0. Techncal report, Japan Automobile Manufacturers Association, Inc., Japan (2022). https://www.jama.or.jp/english/reports/framework.html
17. Will, D., Eckstein, L., Bargen, S.V., Taefi, T.T., Galbas, R.: State of the art analysis for connected and automated driving within the SCOUT project. Montreal (2017)
18. Yang, S., et al.: CHAINS: CHAIN-based fusion safety system framework for intelligent connected vehicle. CHAIN **1**(1), 2–45 (2024). https://doi.org/10.23919/CHAIN.2024.000006
19. Zhao, X., Abdo, A., Liao, X., Barth, M.J., Wu, G.: evaluating cybersecurity risks of cooperative ramp merging in mixed traffic environments. IEEE Intell. Transp. Syst. Mag. **14**(6), 52–65 (2022)

Deep Learning-Based Monocular Depth Estimation Method for Forklift Collision Avoidance

Dong-Ju Kim[1], Chang-Yeop Lee[1], Hyo-Jin Kim[1], and Young-Joo Suh[2](✉)

[1] Institute of Artificial Intelligence, POSTECH, Pohang, South Korea
{kkb0320,lcy8417,hyojinkim}@postech.ac.kr
[2] Graduate School of Artificial Intelligence, POSTECH, Pohang, South Korea
yjsuh@postech.ac.kr

Abstract. This paper proposes a monocular camera-based depth estimation method to prevent accidents between moving heavy equipment, such as forklifts, and workers at manufacturing environments. Traditional depth estimation techniques, which rely on ultrasonic sensors, LiDAR, and stereo cameras, face challenges such as high costs, complex installation processes, and maintenance difficulties. As a result, monocular camera-based approaches have gained increasing research attention. In this paper, we introduce an improved Monocular Depth Estimation (MDE) method aimed at enhancing forklift collision avoidance. Our method incorporates a variant of the Split-Transform-Merge (STM) strategy, traditionally used in deep learning for image classification, to improve depth estimation performance. Conventional MDE techniques generally perform two tasks: regression, which inputs a single image into the network to produce depth outputs, and classification, where discrete depth value classes (i.e., bins) are predefined, and the network outputs the probability of each pixel belonging to a specific bin. However, a significant limitation of these methods arises when high-dimensional feature maps are projected into low-dimensional bins, leading to information loss. To address this, we propose a novel strategy that thoroughly analyzes the relationship between feature maps and bins estimation, applying a variant STM strategy to minimize information loss. To validate the performance of the proposed method, we compared it with a previous MDE method, i.e., AdaBins, and confirmed performance improvements across four evaluation metrics. The experimental results not only confirm enhanced performance in all quantitative metrics but also reveal significant improvements in visual quality.

Keywords: Monocular Depth Estimation · Forklift Safety · Deep Learning

1 Introduction

In manufacturing environments, forklifts are essential equipment for logistics and material handling. However, accidents during forklift operations pose significant risks to worker safety and productivity, ranking among the leading causes of industrial accidents. These accidents are often caused by limited visibility, lack of environmental awareness, and operator inattention. Consequently, the need for effective safety systems to prevent such accidents has become increasingly important [1]. Traditionally, various sensor technologies, including ultrasonic sensors, LiDAR, and stereo cameras, have been employed to enhance forklift safety. Although these sensors effectively detect the depth and location of obstacles in the environment, they suffer from high costs, complicated installation processes, and maintenance challenges. Particularly, the high cost of precise sensors like LiDAR limits their use in smaller manufacturing sites.

On the other hand, monocular camera-based depth estimation offers cost-efficiency and ease of installation, addressing some of these limitations. However, it may lack precision when relying solely on depth maps estimated from images. To overcome these issues, this paper proposes a new strategy to improve monocular camera-based depth estimation, making it more suitable for practical use in manufacturing settings. This study demonstrates that a deep learning-based depth estimation technique using a monocular camera can accurately estimate 3D depth information without the need for expensive equipment [2]. This technique has the potential to be effectively integrated into forklift safety management systems, serving as a key technology to prevent collisions by estimating real-time distances to obstacles or workers. Additionally, it offers significant cost reductions compared to conventional stereo cameras or LiDAR systems.

This paper is structured as follows: Sect. 2 reviews existing research on monocular camera-based depth estimation techniques, Sect. 3 describes the proposed depth estimation technique, Sect. 4 presents the performance verification using the Karlsruhe Institute of Technology and Toyota Technological Institute (KITTI) dataset, and Sect. 5 concludes the study.

2 Related Works

With advances in deep learning technology, research on Monocular Depth Estimation (MDE) has intensified. MDE tasks are broadly categorized into two main approaches: regression-based methods and classification-based methods.

The regression-based approach was initiated by the work of Eigen et al., who employed CNNs to hierarchically integrate global and local spatial information in a coarse-to-fine manner [3]. Subsequently, Li et al. proposed the Hierarchical Aggregation and Heterogeneous Interaction (HAHI) module to facilitate seamless interaction [4]. Yang et al. designed a CNN encoder and an attention-gated decoder [5], while Ranftl et al. leveraged tokens from various stages using a vision transformer [6]. However, these regression-based approaches faced limitations due to slow convergence rates, as they required direct prediction of an infinite continuous depth space.

To overcome the slow convergence of regression approaches, Fu et al. introduced a method that classifies depth into discrete ranges (referred to as bins) instead of predicting depth in a continuous space, thereby accelerating optimization [7]. They pre-defined bins and estimated depth by calculating the probability of each pixel falling into these bins. However, approximating continuous depth values as discrete ones proved to be a suboptimal solution, leading to artifacts in the results. To address this limitation, Bhat et al. introduced an Adaptive Bins (AdaBins) prediction approach, where the bins are dynamically trained by a neural network, rather than being pre-defined, and this method has formed the foundation for subsequent research [8]. Li et al. integrated local representations from CNN feature maps with global representations extracted from learnable queries via a transformer, projecting them to estimate adaptive bins [9]. Bhat et al. developed a multi-scale decoder layer structure, where each layer's feature map is processed with a splitter, and the number of adaptive bins is incrementally increased by splitting the bins estimated from the previous layer [10].

These studies introduced several methods for estimating adaptive bins and achieved performance improvements [9][10]. However, a common limitation across these approaches is the information loss that occurs when high-dimensional feature maps are simply projected to relatively low dimensions for adaptive bins estimation. This information loss poses a challenge for optimal bin estimation, ultimately hindering accurate depth estimation. In this paper, we propose a novel computational strategy that minimizes this information loss and can be effectively applied in bin estimation.

3 Proposed Method

Previous studies on adaptive bins estimation [9][10] projected high-dimensional feature maps of a particular layer into low-dimensional feature maps for bins estimation, leading to information loss, as shown in Fig. 1. During this process, critical information, such as spatial details and contextual information learned in the high-dimensional feature maps, is lost, which limits accurate depth estimation.

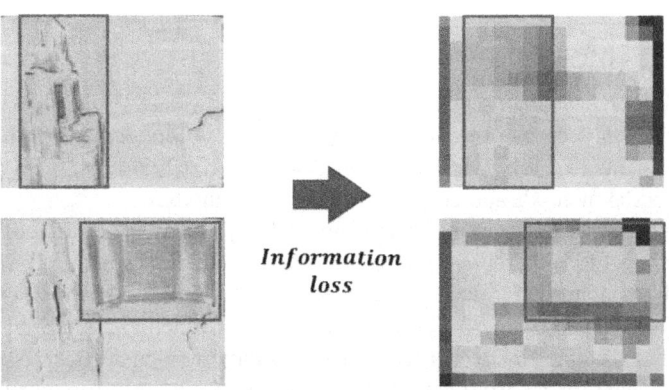

Fig. 1. Information loss due to projection from high dimensions to low dimensions.

To address this issue, this paper proposes a computational strategy that divides the feature map representations into optimal groups for processing, rather than simply projecting high-dimensional feature maps into low-dimensional ones. The proposed strategy modifies the STM strategy, which is commonly used in the image classification domain of deep learning, for application in the MDE domain. The following sections describe the STM strategy and the proposed computational approach.

3.1 Motivation

The STM strategy [11][12], commonly utilized in image classification, is illustrated in Fig. 2. In the split stage, the feature map $X \in \mathbb{R}^{C*H*W}$ is divided into N groups. Here, C denotes the channel, H the height, and W the width. The number of channels in each group $x_i \in \mathbb{R}^{C/N*H*W}$ is obtained by dividing the total number of channels by N groups. Next, in the transform stage, each group undergoes independent operations, providing computational advantages over performing operations on the entire feature map. These independent operations typically involve conventional convolution processes. Finally, in the Merge stage, the N groups, after being processed in the transform stage, are merged to create a new feature map $X' \in \mathbb{R}^{C*H*W}$, with dimensions identical to the original feature map X. This strategy enhances computation speed by dividing the process into groups and executing transform operations in parallel.

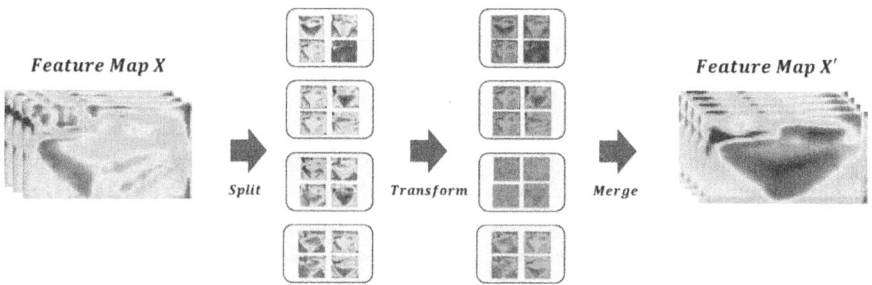

Fig. 2. STM strategy commonly used in image classification domain.

3.2 Proposed Computational Strategy

The computational strategy proposed in this paper extends the STM strategy by introducing a new conversion stage between the transform and merge stages. Initially, the split stage divides the feature map into multiple groups, consistent with the existing STM method. Each group undergoes independent transformations, followed by a conversion stage where the transformed groups are optimally adjusted to estimate a single adaptive bin. Finally, the converted single adaptive bin is merged to form the final adaptive bins in the merge stage. This strategy allows for more precise bin estimation and reduces the complexity of optimization compared to traditional methods, which estimate all bins using the entire feature map.

Figure 3 illustrates the application of the proposed computational strategy to AdaBins, the foundational model for MDE. In the AdaBins approach, a Vision Transformer (ViT) is used to estimate adaptive bins. The process starts by dividing patch embeddings through convolution operations based on patch size (p × p) [13]. After the transformer encoder processes each patch output, referred to as queries, the proposed computational strategy is applied. First, the queries are split into multiple groups during the split stage. Each of these split groups undergoes independent convolution operations in the transform stage (f1, f2, f3, ..., fN). The transformed groups are then converted to their respective optimal positions. Finally, these converted positions are merged to estimate the adaptive bins.

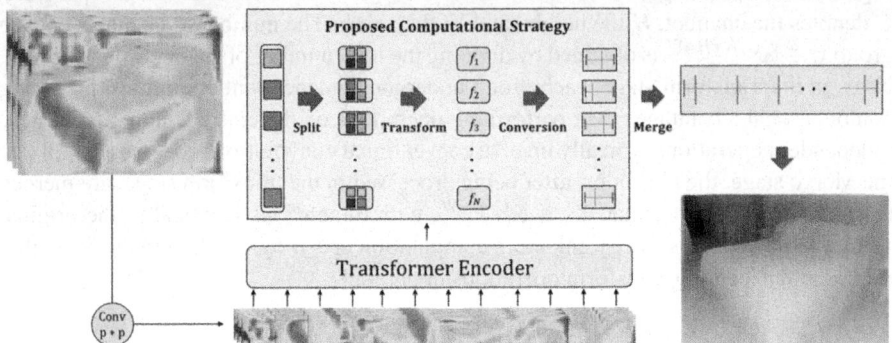

Fig. 3. Example of applying the proposed computation strategy to AdaBins.

In contrast to the traditional process, which involves projecting high-dimensional features into low-dimensional spaces and results in information loss, the proposed method has the advantage of reducing information loss and reducing the difficulty of optimization for estimating adaptive bins by dividing groups of feature maps and performing an independent transformation process for each group. Consequently, the proposed method enables more accurate bin estimation and depth estimation, making it a vital technology for preventing collisions by providing precise depth assessments of obstacles or workers in manufacturing environments.

4 Experiment Result

4.1 Experimental Setup

Since the proposed method focuses on preventing collisions by accurately estimating distances to obstacles or workers, its performance evaluation was conducted using the KITTI outdoor dataset [14], which contains objects such as obstacles and humans.

The KITTI dataset provides stereo images of outdoor scenes and depth maps acquired by LiDAR sensors. The dataset consists of RGB images with a resolution of 1241×376, along with corresponding depth maps. The depth maps were generated by projecting LiDAR points, and they provide depth values up to 80 m. Referring to the standard evaluation protocol by Eigen et al. [3], we used 23,158 images for training and 697 images for testing.

For the loss function of the output depth image, we used the Scale-Invariant (SI) version of log loss, similar to previous studies. AdaBins was used as the baseline model for comparison, and a total of four evaluation metrics were selected to evaluate performance. Equations (1) through (4) define the four-evaluation metrics, where d denotes the ground truth depth values in the images and $\tilde{d}$ denotes the predicted depth values.

Additionally, all experiments were conducted using four Tesla P40 GPUs with a batch size of 8, and the model was trained for 20 epochs.

$$\text{Absolute Relative Error(AbsRel)} : \frac{1}{N} \sum_{i=1}^{N} \frac{|d_i - \tilde{d}_i|}{\tilde{d}_i} \tag{1}$$

$$\text{Root Mean Squared Error(RMSE)} : \sqrt{\frac{1}{N} \sum_{i=1}^{N} d_i - \tilde{d}_i^2} \tag{2}$$

$$\text{Average log error(Log 10)} : \frac{1}{N} \sum_{i=1}^{N} |\log_{10} d_i - \log_{10} \tilde{d}_i| \tag{3}$$

$$\text{Root Mean Squared log Error (RMSE log)} : \sqrt{\frac{1}{N} \sum_{i=1}^{N} \log d_i - \log \tilde{d}_i^2} \tag{4}$$

4.2 Performance Evaluation

Table 1 presents the comparative experimental results of the AdaBins model across four evaluation metrics before and after applying the proposed strategy. First, the proposed strategy improved performance by 0.002 in AbsRel, followed by a 0.008 improvement in RMSE, 0.002 in Log10, and 0.001 in RMSE log. These results indicate that the proposed strategy achieved better performance in all evaluation metrics, quantitatively demonstrating the potential for more precise depth estimation. Additionally, we also confirmed that the number of parameters was reduced by 0.8M, as smaller kernels were applied in parallel during bin estimation.

Table 1. Comparison results of the proposed computation strategy.

Method	AbsRel↓	RMSE↓	Log10↓	RMSElog↓	Parmas↓
AdaBins [8]	0.058	2.360	0.190	0.088	78.9M
AdaBins [8] + Proposed	**0.056**	**2.352**	**0.188**	**0.087**	**78.1M**

As shown in Table 1, not only were the quantitative results improved, but qualitative results were also enhanced, as illustrated in Fig. 4. In Fig. 4, the first column shows the RGB input images, and the second column presents the results from the existing AdaBins model. The third column displays the results after applying the proposed computation strategy to the AdaBins model. In the predicted depth results, areas closer to the camera are colored blue, and areas farther away are colored red.

In the first row, the proposed method accurately predicted the person riding a bicycle, which the existing model failed to detect properly. Similarly, in the second row, the proposed method accurately detected a car, while the existing model struggled. From these visual results, we can confirm that the proposed method is effective in improving depth estimation performance.

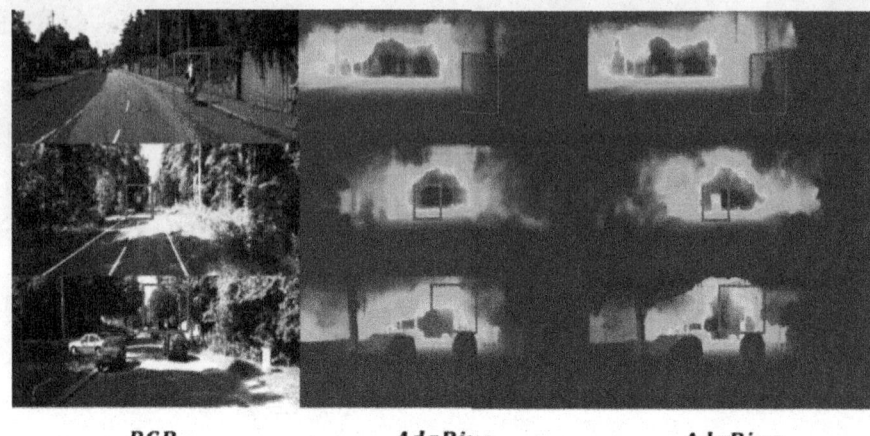

Fig. 4. Experimental results of applying the proposed computation strategy to AdaBins.

5 Conclusion

In this paper, we proposed an effective computational strategy to enhance monocular camera-based depth estimation for the prevention of collision accidents involving moving heavy equipment such as forklifts and workers in manufacturing environments. The monocular camera-based depth estimation method can serve as a crucial technique for collision prevention, and the experimental results demonstrated that the proposed computational strategy improves MDE performance in both quantitative and visual aspects.

The proposed computational strategy is not limited to the specific model used in this experiment, i.e., AdaBins, and it can be generalized and applied to other models with a bin estimation process in the MDE domain. Furthermore, the proposed method holds promise for broader application beyond monocular camera-based depth estimation research, including collision prevention for various types of moving heavy equipment.

Acknowledgment. This research was supported by Basic Science Research Program through the National Research Foundation of Korea (NRF) funded by the Ministry of Education (No.2022R1A6A1A03052954).

References

1. Neuvition,Inc, https://www.neuvition.com/ko/forklift-collision-avoidance-neuvition, last accessed 2024/08/17
2. Masoumian, A., et al.: Monocular depth estimation using deep learning: a review. Sensor, **22**(14), 5353 (2022)
3. Eigen, D., Puhrsch, C., Fergus, R.: Depth map prediction from a single image. in Proc. of the 28th International Conf. on Neural Information Processing Systems (NIPS), pp. 2366–2374, Montreal (2014)

4. Li, Z., Chen, Z., Liu, X., Jiang, J.: DepthFormer: exploiting long-range correlation and local information for accurate monocular depth estimation. ArXiv preprint ArXiv:2203.14211 (2022)
5. Yang, G., Tang, H., Ding, M., Sebe, N., Ricci, E.: Transformer-based attention networks for continuous pixel-wise prediction. In Proc. of the IEEE/CVF International Conf. on Computer Vision (ICCV), pp. 16269–16279, Online (2021)
6. Ranftl, R., Bochkovskiy, A., Koltun, V.: Vision transformers for dense prediction. In Proc. of the IEEE/CVF International Conf. on Computer Vision (ICCV), pp. 12179–12188, Online (2021)
7. Fu, H., Gong, M., Wang, C., B, K, Tao, D.: Deep ordinal regression network for monocular depth estimation. In Proc. of the IEEE Conf. on Computer Vision and Pattern Recognition (CVPR), pp. 2002–2011, Salt Lake City (2018)
8. Bhat, S.F., Alhashim, I., Wonka, P.: AdaBins: depth estimation using adaptive bins. In Proc. of the IEEE Conf. on Computer Vision and Pattern Recognition (CVPR), pp. 4009–4018, Online (2021)
9. Li, Z., Wang, X., Liu, X., Jiang, J.: BinsFormer: Revisiting Adaptive Bins for Monocular Depth Estimation. ArXiv preprint ArXiv:2204.00987 (2022)
10. Bhat, S.F., Alhashim, I., Wonka, P.: Localbins: improving depth estimation by learning local distributions. In Proc. of the European Conf. on Computer Vision (ECCV), pp. 480–496, Tel Aviv (2022)
11. Szegedy, C., Liu, W., Jia, Y., Sermanet, P., Reed, S., Anguelov, D., Erhan, D., Vanhoucke, V., Rabinovich, A.: Going deeper with convolutions. In Proc. of the IEEE Conf. on Computer Vision and Pattern Recognition (CVPR), pp. 7–12, Boston (2015)
12. Szegedy, C., Vanhoucke, V., Ioffe, S., Shlens, J., Wojna, Z.: Rethinking the inception architecture for computer vision. In Proc. of the IEEE Conf. on Computer Vision and Pattern Recognition (CVPR), pp. 2818–2826, Las Vegas (2016)
13. Dosovitskiy, A., et al.: An Image is Worth 16x16 Words: Transformers for Image Recognition at Scale. ArXiv preprint ArXiv:2010.11929 (2020)
14. Garg, R., BG, V.K., Carneiro, G., Reid, I.: Unsupervised CNN for single view depth estimation: Geometry to the rescue. In Proc. of the IEEE Conf. on Computer Vision and Pattern Recognition (CVPR), pp. 567–576, Las Vegas (2016)

Enhancing Multi-user Experience: Optimizing Explanation Timing Through Game Theory

Akhila Bairy[✉] [ID] and Martin Fränzle [ID]

Research Group Foundations and Applications of Systems of Cyber-Physical Systems, Department of Computing Science, Carl von Ossietzky Universität Oldenburg, Oldenburg, Germany
{akhila.bairy,martin.fraenzle}@uni-oldenburg.de

Abstract. As interactive systems increasingly involve multiple users with varying attention patterns and cognitive states, the need for precise explanation timing becomes essential. Our research focuses on minimising cognitive workload by strategically identifying when to deliver explanations that are both timely and effective for all users.

We employ a Markov Decision Process (MDP) framework and use backward Bellman induction to calculate the optimal timing for explanation delivery. This paper utilises a game-theoretic approach to optimising the timing of explanations in multi-user systems, guided by the psychologically validated *SEEV* (Salience, Effort, Expectancy, Value) model. The resulting strategy aims to reduce unnecessary cognitive strain while enhancing user comprehension and trust.

The model's performance is evaluated in several simulated scenarios, demonstrating its ability to adapt explanation timing to different user needs and contexts. Our findings suggest that this approach can significantly improve the user experience in multi-user environments by reducing cognitive load and optimizing information delivery. This work contributes to the broader field of human-computer interaction by providing a scalable method for enhancing multi-user systems through intelligent explanation timing. Future directions involve extending the model to incorporate multi-step explanations for multiple users and expanding its scope to accommodate more complex scenarios.

Keywords: Autonomous Vehicles · Explanation Timing · Reactive Game Theory · Attention Model · Human-Machine-Interaction · User Models

1 Introduction

As autonomous vehicles (AVs) continue to advance, the need for effective communication between the vehicle and its occupants has become increasingly critical. At the heart of this communication is the concept of providing explanations-clear, concise information that helps passengers understand the decisions and actions taken by the *AV*. These explanations are not merely a luxury but a necessity, particularly as vehicles progress through different levels of automation, as defined by the Society of Automotive Engineers (*SAE*).

The *SAE* has established six levels of driving automation [1], ranging from Level 0, where the human driver is fully in control, to Level 5, where the vehicle is completely autonomous and requires no human intervention. As vehicles transition from lower levels of automation (Levels 1–2) to higher levels (Levels 3–5), the role of the human driver shifts from active control to supervisory roles or, eventually, to a mere passenger. This shift in roles creates new challenges, particularly in how the vehicle communicates with its occupants.

At lower levels of automation, drivers are still required to make decisions and take control of the vehicle under certain conditions. However, as vehicles approach full autonomy (Levels 4–5), where the system handles all driving tasks, the timing and necessity of explanations become even more crucial. In these higher levels, occupants need to trust that the vehicle is making safe and appropriate decisions, even in complex or unexpected situations. Timely and well-delivered explanations can enhance this trust by providing insight into the vehicle's reasoning process and its understanding of the driving environment [4,9,13].

The timing of these explanations is also vital. Providing explanations too frequently or at inappropriate times can overwhelm or confuse the occupants, potentially leading to distraction or decreased trust in the system. On the other hand, delayed or absent explanations can leave occupants uncertain or anxious about the vehicle's actions. Therefore, determining the optimal timing for delivering explanations is a key challenge in the development of autonomous vehicle systems. By carefully considering when and how to provide explanations, designers can enhance the overall user experience, ensuring that occupants remain informed, engaged, and confident in the capabilities of autonomous vehicles. The design of an explainable system can also be extended to other safety-critical systems.

In our work, we build a model to determine the optimal time to present an explanation to more than one user in the *AV*. In Sect. 2, we discuss about the existing work in this field and provide an overview of game theory. The model is developed using a reactive game and an attention model (*SEEV*) which is talked about in-depth in Sect. 3. The results obtained are discussed in Sect. 4 and finally we conclude our paper by talking about the future work in Sect. 5.

2 Related Work

Timing is pivotal when delivering explanations in both human and autonomous system interactions, such as with AVs. Shen et al. [14] note that users don't always require explanations, reserving them for critical situations. These explanations can be delivered *before*, *during*, or *after* an action by the vehicle. Despite its importance, timing is often an overlooked aspect in research.

Determining what qualifies as *during an action* is challenging due to varying action durations, leading most studies to focus on explanations before or after actions. Koo et al. [9] found that explanations delivered one second prior to an action increased participants' sense of control and alertness. Du et al. [4] confirmed that pre-action explanations build more trust than post-action ones.

Körber et al. [8] demonstrated that even delayed explanations —such as 14 s after a request for driver intervention— improved user understanding. In essence, while the timing of explanations significantly influences user experience, it remains an underexplored area warranting further research, with studies indicating that pre-action explanations are particularly beneficial for user trust and comprehension [4,9,13].

Building on the understanding that pre-action explanations are beneficial, our previous works focused on determining the optimal timing for providing explanations to a single user. Specifically, we developed models to identify the optimal time to deliver a single explanation [3] and the optimal timing delivery for multi-step explanations [2]. These models employed a game-theoretic approach, and were implemented as reactive games. In this paper, we extend these approaches by utilizing the same game-theoretic framework to determine the optimal timing for providing explanations to multiple users.

Game theory, originally developed by von Neumann and Morgenstern [12], is a mathematical framework used to study strategic interactions between individuals or entities, where the outcome for each participant depends not only on their own decisions but also on the decisions made by others [11]. In essence, it analyzes situations where multiple players make decisions that impact one another, aiming to predict the optimal strategies for all involved.

Within game theory, a particular type of game known as a *reactive game* [5] focuses on scenarios where players must make decisions in response to changing conditions or the actions of other players. In a reactive game, one player's strategy may evolve based on the behaviour of the other player, creating a dynamic interaction. These games are often used to model situations where decisions need to be made in real time, such as in autonomous systems or automated decision-making processes. By using game theory and reactive games, we can design strategies that allow systems to respond effectively to changing environments, ensuring better outcomes in complex, real-world situations.

3 Reactive Game Model

The optimisation of the timing of the provision of explanation for multiple users in an AV setting is done with the help of a reactive game. For the ease of computation, we consider only two users in this paper. Here the game is played between the AV and the mental workload model of two users. There is a correlation between the mental workload and attention paid by the human. Kantowitz, in his paper [7], states that even simple models of attention can accurately predict the mental workload of humans. In our work, we measure the attention of the users using the attention model called $SEEV$ [18]. More about this model is discussed in Sect. 3.1.

3.1 Human Attention Model - *SEEV*

The *SEEV* (Salience, Effort, Expectancy, and Value) attention model [18] is a comprehensive framework designed to elucidate the mechanisms by which individuals allocate their attentional resources in complex environments. Developed primarily within the field of human factors and cognitive engineering, the *SEEV* model integrates four key components that interact to influence attention distribution:

1. Salience (*S*): This component refers to the prominence or distinctiveness of a stimulus within the environment. Salient stimuli are those that stand out due to their unique characteristics, such as brightness, colour, movement, or loudness. The more salient a stimulus, the more likely it is to capture attention.
2. Effort (*Ef*): Effort pertains to the cognitive and physical resources required to shift attention from one stimulus to another. High effort costs can deter individuals from frequently changing their focus, leading to more sustained attention on less demanding tasks or stimuli. This aspect emphasizes the mental workload involved in attention-switching.
3. Expectancy(*Ex*): Expectancy involves the anticipation or likelihood of encountering a particular stimulus based on prior knowledge or experience. When individuals expect certain stimuli to appear in specific contexts or locations, they are more likely to direct their attention there. Expectancy is informed by both short-term and long-term memory and plays a crucial role in guiding attentional focus.
4. Value(*V*): This component addresses the perceived importance or relevance of a stimulus to the individual's goals or tasks at hand. High-value stimuli are those that are deemed critical or beneficial, prompting individuals to allocate more attentional resources towards them. Value is often determined by the immediate or anticipated benefits of attending to a stimulus.

The probability of attention *(P(A))* of a user is calculated using the formula

$$P(A) = S - Ef + Ex \cdot V \ . \tag{1}$$

Salience and effort, connected to the environment's physical characteristics, are collectively known as *bottom-up* factors that impact attention [17]. In contrast, the other two attributes, expectancy and value, are classified as *top-down* factors, as they stem from cognitive processes and an individual's prior experiences and expectations.

The *SEEV* model is particularly influential in applied settings, such as aviation, driving, military operations, and user interface design. It aids in predicting how operators will distribute their attention under varying conditions, thereby informing the design of systems and interfaces that align with natural attentional tendencies. For instance, in aviation, understanding how pilots allocate attention to various cockpit instruments can lead to better cockpit design, reducing the risk of errors and enhancing overall safety. The *SEEV* model was developed by Wickens et al., [18] to determine the attention of pilots in cockpits. Subsequently, *SEEV* model was extended to predict the attention of a driver in traffic situations by Horrey et al. [6] and Wortelen [19].

By incorporating salience, effort, expectancy, and value into a unified framework, the *SEEV* model offers a nuanced understanding of attention dynamics, providing valuable insights for optimizing human performance and designing more efficient and user-friendly systems.

3.2 *SEEV* Model for Two Users

The previous section provided a general overview of *SEEV* model. The different attributes of *SEEV* model are slightly modified for our work. These modifications have been suggested by Wortelen [19] and employed in our previous works as well [2,3].

1. Salience (S): it refers to the degree to which new information of a specific type would capture the human's attention if it were to become available.
2. Effort (Ef): it denotes the amount of physical or cognitive effort required by the individual to process this new information.
3. Expectancy (Ex): it represents the anticipated frequency with which new information will become available, serving as a dynamic variable that estimates the expected time until updated information is received.
4. Value (V): it signifies the anticipated benefit or gain that the individual expects to obtain from the information item.

In this paper, we employ the *SEEV* model to represent two distinct user profiles. This is achieved by systematically varying the expectancy attribute within the *SEEV* framework. Consequently, these variations lead to the emergence of different human emotional states such as anxious, calm, or bored, allowing for a nuanced analysis of user experiences.

The initial values of the expectancies of the users are randomly generated in our model. We generate these random initial values as less than 0.5. For the final values of the expectancies, we use the formula of generating uniformly distributed random number within a specified interval.

$$r = a + (b - a) * rand(N, 1) . \qquad (2)$$

Using Eq. 2, N random numbers can be generated in the interval of *(a,b)* [16]. In our work, for each user, we generate only one random number, i.e., the final expectancy value. Hence the N value in Eq. 2 is one.

The initial and final values of expectancy and the initial and final time values are then fit into an exponential curve using the *fit function* (shown in Eq. 3) in MATLAB [15].

$$f = fit(x, y, \text{'exp1'}) \qquad (3)$$

In Eq. 3, x represents the array of initial and final expectancy values; y represents the array of initial and final time values. The initial time value is

1 and the final time value depicts the time at which the scenario occurs. This final time value is also manipulated to get different expectancy values. *exp1* tells us the type of fit function used. *exp1* generates an exponential curve using a single-term exponential which is shown in Eq. 4.

$$f(x) = a * exp(b * x) \tag{4}$$

3.3 Model Implementation

In any situation, especially safety critical scenarios, the timing of the explanation is crucial, as it needs to balance the need for user understanding with the avoidance of unnecessary distraction or confusion. To better understand the importance of timing, let us consider the following example from [3].

Example 1. At an intersection, an autonomous vehicle v, intends to make a left turn but comes to a stop, even though the traffic light is green and permits an uninterrupted left turn. v has detected the presence of an approaching emergency vehicle and stops to yield the right of way. Although v is equipped to explain its actions to the occupants, it must carefully decide whether and when to provide this explanation.

In the given scenario, if the explanation is provided too early —such as before the users are aware that an intersection is approaching and a left turn is planned— the information might be overlooked or dismissed. In this case, the cognitive effort required to process the explanation would be wasted, as it fails to connect with the users' current understanding of the situation. On the other hand, if the explanation is delivered too late —even by a small margin— or not at all, the users may feel compelled to engage in their own attention strategies. They might begin scanning the environment to deduce why the vehicle stopped at a green light, which would increase their cognitive workload as they actively seek an explanation.

Thus, minimizing cognitive workload depends heavily on the precise timing of the explanation. Delivering the explanation at the optimal moment ensures that it is relevant and informative, preventing unnecessary mental effort while keeping the users well-informed and at ease.

In our approach, we concentrate on identifying the optimal timing for providing explanations within the specific context described in the Example 1 Scenario. In this scenario, because the area of interest is fixed, the effort factor of the *SEEV* model does not vary. Additionally, due to the brief duration of the scenario, we can also treat the salience as a constant throughout. As a result, both Salience and Effort, and the difference between them, can be represented as constants (c) in Eq. 1. This allows us to calculate the probability of attention using only the top-down factors, which is expressed as:

$$P(A) = Ex \cdot V + c \ . \tag{5}$$

The modified *SEEV* model shown in Eq. 5 has been applied as a dynamic factor that influences workload within a Markov decision process game, where the decisions revolve around the timing of explanation presentation. We utilized MATLAB [10] to calculate the reactive strategy for delivering explanations. The SEEV-based game begins n seconds before the scenario starts and concludes when the scenario ends.

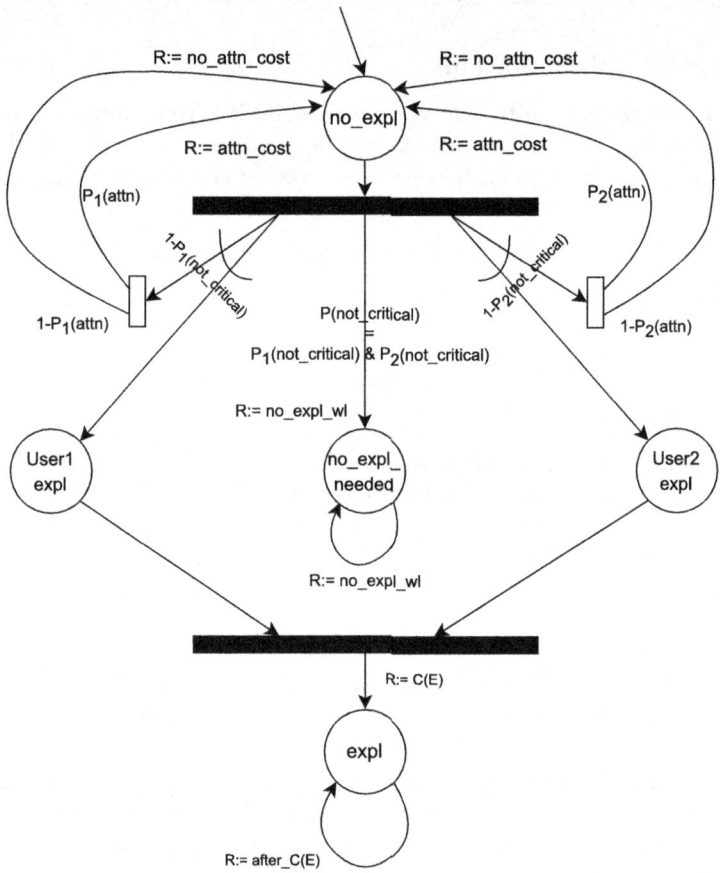

Fig. 1. State diagram of the strategic player with parallel processes

Figure 1 illustrates a state diagram depicting the various states and transitions for the strategic player, incorporating parallel processes. At the start of the game, the strategic player starts by being in the *no_expl* state, which indicates the absence of explanation. From here on, at every time step (each time step is of one-second duration) the strategic player can choose to go to one of the following states based on various probabilities

1. *no_expl_needed* state: This state is reached only with the joint probability of both the users claiming that the situation is not critical (i.e. $P(not\_critical)$).

Once in this state, the player continues to stay in this state until the end of the scenario.
2. *no_expl* state: When both the users consider the situation to be critical (i.e. $(1 - P_i(not\_critical)$ where $i = 1, 2$), this state can be reached with different rewards based on whether the user is paying attention or not. The reward structure is explained in detail in Sect. 4.
3. *expl* state: This state is reached when the strategic player decides to give an explanation, whether optimal or not. The latter is clarified by the reward. It is important to note that even if one of the users feels the need for an explanation, both the users receive it. Once the users receive the explanation, the strategic player stays in this state until the end of the scenario.

4 Results and Discussion

The objectives of the reactive game introduced in this paper were twofold: first, to pinpoint the optimal timing for delivering an explanation to two users that minimizes the cognitive workload of both, and second, to identify the minimum expected cognitive workload for both the users across all possible presentation strategies.

Table 1. MDP Reward Structure

S	S'	Probability	R
no_expl	no_expl	P(not_critical) · P(attn)	attn_cost
no_expl	no_expl	P(not_critical) · (1-P(attn))	no_attn_cost
no_expl	expl	P(not_critical)	C(E)
no_expl	no_expl_needed	P(not_critical)	no_expl_wl
expl	expl	1	after_C(E)
no_expl_needed	no_expl_needed	1	no_expl_wl

To achieve these goals, it was necessary to assign rewards to the various state transitions (based on Fig. 1), as outlined in Table 1. The exact reward/cost values assigned to the various probabilities are shown in Table 2.

Table 2. Reward/Cost Values

Reward in Figure	Reward/Cost Value
attn_cost	0.4
no_attn_cost	0.2
C(E)	0.3
after_C(E)	0.1
no_expl_wl	0.0

Since this is a finite horizon model, backward Bellman induction is used to calculate the minimum cognitive workload induced by the attention strategy at any given point in time. The formula to compute the minimum workload (min_wl) is provided in Eq. 6.

$$min\_wl_n^k = P_i(not\_critical) \cdot no\_expl\_wl + (1 - P_i(not\_critical)) \cdot expl\_wl_n^k \quad (6)$$

As explained before, some instances may not require an explanation if the user has already assessed the surroundings or if the situation resolves itself. $P_i(not\_critical)$, where $i = 1, 2$; accounts for such situations by considering the probability of no critical situation arising for the two users. In situations where no explanation is required, a constant no_expl_wl is employed. The workload associated with providing an explanation, denoted as $expl\_wl_n$, depends on the time when the scenario occurs (k) and the current time (n). [3] shows the formula for this, which is given in Eq. 7.

The $expl\_wl_n^k$ value represents the minimum between the cost of providing an explanation and the cost of not providing one at time n, given a total scenario duration of k. This minimum value reflects the strategic choice of the explanation mechanism, which aims to minimize the expected cognitive workload. The cost of providing an explanation includes the direct cost of the explanation, $C(E)$, and any additional costs incurred after the explanation is given, denoted as $after\_C(E)$. On the other hand, the cost of not providing an explanation varies depending on the probability of attention, which is calculated using the backward Bellman recursion. The probability of attention for the two users, $P_i(attn)$ where $I = 1, 2$, is derived from the *SEEV* model.

$$expl\_wl_n^k = min \begin{cases} C(E) + (k-n) \cdot after\_C(E), \\ P_i(attn)_n \cdot (min\_wl_0^{k-n} + attn\_cost) \\ +(1 - P_i(attn)_n) \cdot (min\_wl_{n+1}^k + no\_attn\_cost) \end{cases} \quad (7)$$

When attention is being paid, $expl\_wl$ represents the workload cost associated with following an attention direction ($attn\_cost$) combined with the backward recursion of the minimum workload, adjusted for the remaining horizon, $k - n$. If no attention is being paid, $expl\_wl$ represents the cost of not paying attention (no_attn_cost) combined with the backward recursion value of the minimum workload.

Equations 6 and 7 are mutually recursive. Using these equations, the optimal explanation timing for each user is determined. These optimal times are then compared and the explanation is delivered to the user who requires it sooner.

Figures 2, 3, and 4 show various graphs, where the scenario occurs at 10 s, with varying expectancy curves for users 1 & 2. Figure 2 shows two graphs where user 1 requires an explanation sooner than user 2. In the two graphs in Fig. 3, user 2 needs an explanation earlier than user 1. In Fig. 4, both users require an explanation at the same time.

Fig. 2. User/addressee 1 requires an explanation earlier

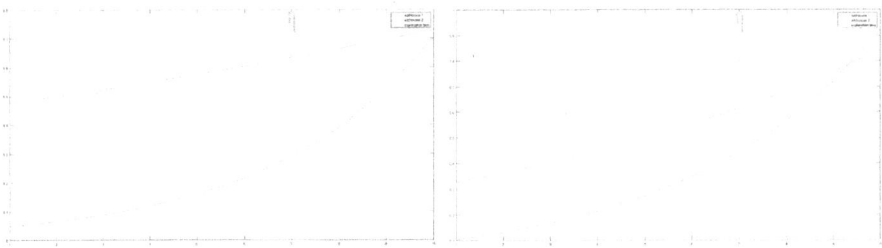

Fig. 3. User/addressee 2 requires an explanation earlier

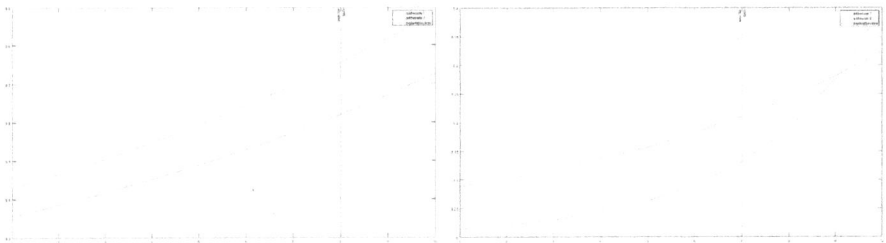

Fig. 4. Both users require an explanation at the same time

As shown in all the figures above, the explanation is typically provided 2 s or 3 s before the scenario concludes. The experiment was repeated at various scenario end times. If the scenario concludes in 2 s or less, then the model suggests that providing an early explanation does not result in a significant reduction in workload. This finding is consistent with our previous results regarding explanation timing for a single user [3].

Since the model is implemented in MATLAB and relies on backward induction, even a one-second increase in the scenario duration leads to a significant rise in computational demands. When the scenario duration exceeds 25 s, the computation time for determining the optimal explanation timing surpasses 2.5 s. Consequently, for longer scenarios, MATLAB proves to be an inefficient platform for implementing this model.

5 Conclusion

In this study, we aimed to determine the optimal timing for providing an explanation in autonomous systems involving multiple users, using a game-theoretic approach. Our findings demonstrate that explanation timing strategies can be effectively adapted to accommodate diverse needs of different users. While our models provide valuable insights, they are constrained by the assumptions made for the *SEEV* model and the reward structure and values shown in Tables 1 & 2. These rewards/costs are currently based on educated estimates intended to demonstrate the technology, but they lack empirical psychological validation.

Future research should focus on applying these models to more diverse real-world scenarios and refining them for broader applications. As an immediate next step, we plan to investigate the outcomes of implementing multi-step explanations for multiple users. Overall, our work contributes to the growing body of literature on human-*AV* interaction, offering practical guidelines for improving user experience in autonomous systems.

Acknowledgments. The research here has been supported by Universität Oldenburg within the RTG Social Embeddedness of Autonomous Cyber Physical Systems (SEAS) and by Deutsche Forschungsgemeinschaft under grant no. DFG FR 2715/5-1 "Konfliktresolution und kausale Inferenz mittels integrierter sozio-technischer Modellbildung" This research was also supported by the Innovation Campus for Future Mobility (https://www.icm-bw.de/) and by the Helmholtz Association within the Core Informatics project.

Disclosure of Interests. The authors have no competing interests to declare that are relevant to the content of this article.

References

1. Taxonomy and definitions for terms related to driving automation systems for on-road motor vehicles J3016_202104. Technical report (2021). https://web.archive.org/web/20211220101755/https://www.sae.org/standards/content/j3016_202104/
2. Bairy, A., Fränzle, M.: Optimal explanation generation using attention distribution model. In: Ahram, T., Taiar, R. (eds.) Human Interaction and Emerging Technologies (IHIET-AI 2023): Artificial Intelligence and Future Applications. AHFE (2023) International Conference. AHFE Open Access. AHFE International, USA (2023). https://doi.org/10.54941/ahfe1002928
3. Bairy, A., Fränzle, M.: Efficiently explained: leveraging the SEEV cognitive model for optimal explanation delivery. In: Praetorius, G., Sellberg, C., Patriarca, R. (eds.) Advances in Human Factors of Transportation. AHFE (2024) International Conference. AHFE Open Access, vol. 148. AHFE International, USA (2024). https://doi.org/10.54941/ahfe1005221
4. Du, N., et al.: Look who's talking now: implications of AV's explanations on driver's trust, AV preference, anxiety and mental workload. Transp. Res. Part C: Emerg. Technol. **104**, 428–442 (2019)

5. Gutierrez, J., Harrenstein, P., Wooldridge, M.: From model checking to equilibrium checking: reactive modules for rational verification. Artif. Intell. **248**, 123–157 (2017)
6. Horrey, W.J., Wickens, C.D., Consalus, K.P.: Modeling drivers' visual attention allocation while interacting with in-vehicle technologies. J. Exp. Psychol. Appl. **12**(2), 67–78 (2006)
7. Kantowitz, B.: Attention and mental workload. Proc. Hum. Fact. Ergon. Soc. Ann. Meet. **44**, 3–456 (2000). https://doi.org/10.1177/154193120004402121
8. Koerber, M., Prasch, L., Bengler, K.: Why do I have to drive now? Post Hoc explanations of takeover requests. Hum. Factors **60**(3), 305–323 (2018). https://doi.org/10.1177/0018720817747730
9. Koo, J., Shin, D., Steinert, M., Leifer, L.: Understanding driver responses to voice alerts of autonomous car operations. Int. J. Veh. Des. **70**, 377 (2016). https://doi.org/10.1504/IJVD.2016.076740
10. MATLAB: version 9.14.0 (R2023a). The MathWorks Inc., Natick, Massachusetts (2023)
11. Myerson, R.B.: Game Theory: Analysis of Conflict. Harvard University Press (1991). http://www.jstor.org/stable/j.ctvjsf522
12. von Neumann, J., Morgenstern, O.: Theory of Games and Economic Behavior (60th-Anniversary Edition). Princeton University Press (2007). http://press.princeton.edu/titles/7802.html
13. Ruijten, P.A.M., Terken, J.M.B., Chandramouli, S.: Enhancing trust in autonomous vehicles through intelligent user interfaces that mimic human behavior. Multimodal Technol. Interact. **2**(4), 62 (2018)
14. Shen, Y., et al.: To explain or not to explain: a study on the necessity of explanations for autonomous vehicles (2020). arXiv:abs/2006.11684
15. The MathWorks Inc.: Mathworks Documentation for Exponential Models. https://de.mathworks.com/help/curvefit/exponential.html
16. The MathWorks Inc.: Mathworks Documentation for Uniformly Distributed Random Numbers. https://de.mathworks.com/help/matlab/ref/rand.html
17. Wickens, C.: Noticing events in the visual workplace: the SEEV and NSEEV models. In: The Cambridge Handbook of Applied Perception Research, pp. 749–768. Cambridge Handbooks in Psychology, Cambridge University Press (2015). https://doi.org/10.1017/CBO9780511973017.046
18. Wickens, C., Helleberg, J., Goh, J., Xu, X., Horrey, W.: Pilot task management: testing an attentional expected value model of visual scanning. Savoy, IL, UIUC Institute of Aviation Technical Report (2001)
19. Wortelen, B.: Das Adaptive-Information-Expectancy-Modell zur Aufmerksamkeitssimulation eines kognitiven Fahrermodells. Ph.D. thesis, Carl von Ossietzky Universität, Oldenburg, Germany (2014)

User-Intimate and Trustworthy Ontology-Based Requirement Engineering Methodology: A Case Study on Connected and Autonomous Electrical Vehicles

Alper Kanak[1,2(✉)], Ali Serdar Atalay[3], Oğuzhan Herkiloğlu[3], Ahu Ece Hartavi Karcı[4], Elif Toy Aziziaghdam[5], and Salih Ergün[1,2]

[1] Ergünler R&D Co. Ltd., Isparta, Turkey
{alper.kanak,salih.ergun}@erarge.com.tr
[2] Ergtech Sp.Z.O.O, Warsaw, Poland
{alper.kanak,salih.ergun}@ergtech.eu
[3] AI4SEC OÜ, Tallinn, Estonia
{ali.atalay,oguz}@ai4sec.eu
[4] University of Surrey, Guildford, UK
a.hartavikarci@surrey.ac.uk
[5] Otokar AŞ, Sakarya, Turkey
etoy@otokar.com.tr

Abstract. This paper introduces a conceptual framework motivated by the growing need for a systematic methodology to accurately identify requirements and improve the design of cyber-physical systems (CPS). The proposed approach focuses on addressing the needs of users and stakeholders by presenting an ontology-based requirements engineering methodology that integrates a comprehensive security, privacy, and safety threat model. This model is adaptable to any CPS design, particularly in the context of intelligent transportation systems and cyber-physical infrastructure. The methodology emphasizes a user-centered, system-oriented approach from the outset, leveraging the "X-by-design" principle, ensuring that critical requirements like security, privacy, and safety are embedded throughout the design process, rather than being addressed retroactively. As a case study, the paper demonstrates how this model is applied in a research project focused on the use of connected and autonomous electrical vehicles (CAEVs) in smart transportation and mobility systems. This approach offers a holistic solution to CPS design challenges by considering all key requirements from the very beginning.

Keywords: User-Intimate Requirements Hierarchy Resolution Framework (UI-REF) · threat model · cyber security · privacy · safety · ontology-based requirement engineering (OBRE) · connected and autonomous electrical vehicles

1 Introduction

The deployment of connected and autonomous electric vehicles (CAEVs) within intelligent transportation systems (ITS) represents a transformative shift in modern transportation. These systems rely on a network of interconnected devices, sensors, communication protocols, and cloud services, creating a complex cyber-physical environment. While CAEVs promise enhanced safety, efficiency, and user convenience, they also introduce new vulnerabilities that can be exploited by adversaries. The cybersecurity and privacy of these systems are paramount, requiring a proactive approach that integrates advanced requirement engineering, threat modeling, and trust management.

In CAEV and ITS environments, where numerous subsystems and stakeholders are involved, there is a strong need to apply ontology-based approaches in the design phase, as they offer a way to ensure that all requirements are captured comprehensively. Ontology-based requirement engineering (OBRE) [1], as proposed in this paper, employs formalized structures to represent knowledge, requirements, and relationships between different elements of the system. OBRE provide a semantic foundation that supports the integration of domain-specific knowledge, threat models, and stakeholder objectives, resulting in a more accurate and consistent requirement set. User-Intimate Requirements Hierarchy Resolution Framework (UI-REF) is proposed as a requirement engineering methodology [2] and used as the basis of the methodology presented in this paper. This approach forms the basis of the X-by-design paradigm which is a design philosophy that aims to embed various critical attributes—such as security, privacy, safety, resilience, and reliability—into systems from the outset. Rather than treating these attributes as afterthoughts or add-ons, X-by-Design ensures that they are foundational aspects of system architecture and development.

OBRE tackles not only the technical requirements and the design of the system architecture but also the users' and stakeholders' needs. The user-intimated requirements elicitation, i.e. UI-REF, addresses these challenges by incorporating user input, system specifications, and security best practices into a cohesive strategy. UI-REF goes beyond conventional requirement gathering by actively involving stakeholders throughout the process, using ontology-based techniques to model trust, privacy, and safety requirements. Additionally, the integration of threat models such as STRIDE and LINDDUN ensures that the framework can identify and address a comprehensive range of potential attack vectors and privacy risks.

In the realm of cybersecurity and privacy engineering for ITS and CAEVs, a robust and systematic approach to requirements gathering is paramount. As these systems become increasingly complex, autonomous and interconnected, traditional requirement elicitation methods are often insufficient to address the nuanced security, privacy, and safety challenges. This calls for a sophisticated framework that not only captures functional and non-functional requirements but also integrates threat modeling, and OBRE to ensure holistic security and privacy solutions. The so-called user-intimated requirements elicitation emerges as a critical methodology in this space, bridging the gap between user expectations, system functionalities, and trust objectives.

Trust models are foundational in establishing secure interactions within CAEVs and ITS. These models define the relationships, dependencies, and trustworthiness between system components, entities, and users. They ensure that only authenticated and authorized entities can interact with sensitive vehicle and network data. Trust models form the basis for access control, cryptographic operations, and secure communication, playing a pivotal role in protecting the integrity and confidentiality of data in complex vehicular networks. In connected and autonomous vehicle ecosystems, not only cybersecurity and privacy but also safety is critical and all related requirements are deeply intertwined towards fostering holistic trustworthiness. Cybersecurity ensures that vehicle systems are protected against malicious attacks, while privacy focuses on safeguarding the personal and sensitive information of drivers and passengers. Safety, on the other hand, is concerned with preventing any operation that may cause physical harm. As noted in this paper, a comprehensive requirements elicitation framework must address these three pillars in a balanced manner, ensuring that security and privacy do not come at the expense of operational safety.

The proposed methodology incorporates the UI-REF and OBRE by relying on security, privacy and safety threat modeling. Threat modeling is indispensable in CAEV-utilized ITSs as it presents a structured approach for identifying and assessing potential threats in a system. STRIDE (Spoofing, Tampering, Repudiation, Information Disclosure, Denial of Service, and Elevation of Privilege) [3] is commonly used to evaluate security threats. In parallel, LINDDUN (Linkability, Identifiability, Non-Repudiation, Detectability, Disclosure of Information, Unawareness, and Non-Compliance) [4] focuses on privacy-related threats. Both methodologies provide systematic ways to identify vulnerabilities and define corresponding mitigation strategies, making them indispensable in the development of resilient CAEV systems.

The paper is organized as follows. Section 2 gives an overview of the current state-of-the-art. Section 3 presents the proposed User-intimated and Trustworthy Ontology-based Requirement Engineering Methodology (REM in short). Section 4 presents a use case based on the gathered experiences on ongoing research projects ESCALATE and YEHO for the sake of proof of concept. Section 5 concludes the paper.

2 Literature Review

The integration of CAEVs into ITS necessitates a comprehensive understanding of user-intimated requirements and x-by-design philosophy. User-intimated requirements elicitation is used for capturing the needs and expectations of users in the context of ITS. UI-REF or in general user-intimated requirements elicitation frameworks facilitate the identification of user requirements through participatory design and iterative feedback mechanisms, ensuring that the systems developed are user-centered and contextually relevant [5, 6]. Recent studies emphasize the importance of involving end-users in the design process to enhance the usability and acceptance of cyber technologies [7]. Natsiavas et al. [8] present a comprehensive methodology for secure and interoperable health data exchange, emphasizing the importance of stakeholder involvement in identifying barriers and facilitators for health information technology adoption. Their methodology aligns with best practices in requirements engineering, highlighting the need for effective requirements elicitation and validation processes. Similarly, Rizk et al. [9] discuss the significance of crowdsourcing in requirements elicitation for eLearning systems. They argue that user involvement is a critical success factor in software development projects, and traditional approaches often fail to engage a diverse user base. Their study highlights the limitations of conventional methods and advocates for innovative approaches that leverage crowdsourcing to gather user insights effectively. The need for continuous requirements elicitation has become increasingly apparent, especially in dynamic environments where user needs evolve. Oriol et al. [10] propose a framework that combines user feedback and monitoring to support ongoing requirements elicitation. This approach addresses the challenges faced by requirements engineers in maintaining alignment with user needs throughout the software lifecycle. The relationship between user-oriented requirement engineering and technology acceptance is underscored by the need for user-centered design principles [12]. Ahmad et al. [11] propose a framework for engineering human-centered artificial intelligence-based software systems, emphasizing the importance of incorporating user perspectives in the development process. This focus on human-centered design is critical for fostering user acceptance of technology. Moreover, the conceptualization of privacy concerns is crucial in requirement engineering. Peras & Mekovec [13] discuss the privacy concerns of cloud users, emphasizing that despite the implementation of privacy protection mechanisms by cloud service providers, privacy remains a significant barrier to adoption. This highlights the need for requirement engineering frameworks to address user privacy concerns explicitly, integrating privacy features that align with user expectations and regulatory standards.

OBRE has emerged as a significant approach in the field of software engineering, particularly for enhancing the quality and consistency of requirements elicitation, specification, and management. Ontologies serve as formal representations of knowledge within a specific domain, providing a shared vocabulary and a structured framework for capturing domain concepts and relationships. Recent studies emphasize that the use of ontologies can significantly enhance the requirements engineering process by improving communication among stakeholders and ensuring a common understanding of requirements [14, 15]. For instance, Alsanad et al. [15] highlight that ontologies facilitate the analysis of domain knowledge, which is crucial for maintaining consistency in software requirements and enhancing collaboration among development teams. Moreover, ontologies can help in managing the complexity of requirements by providing a clear structure that separates domain knowledge from operational knowledge. This separation allows for the reuse of existing ontologies across different projects, thereby streamlining the requirements engineering process [16].

The integration of security, safety and privacy threat models into various domains has become increasingly important as technology evolves, and cyber threats become more sophisticated. STRIDE [3] and LINDDUN [4] are two prominent frameworks used for identifying and categorizing threats in software systems. STRIDE provides a systematic approach to threat modeling by using Data Flow Diagrams (DFDs) to visualize the interactions within a system and identify potential threats [17, 18]. Recent studies have highlighted the effectiveness of STRIDE in various contexts, including application security planning and cloud computing. For instance, Tedyyana et al. [18] utilized STRIDE to categorize threats in a public service complaint application, demonstrating its applicability in real-world scenarios. On the other hand, LINDDUN focuses specifically on privacy threats, addressing concerns such as data protection and user privacy in systems.

The integration of LINDDUN with other security models, such as STRIDE, has been proposed to create a comprehensive threat modeling approach that encompasses both security and privacy concerns [19]. Safety threat models are essential for identifying and mitigating risks in various domains, including healthcare and transportation. For example, Formosa et al. [20] developed a modeling approach to predict vehicle-based safety threats, emphasizing the importance of understanding driver variability and its impact on conflict prediction. Similarly, the SEIPS model has been adapted to assess safety threats in surgical environments, highlighting the need for a systems-based classification scheme to improve patient safety [21]. In the context of smart grids, Castiglione et al. [22] introduced a hazard analysis methodology that incorporates both safety and security modeling to analyze the impact of cyber threats on smart grid infrastructures. This approach underscores the critical need to identify attack paths that could lead to safety violations, thereby ensuring the resilience of essential services.

3 User-Intimated and Trustworthy Ontology-Based Requirement Engineering Methodology (REM)

The requirement engineering and design procedure relies on a pyramid model (Fig. 1) that makes use of (i) OBRE to better understand the semantic relations among the requirements of system components and the baseline assessment criteria that are needed to evaluate the success criteria of targeted user needs; (ii) Privacy- security and safety threat modelling to comprehend the potential gaps and weaknesses of the overall design; and (iii) UI-REF to particularize the user-intimate requirements.

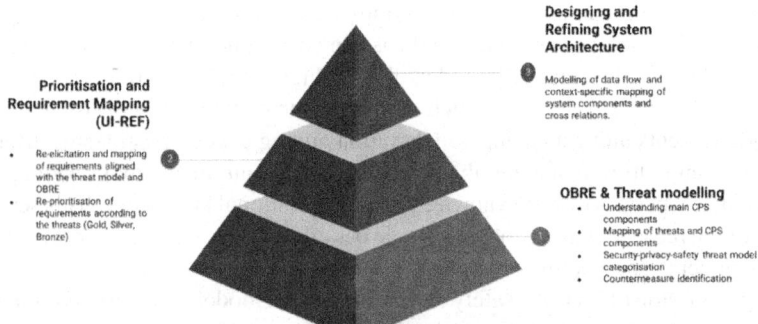

Fig. 1. Pyramid model for REM

At the lowest layer of the REM pyramid, OBRE and threat modelling are applied to understand the functions of main CPS components, mapping of threats and CPS components, security-privacy-safety threat model categorization and countermeasure identification. Then, in the middle layer, UI-REF is applied to investigate the end user and stakeholders' needs by re-eliciting and mapping requirements aligned with the threat model and OBRE. Here, (re-)prioritisation of requirements according to the level of severity and likelihood (Gold, Silver, Bronze) can be applied for more effective project management. Finally, in the top layer modelling of data flow and context-specific mapping of system components and cross relations are used to come up with a high-level system architecture. This architecture can be detailed in further steps by applying the same methodology depending on the level of complexity and size of the CPS.

3.1 Ontology-Based Requirement Engineering

OBRE is an approach that utilizes ontologies to improve the process of capturing, modelling, and managing requirements in software engineering. Ontologies are formal and explicit specifications of concepts, entities, and their interrelationships within a specific domain. In the context of requirement engineering, ontologies provide a structured and standardized way to represent domain knowledge, making it easier to understand and manage complex requirements.

Requirement elicitation (similar to UI-REF) is handled in two ways: (i) Domain Understanding: An ontology, addressing the requirements and Key Performance Indicators (KPIs) related to the vehicle, infrastructure, economic, sustainability, modularity, communication, security, privacy, legal, end-user, standards and pilot needs, helps in understanding the domain of the offered solution space defining relevant terms and their relationships. (ii) Stakeholder Collaboration: An ontology that provides a common vocabulary, enabling effective communication between stakeholders with different backgrounds, is useful to model the end user and pilot needs.

The ontological approach helps the system analysts see the big picture and the targeted specification and associated requirements to narrow down the project scope to a more realistic level. The ontology-based requirement engineering helps the consortium to enable: (i) Better semantic clarity (Ontologies add semantic clarity to requirements, reducing ambiguity and ensuring a common understanding among stakeholders); (ii) More accurate classification and categorization (Ontologies enable the classification of requirements based on their domain, functionality, or other criteria, making it easier to manage complex requirements sets); (iii) Holistic dependency analysis (Ontologies can represent dependencies between requirements to better understand the changes).

OBRE helps the analysts improve: (i) Consistency Checking: Ontologies allow automated consistency checks to ensure that requirements do not conflict with each other; and (ii) Completeness Analysis: Ontologies can be used to check if all necessary aspects of the domain are covered by the requirements. Another advantage of OBRE is to better manage the requirement engineering life cycle that is composed of: (i) Traceability: Ontologies facilitate traceability between requirements, design, and implementation, ensuring that each requirement is addressed in the final product; and (ii) Change Management: Ontologies help in understanding the impact of requirement changes by visualizing how changes affect other related requirements and elements.

OBRE presents a structured way of gathering knowledge about the system components and reusing that knowledge in similar situations. Knowledge reusability helps the consortium mainly in two dimensions: (i) Reusable Knowledge Components: Ontologies can include reusable knowledge components, making it easier to leverage existing domain knowledge for different pilots, or even for new projects; and (ii) Standardization: Ontologies provide a standardized way to represent domain knowledge, promoting consistency and reusability across projects.

The proposed OBRE methodology can be applied by implementing the following procedure: **Step-1** Domain Analysis and Ontology Development: Identify key domain concepts, define relationships, and develop an ontology schema using formal languages like OWL. Validate the ontology with stakeholders. **Step 2** Requirement Elicitation Using the Ontology: Use the ontology to guide requirement capture, ensuring completeness and consistency. Map requirements to relevant ontology elements. **Step 3** Formalization of Requirements: Translate requirements into logical rules and constraints using ontology languages. Validate consistency and completeness with reasoning tools. **Step 4** Ontology-Based Requirement Analysis: Detect conflicts and redundancies through automated reasoning, and ensure traceability between high-level and low-level requirements. **Step 5** Requirement Validation and Verification: Validate requirements with stakeholders and automate verification to ensure compliance with defined rules. **Step 6** Implementation Support and Traceability: Maintain links between ontology elements and system components, updating as the system evolves. **Step 7** Tool Integration and Automation: Use ontology modeling tools (e.g., Protégé) and integrate with requirement management systems to automate updates and maintain coherence.

3.2 Threat Modelling

Threat modeling enables the identification of potential threats and corresponding security mitigations that a cyber-physical system may face. Essentially, modeling tools and frameworks provide a comprehensive overview of the system's structure, outlining possible threats, the actors who could exploit them, and the methods they might use. Addressing threats discovered later in a system's lifecycle can be costly, making early threat identification crucial for reducing remediation expenses. To achieve this, threat modeling is integrated early in the development cycle. Depending on the project's needs, various modeling methods can be applied—each tailored to address specific aspects of the process. Often, multiple approaches are combined to ensure comprehensive threat coverage. The proposed threat modelling approach is based on the algorithm below (Algorithm 1) and provides a structured approach to applying the STRIDE, LINDDUN, and safety threat models for identifying vulnerabilities in CAEVS and ITS.

Algorithm 1. Algorithmic approach for threat modelling

Step 1: Initialize System Model
1.1. Gather system architecture details
1.2. Define the system boundary, external entities, and subsystems.
1.3. Identify assets, actors, data stores, and communication channels.
Step 2: Generate Data Flow Diagram (DFD)
2.1. Create a DFD for the system to represent data sources, data sinks, processes, and flows.
2.2. Annotate the DFD with component labels, flow types and trust boundaries.
Step 3: Apply STRIDE Threat Model (Security)
3.1. For each component, data flow, and data stored in the DFD:
-S: Check for potential Spoofing attacks (e.g., spoofing a vehicle's ID or sensor signal).
-T: Check for Tampering of data or processes (e.g., modifying vehicle commands).
-R: Evaluate Repudiation risks (e.g., lack of logging in critical safety functions).
-I: Look for Information disclosure vulnerabilities (e.g., eavesdropping on V2X).
-D: Identify Denial of Service (DoS) threats (e.g., jamming of wireless signals).
-E: Check for Elevation of Privilege scenarios (e.g., unauthorized access to vehicle).
3.2. Document identified STRIDE threats and their affected components.
Step 4: Apply LINDDUN Threat Model (Privacy)
4.1. For each data flow and data stored in the DFD:
- L: Identify Linkability risks (e.g., tracking driver behavior across trips).
- I: Check for Identifiability risks (e.g., linking vehicle ID to driver identity).
- N: Evaluate Non-Repudiation issues (e.g., inability to deny actions in data logs).
- D: Assess Detectability threats (e.g., unauthorized observation of vehicle status).
- D: Look for Disclosure of Information (e.g., leaking private data during communication).
- U: Identify Unawareness risks (e.g., lack of user knowledge on data collection).
- N: Evaluate Non-Compliance with privacy regulations (e.g., GDPR, CCPA).
4.2. Document identified LINDDUN threats and their affected data flows.
Step 5: Apply Safety Threat Model (Functional Safety and System Safety)
5.1. For each system component, data flow, and process:
- Identify potential safety-related hazards (e.g., unexpected braking, steering failures).
- Evaluate potential causes and scenarios for each hazard.
- Analyze possible cascading failures
- Consider environmental factors that may trigger unsafe conditions.
5.2. Prioritize safety threats based on risk assessment (severity, exposure, controllability).
5.3. Document identified safety threats and corresponding risk levels.
Step 6: Consolidate Threats and Map Vulnerabilities
6.1. Merge identified threats from STRIDE, LINDDUN, and Safety models.
6.2. Categorize threats based on the type (security, privacy, safety) & impacted components.
6.3. Map threats to specific vulnerabilities in components, data flows, and processes.
Step 7: Suggest Mitigation Strategies
7.1. Propose security controls and countermeasures for STRIDE threats
7.2. Recommend privacy-preserving techniques for LINDDUN threats
7.3. Suggest safety measures for identified hazards
Step 8: Document and Review with Stakeholders
8.1. Prepare a threat analysis report with identified vulnerabilities, risk levels, mitigations.
8.2. Review the report with security, privacy, and safety experts to validate findings.
8.3. Iterate the analysis based on feedback and updates to the system model.
Step 9: Update and Maintain the Threat Model
9.1. Update the threat model as the system evolves.
9.2. Perform periodic re-assessment to identify emerging threats and refine mitigations.

3.3 UI-REF

UI-REF serves as a comprehensive approach to various aspects of requirements elicitation, conflict resolution, ranking, and usability and impact assessment. An overview of the key components and principles of UI-REF is as follows: 1. Holistic Requirements Elicitation: UI-REF aims to comprehensively capture all relevant requirements for a system. It considers the diverse needs and contexts of users to ensure that no critical requirements are overlooked. 2. Conflict Resolution: UI-REF provides mechanisms to identify and resolve conflicts among different requirements or stakeholder interests. This ensures that the final set of requirements is coherent and can be effectively implemented. 3. Ranking and Prioritization: UI-REF categorizes requirements into priority levels. The framework supports a descending-order prioritization approach to determine which requirements should be implemented first. This helps in focusing on the most critical and high-impact features or functionalities. The requirements are ranked as follows: (i) MUST-have/Mandatory—These are those core design features which are perceived by the majority of a user group as offering the most needed affordances that can be provided by the target system. (ii) SHOULD-have/Desirable—These are those features that are the lower priority design features to be provided by the system within the project scope (within the limits of resources and technological constraints). (iii) NICE-to-have/Optional—These are those features that are highly contextualised to particular (sub)-sectors of the user group and are less common, and/or possibly more controversial and conflictual category. 4. Relationships-Based Usability and Impact Assessment: UI-REF incorporates methods for assessing usability and the potential impact of requirements. It considers how each requirement affects the overall user experience and the achievement of key performance indicators. 5. Ontologically-Guided Models (i.e. OBRE): UI-REF utilizes ontological models to ensure a clear and structured representation of requirements. This helps in formalizing and documenting requirements in a standardized manner. 6. Techniques and Tools: UI-REF employs a variety of techniques and tools, including online self-reporting, card sorting, laddering, nested video interviews, cognitive probes, and task walk-throughs. These techniques are used to gather user input, understand their needs, and clarify requirements. 7. Contextualization: UI-REF emphasizes the importance of understanding the context in which requirements will be applied. It takes into account the specific use contexts or scenarios in which the offered solutions will be used. 8. User Involvement: UI-REF promotes user involvement throughout the requirements elicitation process. Users play a crucial role in specifying their needs, confirming requirements, and prioritizing them. 9. Co-Design: UI-REF supports co-design approaches, where users actively participate in shaping the design of the solution. It enables users to specify context-aware requirements, including those related to privacy, security, and usability.

UI-REF is implemented through a layered approach (Fig. 2) starting with domain analysis and prototypical scenario identification. The UI-REF compliance verification is based on a natural consequence of the fact that the users in stating their requirements cannot be expected to be either exhaustive or factor in technology, market and practice constraints (State-of-the-art: SoA, State-of-Market: SoM, State-of-Practice: SoP) and trends of which they are not necessarily expected to be fully aware. Further, users are expected to articulate their own perceived requirements which may or may not be complete and may be incompatible with other users' requirements or project resources or in conflict with the technological and/or market imperatives and trends.

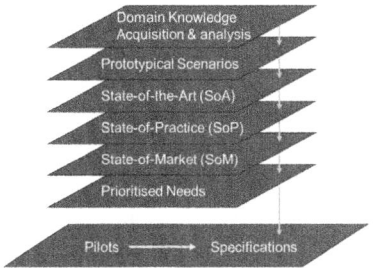

Fig. 2. UI-REF Layers

The insights arising from examining the dynamics of the confluence of market-pull and technology-push forces have a positive impact on the development and adoption of technologies and innovations. Such pull and push factors representing constraints and affordances invoked as requirements filters and augmenters are understood by performing respectively an SoA, SoP and SoM analysis, where the State-of-X is represented by the latest update on the state of current modus operandi, gaps, and, available enabling and emergent innovations from the viewpoint of X. This approach serves developer teams to indicate the best point of departure for the innovation developments to be achieved within a project, as well as clarify what technologies are needed by the project possibly as COTS that shall not actually be available from external technology in time to be integrated into the project and, therefore, must be developed by the project within the allocated resources. Tradeoffs must be considered if there is a project need from external technology that cannot itself be fully met promptly to enable the project to deliver the full list of user-demanded functionalities within given resources.

4 Proof of Concept

To prove the concept, the offered REM is applied in two research projects: (i) YEHO: A national project funded by the Scientific and Technological Research Council of Turkiye, focusing on the design and implementation of a 6-m autonomous bus, namely e-Centro by Otokar; (ii) ESCALATE (Powering European Union Net Zero Future by Escalating Zero Emission Heavy Duty Vehicles and Logistic Intelligence) focusing electrical trucks and their use in long-haul logistics.

The main concepts of the designed ontology relies on the ESCALATE project and they are modelled (may have either a 'subject' or 'object' role) in two classes: (i) Requirements (Vehicle_Requirements, Infrastructure_Requirements, Economic_Requirements, Sustainability_Requirements, Safety_Requirements, Security_Requirements, Communication_Requirements, Modularity_Requirements, End_User_Requirements, and Pilot_Requirements); ii) Assessment Criteria (Economic_Criteria, End_User_Assessment_Criteria, Pilot_Evaluation_Criteria, Infrastructure_Assessment_Criteria, Modularity_Assessment_Criteria, Safety_Criteria, Sustainability_Criteria, Vehicle_Cost_Assessment_Criteria, Vehicle_Perfromance_Assessment_Criteria, Laws, Regulations, Standards). The snapshot of the designed OBRE can be seen in

Fig. 3. The open source Protégé Ontology tool is used to design the main classes (column: [a] below) associated with the concepts linked with the requirements and assessment criteria listed above. The properties, which are so-called 'predicates' (column: [b] below), are also listed to present semantic relations between these concepts.

Main classes may either be the 'subject' or 'object' that are semantically linked through the 'predicates'. For instance, "subject: Economic_Requirements; predicate: assessed_in_terms_of_economic; object:Economic_Criteria" is a subject-predicate-object triple representing the following ontological statement: "Economic Requirements MUST/SHOULD be assessed in terms of Economic Criteria". e-Centro Autonomous is a battery EV and may be subject to serious security, safety and privacy threats. To address these threats, ENISA cyber threat taxonomy is used [23]. Safety threats are considered as the consequences of security weaknesses. For instance, an intrusion into the CAN-bus may block the Electronic Control Units and the braking system which may result in serious accidents. According to the STRIDE threat model, Table 1 presents some example cyber security threats that may lead to safety risks as well as countermeasures and priority ranks.

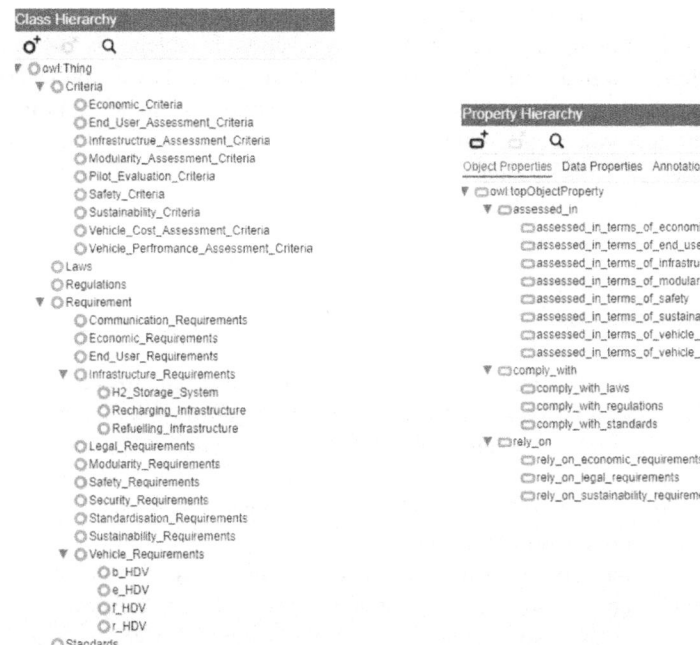

[a] Classes associated with high-level project super-concepts that can either be a 'subject' or 'object'

[b] Properties or 'predicates' representing the semantic relationship between any 'subject'-'object' pair

Fig. 3. A snapshot from the Protégé Ontology presenting the list of main classes and properties of the ESCALATE's requirements space

Table 1. Some examples of security threats, safety risk level and offered countermeasures (not limited to)

STRIDE Categories	Priority Ranking	Safety Risk	Threat Type	Countermeasure Description
S	High	High	Authentication failures	Storage of vehicle data; Authorisation of main system and framework Users; User Authentication; Users' anonymity; Consensus mechanism (Zero-knowledge Proof) and protocol; Hardware Security Module integration for Privacy Preservation, Multifactor Authentication, one-time-passwords
T	High	Medium	Integrity disruption	Integrity of vehicle, mobility data and passenger information
R	High	Low	Non-Repudiation	Non-Repudiation by Timestamping Non-repudiation of users' actions
I	High	Low	Confidentiality violation	ID Document Personal Data Presentation; ID Document Presentation of Proof of Address; Confidentiality of transactions; Privacy notice; Transaction Data Encryption; Confidentiality of Users; Data Masking for Avoiding Disclosure of Personal Data; Confidentiality of users' access privileges
D	Medium	High	Availability disruption	Availability of the vehicle, main framework and mobile application even under cyber-attacks Global Access via the vehicle interface even under Cyber-Physical Attacks
E	High	High	Unauthorized access	Protecting Private Authorization Information Authorization of main framework Users
TDE	Medium	High	Firewall protection failure	Network elements can be configured in a selectively permeable structure during communication with the device
DE	Medium	Medium	Blacklisting vs Whitelisting	Network elements can be configured in a selectively permeable structure during communication with the device
DE	Medium	High	Intrusion	Detection and prevention of security attacks

(*continued*)

Table 1. (*continued*)

STRIDE Categories	Priority Ranking	Safety Risk	Threat Type	Countermeasure Description
SIE	High	High	Weak Password Policies	Critical elements must store passwords within themselves through cryptographic processes, and when creating these passwords, they must contain at least 1 uppercase, 1 lowercase letter, 1 special character and 1 number
RE	Medium	High	False System Monitoring	Admin processing logs within a certain period
S	High	Low	Know Your Customer/User	Access Requirements for Individuals ID Document Personal Data Presentation ID Document Presentation of Proof of address
TI	High	High	Weak Encryption	Protecting Private Authorisation Information Data Masking for Avoiding Disclosure of Personal Data
IE	Medium	High	Problems with Role-based Access Control	Confidentiality of users' access privileges
STRIDE	High	High	Penetration attacks,	Penetration Testing by a 3rd party institution
STRIDE	High	Medium	Threat Modelling errors	Revisiting STRIDE and LINDDUN, repeat REM
STDE	Medium	High	Anomaly Detection Failures	Confidentiality of transactions
D	Low	High	Physical access to critical zone	Demilitarized Zones, strong authentication and authorization policies
SIE	Medium	High	Password Hashing errors	Secure storage on device
RE	Medium	Medium	Inaccurate Logging and Auditing	Local gateway to connect the vehicle to company servers

Similarly, privacy threat modelling is also applied by following the LINDDUN threat model. Table 2 presents the recommended countermeasures against privacy threats.

Table 2. Recommended countermeasures based on the privacy threat model

LINDDUN Categories	Priority Ranking	Countermeasure Type	Countermeasure Description
LINDDUN	Medium	Program Management	Data Processing Agreement with the data provider w.r.t. GDPR
Linkability	Medium	Security Assessment	Confidentiality of users' access privileges; Access Requirements for Individuals; Protecting Private Authorization data
Unawareness	Medium	Training	Consent to the processing of personal data, Privacy notice
LINDDUN	High	Configuration Management	Consensus mechanism (Zero-Knowledge Proof)
LINDDUN	High	Contingency Planning	Resiliency by backup plan in case of a cyber-attack
LINDDUN	High	Incident Response	Detection and Prevention of Network Security Attacks and System Violation
LINDDUN	Medium	Maintenance	Maintainability of the database by considering flowing data
Disclosure	High	Password Rules	Length of PIN code
Disclosure	Medium	Personnel Security	ID Document Personal Data Presentation ID Document Presentation of Proof of Address
Disclosure	High	System and data Integrity	Best-suited consensus protocol; Hardware Security Module integration for Privacy Preservation
LINDDUN	Medium	Using a Standard	Encryption algorithms should follow standards Timestamp format should follow standards
Linkability	Medium	Access Control Mechanism	Confidentiality of users' access privileges; Access Requirements for Individuals; access without user's involvement
Identifiability	High	Anonymity System	Users' anonymity
LINDDUN	High	Backups	Storage of passenger data in a secure environment
LINDDUN	Medium	Emissions Security	Overall Framework Security against Cyber-Physical Attacks
Disclosure	High	Encryption	Protecting Private Authorization Information
LINDDUN	Medium	Error Control)	Fault Tolerance in case of a System Failure
Detectability	Medium	Logging	Anonymisation of passenger-related data in a secure environment
Disclosure	High	Memory Protection	Protecting Private Authorization Information, Cryptographic Key Generation

(*continued*)

Table 2. (*continued*)

LINDDUN Categories	Priority Ranking	Countermeasure Type	Countermeasure Description
Linkability	Medium	Message Digest	Cryptographic Key Generation
Disclosure	High	Penetration Testing	Protection of the main framework against Reverse Engineering (needs to be done by a 3rd party)
Disclosure	High	Sanitize Input	Data Masking for Avoiding Disclosure of Personal Data
LINDDUN	High	Secure Network Communication	Main Framework (Cloud Environment) Sensitive Data Protection; Local gateway to connect to the vehicle with/without internet access; Confidentiality of transactions; Non-repudiation of main framework users' actions

5 Conclusion

This paper presents a user-intimate and trustworthy ontology-based REM that incorporates UI-REF and security, privacy and safety threat models. REM provides a comprehensive methodology for addressing the complex security, privacy, and safety challenges inherent in CAEVs and ITS. By incorporating user input, system specifications, and advanced threat models like STRIDE and LINDDUN, this approach ensures that all functional and non-functional requirements are captured systematically. The integration of trust, security, safety, and privacy principles into a cohesive framework establishes a robust foundation for developing resilient and secure systems, where cybersecurity, privacy, and safety are balanced from the outset to foster holistic trustworthiness. REM concept is presented by focusing on two research projects where OBRE, UI-REF and threat models are effectively used. For instance, OBRE is applied to heavy-duty vehicles, and electrical trucks in the ESCALATE project where long-haul logistics and fleet operations are targeted. The threat model exercise is presented for a 6-m autonomous bus in the YEHO project dealing with security, safety and privacy concerns and regarding countermeasures. Future research can focus on enhancing the proposed methodology by integrating machine learning for dynamic threat adaptation, aligning cross-domain ontologies for interconnected systems, and improving scalability for large-scale deployments. Combining OBRE with formal verification techniques, developing multi-domain risk assessment frameworks, and incorporating human-centric threat modeling are additional avenues. Other directions include automating conflict resolution using ontological reasoning, real-time compliance monitoring, creating domain-specific ontologies, and addressing ethical considerations to ensure transparency and accountability in system design.

Acknowledgement. This work is partly conducted in two research projects: (i) YEHO (Research and Innovation Act for trustworthy and Novel Connected and Autonomous Special and General-Purpose Vehicles), a national project funded under Turkiye's TEKNOHAMLE Programme with project ID: 101097267 and (ii) ESCALATE (Powering European Union Net Zero Future by

Escalating Zero Emission HDVs and Logistic Intelligence) a Horizon Europe project under the Grant Agreement No. 101096598.

References

1. Dobson, G., Sawyer, P.: Revisiting ontology-based requirements engineering in the age of the semantic web. In Proceedings of the International Seminar on Dependable Requirements Engineering of Computerised Systems at NPPs, pp. 27–29, (2006, November)
2. Badii, A., Fuschi, D.L.: User-Intimate Requirements Hierarchy Resolution Framework (UI-REF) in Work-flow Design for 3D Media Production & Distribution, Proceedings e report: 171 (2008)
3. Khan, R., McLaughlin, K., Laverty, D., Sezer, S.: STRIDE-based threat modeling for cyber-physical systems. In 2017 IEEE PES Innovative Smart Grid Technologies Conference Europe (ISGT-Europe), pp. 1–6. IEEE (2017, September)
4. Wuyts, K., Sion, L., Joosen, W.: Linddun go: a lightweight approach to privacy threat modeling. In 2020 IEEE European Symposium on Security and Privacy Workshops (EuroS&PW), pp. 302–309. IEEE (2020, September)
5. Wilson, T.: Multifractality in stride-to-stride variations reveals that walking involves more movement tuning and adjusting than running. Frontiers Network Physiol, **3** (2023)
6. Baumgartner, J., Gusmer, R., Hollman, J. H., Finnoff, J.T.: Increased stride-rate in runners following an independent retraining program: a randomized controlled trial. Scandinavian J. Med. Sci. Sports, **29**(11), 1789–1796 (2019)
7. Sutcliffe, A.: User-oriented requirements engineering. In Usability-and Accessibility-Focused Requirements Engineering: First International Workshop, UsARE 2012, Held in Conjunction with ICSE 2012, Zurich, Switzerland, June 4, 2012 and Second International Workshop, UsARE 2014, Held in Conjunction with RE 2014, Karlskrona, Sweden, August 25, 2014, Revised Selected Papers 1, pp. 11–33, Springer International Publishing (2016)
8. Natsiavas, P., Rasmussen, J., Voss-Knude, M., Votis, K., Coppolino, Komnios, I.: Comprehensive user requirements engineering methodology for secure and interoperable health data exchange. BMC Med Inf Decis Making, **18**(1) (2018)
9. Rizk, N.M., Mervat, H., Ahmed, M., Eman, S.: Creels: crowdsourcing based requirements elicitation for elearning systems. Int. J. Adv. Comput. Sci. Appl., **10**(10) (2019)
10. Oriol, M., Stade, M., Fotrousi, F., Nadal, S., Varga, J., Seyff, N., Schmidt, O.: Fame: supporting continuous requirements elicitation by combining user feedback and monitoring. 2018 IEEE 26th International Requirements Engineering Conference (RE) (2018)
11. Ahmad, K., Abdelrazek, M., Arora, C., Baniya, A.A., Bano, M., Grundy, J.: Requirements framework for engineering human-centered artificial intelligence-based software systems (2023)
12. Mahurkar, A.: A note on end-user requirements elicitation for electronic medical records implementations. Health Econ. Manage. Rev. **3**(4), 74–82 (2022)
13. Peras, D., Mekovec, R.: A conceptualization of the privacy concerns of cloud users. Inf. Comput. Secur., **30**(5), 653–671 (2022)
14. Bencharqui, H., Haidrar, S., Anwar, A.: Ontology-based requirements specification process. E3S Web Conf., **351**, 01045 (2022)
15. Alsanad, A.A., Chikh, A., Mirza, A.A.: A domain ontology for software requirements change management in global software development environment. IEEE Access **7**, 49352–49361 (2019)

16. Belani, H., Šolić, P., Perković, T.: Towards ontology-based requirements engineering for iot-supported well-being, aging and health. 2022 IEEE 30th International Requirements Engineering Conference Workshops (REW) (2022)
17. Sion, L., Yskout, K., Landuyt, D.V., Joosen, W.: Solution-aware data flow diagrams for security threat modeling. Proceedings of the 33rd Annual ACM Symposium on Applied Computing (2018)
18. Tedyyana, A., Ratnawati, F., Syam, E.: Threat modeling in application security planning citizen service complaints. Indonesian J. Electric. Eng. Comput. Sci. **28**(2), 1020 (2022)
19. Yeng, P.K., Wolthusen, S.D., Yang, B.: Comparative analysis of threat modeling methods for cloud computing towards healthcare security practice. Int. J. Adv. Comput. Sci. Appl., **11**(11) (2020)
20. Formosa, N., Quddus, M., Ison, S., Timmis, A.: A new modeling approach for predicting vehicle-based safety threats. IEEE Trans. Intell. Transp. Syst. **23**(10), 18175–18185 (2022)
21. Adams-McGavin, R.C., Jung, J.J., Dalen, A.S.H.M., Grantcharov, T. P., Schijven, M. P.: System factors affecting patient safety in the or. Ann. Surg., **274**(1), 114–119 (2019)
22. Castiglione, L.M., Hau, Z., Get, P., Co, K.T., Muñoz-González, L., Teng, F., Lupu, E.: Hagrid: security aware hazard analysis for smart grids. 2022 IEEE International Conference on Communications, Control, and Computing Technologies for Smart Grids (SmartGridComm) (2022)
23. ENISA Threat Landscape, https://www.enisa.europa.eu/publications/enisa-threat-landscape-2024 (2024)

Intelligent Transportation Management and Traffic Flow Analysis

AI and Leadership in Startup Innovation and Disruption: The Case of Mobility as a Service (MaaS)

Michele Vincenti[✉]

University Canada West, Vancouver, BC V6Z 0E5, Canada
Michele.Vincenti@ucanwest.ca

Abstract. Incorporating AI into Mobility as a Service (MaaS), the evolving transportation landscape shows promise in improving service efficiency and sustainability. However, the rapid growth of progress poses challenges that demand skilled leadership. By examining MaaS implementations and upcoming new technologies, this article demonstrates how AI can elevate user experiences and operational effectiveness. It also proposes leadership qualities to encourage innovation to effectively manage technological integration in the industry. Looking ahead, leaders will be influential only in steering AI integration within MaaS while fostering a culture of innovation and accountability. This article underscores the significance of learning and adaptability among leaders to ensure the growth of AI-powered MaaS ecosystems, ultimately developing sustainable urban mobility solutions. Mobility as a Service (MaaS) provides efficient and effective transportation options to offer users personalized mobility solutions. This approach plays a role in enhancing transportation efficiency and sustainability through advanced technology. By integrating transport modes into a platform, MaaS simplifies access to booking and payment services, resulting in flexible, personalized, on-demand travel experiences for users and improving sustainability. This article also gives an overview of the industry status and the obstacles AI-driven MaaS startups face. It advises industry leaders on the skills and capabilities they need to navigate the future of leadership in the AI-powered MaaS ecosystem. Leaders will be successful only if they embrace strategic planning, and the article guides mastering it.

Keywords: AI-driven MaaS · Leadership in startups · Mobility as a service · Innovation management

1 Introduction

Mobility as a Service (MaaS) combines various modes of transportation to provide seamless, efficient, and tailored mobility services to users. This integration is crucial for modern transportation as it seeks to improve the efficiency and sustainability of urban mobility by leveraging technological advancements. Through one interface, MaaS enhances user experience by providing streamlined access to multiple transport services, including booking and payment systems. The result is flexibility, personalization, on-demand availability, and seamless user experience.

Recent environmental movements have underscored the importance of reducing the impact of fuels and promoting sustainable transportation practices. The fluctuating nature of oil prices due to geopolitical factors has further highlighted the need for alternative transportation solutions.

Cities worldwide have been exploring approaches to transportation that prioritize efficiency and sustainability over models.

Technology played a crucial role in accelerating the popularity of MaaS, first with the implementation of the Internet of Things (IoT) and then with the arrival of Artificial Intelligence (AI).

In 1999, Kevin Ashton, a co-founder of the Auto-ID Center at MIT, created the term "Internet of Things" because of his work on radio-frequency identification (RFID) tags for connecting objects to the internet for tracking and management purposes [1].

The Internet of Things (IoT) is a network of interconnected devices with sensors, software, and technologies. These devices can share data, enabling them to interact with their surroundings and other devices without intervention. IoT is applied in homes, healthcare, transportation, and other areas to facilitate data integration and automated control in daily tasks and industrial settings [2].

A new revolutionary player in technology is Artificial Intelligence (AI). Artificial Intelligence is a branch of computer science focused on developing systems for tasks requiring intelligence. These tasks include learning, reasoning, problem-solving, perception and language comprehension. AI spans from specialized systems handling tasks to potential future systems that could mimic human cognitive abilities [3].

In history, innovation has always been the catalyst for economic improvements. Nicolay Kondratieff, in 1984, developed the theory of the long wave cycle. He discovered the existence of long-term economic cycles that last approximately 50–60 years. These cycles are characterized by rapid economic growth followed by a slowdown. Kondratieff identified these cycles by analyzing price behaviors, interest rates, and other financial indicators across several countries.

Kondratieff's theory suggests that technological innovations drive these waves and increase productivity and economic output. Each wave typically ends with a period of crisis or depression, followed by a new cycle of growth driven by new technological advancements [4].

When we consider the evolution of IoT and AI in MaaS applications, it is evident that today, technological progress presents business opportunities for entrepreneurs and contributes to economic advancement.

The forthcoming leadership landscape in AI-powered MaaS will require a mix of expertise, ethical insight, and innovative strategic thinking. Leaders must leverage AI's capabilities to adopt innovation while effectively addressing the complexities of integrating these technologies into private and public transportation services. Promoting innovation guarantees that leadership remains responsive and forward-thinking when incorporating AI into decision-making processes.

This article aims to equip entrepreneurs within the MaaS sector with the knowledge to lead their teams. This article also gives an overview of the industry status and the obstacles AI-driven MaaS startups face. It advises industry leaders on the skills and capabilities they need to navigate the future of leadership in the AI-powered MaaS ecosystem.

Startup leaders recognize the importance of AI in MaaS for delivering personalized travel recommendations using user profiles and real-time data, and at the same time ensuring services meet individual preferences and needs. Strategic planning and management's ability to handle complex processes with a flexible and responsive team are essential for achieving these goals [5].

Sustainability is a key objective in MaaS, with AI playing a vital role in enhancing efficiency and the user experience by transitioning from rigid to user-centric transportation systems. Concerns such as data privacy, system compatibility, and robust solutions are addressed alongside updated regulations to reduce urban congestion and pollution [6].

The electric vehicle (EV) market is experiencing rapid growth, with global sales hitting 13.2 million units in 2023 due to increasing consumer interest and investments in battery technology [7]. The shift toward battery electric vehicles (BEVs) is significant, with over 40 million units worldwide, driven by consumer preferences and advancements in battery technology [8].

Additionally, there is a noticeable increase in the adoption of electric public transportation. The U.S. had 6,147 buses by the end of 2023, with cities like São Paulo and Bogotá planning significant expansions of their electric bus fleets by 2030. European cities like Amsterdam and Rotterdam are incorporating electric buses to achieve environmental targets, emphasizing the role of EVs in reducing oil dependency and promoting the decarbonization of transportation [9, 10].

The concept of Electric Mobility as a Service (eMaaS) expands on mobility as a Service (MaaS) by not incorporating modes seamlessly but also emphasizing sustainability through the integration of electric mobility systems and shared electric mobility services (SEMS). To make this integration successful, it is crucial to develop system architectures that support the inclusion of electric mobility within the broader MaaS framework [11].

2 Startup Dynamics in MaaS

The landscape of MaaS startups is evolving rapidly, and a range of service models and operational approaches are being explored. These startups are classified based on their level of integration and the scope of services they provide, from route planning to mobility solutions that combine payment and ticketing on a unified platform [12].

A good example of implementation in public transportation in the United States is Parkway Autonomous Operations. The system developed by the company is designed to adapt based on community usage, fairness, cost-effectiveness, safety, and sustainability. This flexibility ensures that the solutions offered by Parkway Autonomous remain relevant and efficient over time. Their product is tailored to meet the needs of various public transportation agencies [13].

These strategic benefits set Parkway Autonomous apart in the public transportation industry by delivering sustainable and flexible solutions that address the changing demands of modern cities.

Moovel provides another example of a startup that has found solutions to the initial challenges. These challenges are not unique to this company, but they are common among Mobility as a Service (MaaS) startups. During their launch phase, they encountered obstacles: (a) Competition and Market Saturation, (b) Operational Challenges, (c) Financial Viability and (d) Technological and Regulatory Barriers.

(a) Competition and Market Saturation: The MaaS market is fiercely competitive, with startups competing for market presence. This stiff competition makes it challenging for companies like Moovel to distinguish themselves unless they offer unique or superior services [14].
(b) Operational Challenges: Managing a MaaS platform involves handling transport services, user interfaces, payment systems and city infrastructure. Moovel may have faced difficulties integrating these elements to ensure a user experience, but in the end, they have successfully integrated [15].
(c) Financial Viability: like new ventures, maintaining financial well-being is crucial. Moovel had to secure revenue through partnerships, user fees, and other channels to sustain its operations. Any shortfall in funding or revenue generation could have hindered long-term sustainability [15].
(d) Technological and Regulatory Barriers: Implementing MaaS solutions requires navigating frameworks and technological obstacles. These challenges could involve issues related to data privacy, local transportation regulations and the technical integration of modes of transportation that might have impeded Moovel's growth or operational efficiency. Also, they successfully solved the challenges here [15].

3 Leadership in MaaS Startups

After examining startup dynamics, we can explore the skills and competencies that leaders must develop to lead the MaaS startup's team.

In facing the obstacles encountered by Moovel, here's how leadership can tackle each of these hurdles:

(a) Dealing with Competition and Market Saturation: Leaders should focus on innovation to set their product or service apart and navigate a new market. Focus on innovation may involve identifying target markets, enhancing the customer experience, or emphasizing selling points that distinguish the startup. Forming partnerships and alliances can also play a role in expanding market presence and gaining a competitive edge [16].
(b) Overcoming Operational Challenges: Tackling operational issues requires implementing innovative strategies to optimize processes, manage the supply chain effectively, and ensure quality control. Leaders may need to invest in technologies or procedures that enhance efficiency and productivity. Moreover, adopting approaches that can respond to changes in the business landscape is essential for maintaining adaptability [17].
(c) Financial Sustainability: Making sure a startup remains financially stable is crucial. Strategic financial management may involve securing funding, improving cash flow management, and making economic choices promoting long-term growth. Creative funding options, like crowdfunding, forming partnerships or seeking venture capital, can bring in the needed capital while reducing risks [18].

(d) Navigating Regulatory Challenges: Successfully overcoming regulatory obstacles requires a proactive and well-informed approach. Startups must stay updated on tech trends and regulatory changes by investing in research and development, advocating for their interests, or adjusting their business strategies to comply with laws. Leaders must have foresight in predicting trends and challenges, strategically using them as opportunities rather than obstacles [19].

Amabile and Khaire (2008) stress the significance of fostering creativity within companies based on insights from discussions with leaders at firms like Google and IDEO. They highlight that leadership should focus on generating ideas and nurturing employee creativity. This involves encouraging viewpoints among employees, leveraging technology for collaboration, and managing processes effectively to encourage innovation without hindering it [20].

In all these scenarios, leadership must be responsive and proactive, employing forward-looking approaches to navigate challenges and seize opportunities for expansion and advancement.

The initial suggestion for leaders to enhance their strategic management skills in fostering innovation is understanding their domain. This knowledge guides the processes and innovation efforts effectively, enabling leaders to make informed decisions while earning the admiration and trust of their team members, who are likely experts in their respective fields. The second recommendation is that leaders should adeptly utilize influence tactics tailored to meet the requirements of creative individuals within organizational contexts. Selecting a tactic based on followers and situational dynamics is crucial rather than applying a one-size-fits-all-all approach. The third recommendation underscores the importance of cultivating a workplace culture that nurtures creativity through relationships and proper allocation of resources, fostering an environment where novel ideas can be nurtured and put into practice efficiently [21].

4 AI-Driven Innovations and Business Models in MaaS

After exploring the essential leadership competencies required in mobility as a Service (MaaS), it is time to delve into the opportunities that AI can bring forth.

Rajabi et al. (2023) proposes a knowledge-based AI framework for Mobility as a Service (MaaS) that improves data integration and customization by combining various data sources, including mobility experts and environmental factors and at the same time emphasizes continuing learning to improve adaptability. The framework also suggest using a dataset as a proof of concept to improve the transparency and trust in the automated decisions [5]. Michelin is an example of company that successfully used AI to optimize last-mile delivery, reduce costs, and improve customer satisfaction by scheduling delivery slots and minimizing failed deliveries, which also has appositive effect on lowering the consumption and CO_2 emissions [22].

(MaaS) reflects a dedication to enhancing efficiency, safety and sustainability within fleet management, illustrating how technology can significantly transform industry norms.

Schweiger (2023) describes how integrating AI in public transportation improves efficiencies, safety, and customers satisfaction through various applications. AI improves surveillance by analyzing CCTV footage to manage crowd sizes and detect safety concerns. AI helps to adopt proactive maintenance to predict equipment failures analyzing sensor data to reduce downtime and maintenance costs [23]. AI also optimizes traffic ad routes by adjusting schedules based on real-time traffic conditions, improving punctuality, and reducing traffic congestion [24].

Furthermore, IoT technology supports smoother toll collection and ticketing processes through technologies such as RFID tags and sensors integrated into vehicles and infrastructure. These technologies streamline payments while reducing traffic backups at toll booths [25].

AI-driven natural language processing (NLP) systems enhance customer service because of its engagement with the customers, and the ability to gather feedback by using chatbots and virtual assistants. These tools provide real-time information on schedules, fares, and route changes, addressing inquiries effectively and enhancing the customer experience by keeping passengers well-informed [26]. AT&T exemplifies the integration of such technology, where chatbots effectively manage real-time interactions, increasing user satisfaction and improving operational efficiency in mobility services [27].

By harnessing the capabilities of AI, public transport operators improve efficiency and enhance passenger satisfaction while contributing to more sustainable urban mobility solutions. Intelligently integrating these technologies can revolutionize transportation into a considerate, safer, and responsive system that meets the needs of contemporary city residents.

5 Challenges in Leading AI-Driven MaaS Startups

Successfully leading AI-driven MaaS startups requires anticipating needs, overcoming bureaucratic obstacles, ensuring scalable and reliable AI systems, managing the operations in real-time [28]. It is also crucial to address ethical issues by avoiding biases, ensuring transparency in data use, and adhering to the regulatory framework in order to build trust [29].

The development of AI (XAI) models plays a role in sensitive applications to ensure that humans can understand and evaluate the decisions made by AI systems [30].

AI (XAI) means explainable, using transparency, trust, compliance, error reduction and collaborative enhancement as its pillars. The ultimate goal is to bridge the gap between human cognitive processes and AI.

Policymakers are responsible for creating regulations that encourage innovation in AI while safeguarding data privacy. Striking this balance is essential to ensure AI technologies benefit society without compromising rights. Efforts have been made to revise and tailor existing privacy laws, such as GDPR and CCPA, to address AI's challenges, including data usage, consent, and data subject rights [31].

Ethical dilemmas have started to be addressed over the years and remain ongoing. Significant efforts in intelligence include implementing policies such as the Artificial Intelligence Act by the European Union, the US Artificial Intelligence Bill of Rights and the initiatives led by the Global Partnership on Artificial Intelligence (GPAI). These measures set a bar for ensuring AI use by focusing on human control, preventing harm, promoting fairness, and enhancing transparency. They establish values and practical guidelines to guarantee that AI systems operate safely, fairly, and transparently to combat biases in automation effectively [32].

Leading an AI-focused startup involves creating an environment that fosters innovation while mitigating risks associated with AI technology. Cultivating a culture requires attracting and retaining individuals well-versed in cutting-edge AI technologies, nurturing a collaborative atmosphere, and overseeing multidisciplinary teams capable of addressing the intricate challenges of integrating AI into mobility services [33].

Furthermore, ongoing workforce development and leadership training must be continuous. The entire organization needs to keep pace with evolving trends.

Effective leadership guides the company's advancements and ethical considerations in startups centered around AI-driven mobility as a Service (MaaS). Leaders in this domain must know about AI technologies and their practical applications. Managerial competencies are not enough, and leaders must be trained in technology to lead a successful team.

Understanding this information is essential for creating business strategies and promoting a culture of innovation within the company [34].

To equip leaders to cultivate a culture of innovation within an organization, it is essential to highlight the importance of learning and flexibility. Leaders must possess the ability to manage and incorporate technologies effectively. Effective technology management entails participating in training programs focusing on emerging trends and fostering thinking. Promoting communication and collaboration across departments can ignite creative ideas and propel innovation forward. Moreover, leaders should create an environment that embraces risk-taking, where experimental concepts and setbacks are seen as parts of the learning and innovation journey.

6 The Future of Leadership in AI-Powered MaaS Ecosystems

Exploring the leadership landscape in AI-powered MaaS ecosystem means understanding how leadership roles will transform with AI's increasing impact. Leaders must master technological nuances, uphold ethical standards, and improve service quality while maintaining fairness and privacy [35]. Leaders must have skills in utilizing AI to predict consumer behavior and improve transportation solutions, in the same way companies like Amazon and Netflix use behavioral analysis in their businesses [36]. Cultivating a learning environment that emphasizes ethical technology use is necessary, and leaders must be able to motivate their teams to remain updated on AI advancements and moral standards [37]. Soon, advancements in quantum computing will play a critical role in strategic decision-making, requiring leaders to understand quantum technologies to optimize operational efficiency and planning [38].

Transformational leaders inspire creativity by sharing a vision and contagious enthusiasm, nurturing development, and creating an atmosphere that encourages team members to explore innovative ideas and take calculated risks. This leadership style plays a role in uniting the team's efforts with the organization's goals, driving progress, and fostering innovation.

Leaders must craft an innovation strategy that complements and enhances the business objectives. This comprehensive strategy considers factors such as market needs and technological advancements. When formulating these strategies for innovation, leaders must stay familiar with market dynamics and their organization's internal capabilities. These innovation strategies must be harmonized with the business framework to enhance competitiveness and performance [39].

7 Conclusion

In summary, the evolution of Artificial Intelligence in Mobility as a Service (MaaS) brings challenges and prospects for leadership. Influential leaders must leverage AI's potential to improve decision-making processes and operational efficiency while grappling with dilemmas and regulatory complexities posed by emerging technologies. They should foster a culture that promotes thinking and ethical conduct, ensuring the utilization of AI tools. This adaptive approach will be pivotal for sustaining growth in AI-driven MaaS ecosystems, underscoring the necessity for leaders with understanding and strategic perspectives.

This article emphasizes the significance of training leaders to navigate these challenges, emphasizing the importance of leadership in unlocking the capabilities of AI to revolutionize transportation.

References

1. Ashton, K.: That 'Internet of Things' Thing. RFID J. (2009) Available at: http://www.rfidjournal.com/articles/view?4986
2. Ishak, M.K., Bhatti, M.K.L., Khan, I., Kim, K.-I.: A comprehensive review of Internet of Things: TECHNOLOGY stack, Middlewares, and Fog/Edge computing interface. Sensors 22(3), 995 (2024). https://doi.org/10.3390/s22030995
3. Rodrigues, M., Silva, R., Borges, A. P., Franco, M., Oliveira, C.: Artificial intelligence: threat or asset to academic integrity? A bibliometric analysis. Kybernetes (2024). https://doi.org/10.1108/K-09-2023-1666
4. Kondratieff, N.D.: The Long Wave Cycle. Richardson & Snyder, New York (1984)
5. Rajabi, E., Nowaczyk, S., Pashami, S., Bergquist, M., Ebby, G.S., et al.: A knowledge-based AI framework for mobility as a service. Sustainability 15(3), 2717 (2023). https://doi.org/10.3390/su15032717
6. Maas, B.: Literature review of mobility as a service. Sustainability, 14(8962) (2022). https://doi.org/10.3390/su14148962
7. S&P Global Mobility.: Electric vehicle trends (2023). Retrieved on April 24, 2024, from https://www.spglobal.com/mobility/en/topic/electric-vehicle-trends.html
8. Ritchie, H.: Tracking global data on electric vehicles. Our World in Data (2024). Retrieved from https://ourworldindata.org/electric-car-sales
9. Sustainable Bus.: Electric bus in public transport (2023). Retrieved from https://www.sustainable-bus.com/electric-bus/electric-bus-public-transport/
10. International Energy Agency: Electric vehicles (2024). Retrieved from https://www.iea.org/energy-system/transport/electric-vehicles
11. Reyes García, J.R., Lenz, G., Haveman, S.P., Bonnema, G.M.: State of the art of mobility as a service (MaaS) ecosystems and architectures—an overview of, and a definition, ecosystem and system architecture for electric mobility as a service (eMaaS). World Electric Vehicle J. 11(1), 7 (2019). https://doi.org/10.3390/wevj11010007

12. Esztergár-Kiss, D., Kerényi, T., Mátrai, T., et al.: Exploring the MaaS market with systematic analysis. Eur. Transp. Res. Rev. **12**, 67 (2020). https://doi.org/10.1186/s12544-020-00465-z
13. StartUs Insights.: Mobility startups (2024). Retrieved from https://www.startus-insights.com/innovators-guide/mobility-startups/
14. StartUs Insights. (n.d.). 5 top mobility as a service startups out of 730. Retrieved April 24, 2024, from https://www.startus-insights.com/innovators-guide/5-top-mobility-as-a-service-startups-out-of-730/
15. Moovel US. (2024). Retrieved April 24, 2024, from https://www.moovelus.com/
16. Calvino, F., Criscuolo, C.: Digital innovation and entrepreneurship: a review of challenges in competitive markets. J. Innov. Entrepreneurship, **8** (2019), Article 11. https://innovation-entrepreneurship.springeropen.com/articles/https://doi.org/10.1186/s13731-019-0101-4
17. McGrath, A.: Streamlining supply chain management: Strategies for the future. IBM Blog (2024, February 19). https://www.ibm.com/blog/supply-chain-strategy/
18. Jeong, J., Kim, J., Son, H., Nam, D.: The role of venture capital investment in startups' sustainable growth and performance: focusing on absorptive capacity and ven-ture capitalists' reputation. Sustainability **12**(8), 3447 (2020). https://doi.org/10.3390/su12083447
19. OECD: Case studies on the regulatory challenges raised by innovation and the reg-ulatory responses (2020). https://www.oecd-ilibrary.org/sites/70df2cab-en/index.html?itemId=/content/component/70df2cab-en
20. Amabile, T.M., Khaire, M.: Creativity and the role of the leader. Harvard Bus. Rev. **86**(10), 100–109 (2008)
21. Mumford, M.D., Scott, G.M., Gaddis, B., Strange, J.M.: Leading creative people: orchestrating expertise and relationships. Leadersh. Q. **13**(6), 705–750 (2002)
22. Michelin Connected Fleet. (2024, February 1). Fleet and Mobility Trends for 2024 You Need to Know. Retrieved from https://connectedfleet.michelin.com/blog/fleet-and-mobility-trends-for-2024-you-need-to-know/
23. Schweiger, C.: Technology-enabled Mobility: Five predictions for 2023. Intelligent Transport (2023). Retrieved from https://www.intelligenttransport.com/transport-articles/146993/technology-enabled-mobility-predictions-2023/
24. Bowen, E.: Real-world applications of IoT for traffic monitoring. Telnyx (2024, January 8). https://telnyx.com/resources/iot-monitoring-traffic
25. Kapoor, M.: IoT in Transportation: The Role of IoT Solutions in Transforming Mobility. CognitiveClouds (2023, August 11). https://www.cognitiveclouds.com/insights/iot-in-transportation
26. RTS Labs.: AI in logistics: Transforming supply chain efficiency and customer service. RTS Labs (2024, February 23). Retrieved from https://rtslabs.com/ai-logistics-transformation-efficiency-customer-service
27. AT&T. (n.d.). Mobility as a Service—IoT Solutions at AT&T Business. Retrieved from https://www.business.att.com/products/mobility-as-a-service.html
28. Microsoft for Startups.: Simplifying the journey of AI startups through experimentation and scaling. Microsoft for Startups Blog (2023, June 23). Retrieved from https://startups.microsoft.com/blog/simplifying-the-journey-of-ai-startups-through-experimentation-and-scaling/
29. Chen, J., Storchan, V., Kurshan, E.: Beyond Fairness Metrics: Roadblocks and Challenges for Ethical AI in Practice (2021). CoRR, abs/2108.06217. Retrieved from https://ar5iv.labs.arxiv.org/html/2108.06217
30. Shan, R.: The Ethical Algorithm: Balancing AI Innovation with Data Privacy. Datafloq (2024, January 8). Retrieved from https://datafloq.com/read/ethical-algorithm-balancing-ai-innovation-data-privacy/
31. IABAC.: AI And Privacy: Balancing Innovation with Data Protection (2023, July 26). Retrieved from https://iabac.org/blog/ai-and-privacy-balancing-innovation-with-data-protection

32. European guidelines on the responsible use of generative AI in research. (2024). CESAER. https://www.cesaer.org/news/european-guidelines-on-the-responsible-use-of-generative-ai-in-research-694/#:~:text=The%20guidelines%20recommend%20researchers%20to,the%20evaluation%20of%20research%20proposals
33. DataRoot Labs.: AI in Logistics: Emerging Startups, Challenges and Use Cases [UPDATED 2024] (2024, March 11). Retrieved from https://datarootlabs.com/blog/ai-in-logistics-emerging-startups-remaining-challenges-and-new-models
34. Davenport, T., Foutty, J.: AI-Driven Leadership. MIT Sloan Management Review (2018, August 10). Retrieved from https://sloanreview.mit.edu/article/ai-driven-leadership/
35. Wassouf, W.N., Alkhatib, R., Salloum, K., et al.: Predictive analytics using big data for increased customer loyalty: Syriatel telecom company case study. J. Big Data **7**, 29 (2020). https://doi.org/10.1186/s40537-020-00290-0
36. Froot, K.: Predicting Performance Using Consumer Big Data (2021, August 18). Retrieved from https://scholar.harvard.edu/files/kenfroot/files/Predicting_Performance_Using_Consumer_Big_Data-Aug18.2021.pdf
37. World Economic Forum: How artificial intelligence will transform decision-making (2023, September). Retrieved from https://www.weforum.org/agenda/2023/09/how-artificial-intelligence-will-transform-decision-making/
38. McKinsey and Company: Steady progress in approaching the quantum advantage (2023). Retrieved from https://www.mckinsey.com/capabilities/mckinsey-digital/our-insights/steady-progress-in-approaching-the-quantum-advantage
39. López, D., Oliver, M.: Integrating Innovation into Business Strategy: Perspectives from Innovation Managers. Sustainability, **15**(6503) (2023). https://doi.org/10.3390/su15086503

Research on Traffic Analysis Zones Division Model for New Roads Based on Complex Network Theory

Zhiyong Wen(✉), Xiaoxiong Weng, and Bangquan Xie

School of Civil Engineering and Transportation, South China University of Technology, Guangzhou510641, China
ctwenzy@mail.scut.edu.cn

Abstract. In this research, we present a traffic analysis zones division model based on complex network theory, which is crucial for conducting feasibility studies for new road projects. The model is designed with three fundamental components: a multi-layer network, feature extraction, and a two-step division method. The highway network is represented using the multi-layer approach that includes the travel cost network, road structural network, and Origin-Destination network. Additionally, to measure the similarity between nodes, both structural attributes and clustering attributes are considered as eigenvalues in the similarity measurements. Furthermore, the two-step division method comprises Louvain Community Detection algorithm for coarse division and modified clustering algorithm for fine division. With the proposed model, the nodes of highway can be divided into traffic analysis zones. Through the case study of the Guangdong-Hong Kong-Macao Greater Bay Area, the results of the model demonstrate a significant improvement in three evaluation metrics when compared to other methods, highlighting the superiority of the proposed model.

Keywords: Traffic Analysis Zones · Complex Network · New Construction Road · Feasibility Study

1 Introduction

Traffic analysis zones (TAZs) are defined as that the traffic area is divided into several geographical units according to the needs of traffic analysis and traffic demand prediction model. It is the smallest spatial unit to analyze the travel and distribution of residents and vehicles. TAZs division is an important step to evaluate the traffic volume of new highway. The purpose of traffic engineering is the functional design of physical infrastructure and the management of traffic. For the improvement of the highway network, new road sections are being built. Transportation modelling allows the prediction of the impacts for constructing a new road to assess the suitability of traffic supply and demand in the phase of evaluation. In the feasibility study phase. The well-known four-step model for travel-demand is ubiquitously used due to institutional and financial requirements. Traffic demand is a collection of individual trips, and each trip is the result of multiple choices made by users, it's a challenge for comprehensively studying the entire network with the stochastic, dynamic, and diverse traffic. For facilitating traffic investigation and analysis, researchers have proposed the concept of TAZs, which involves dividing the study area into distinct zones. As in Lu H.P.'s research [1], the necessity of TAZs as the initial step in the traditional four-step procedure for travel demand forecasting models was emphasized. Significantly, a well-defined TAZs plays a pivotal role as they are used in all phases of four-step model.

In the prior researches, TAZs are divided based on the administrative areas that allows for future traffic volume forecasting in conjunction with regional economic considerations, but it suffers from the drawback that individual characteristics within a zone are not always homogenous. On the other hand, the quantitative method of TAZs is lacking. To address these limitations, we propose the TAZs Division model based on complex network theory, and the model is driven by data. The contributions of this research are as follows: Multi-layer complex network for highway was proposed and applied to generate multiple attributes of nodes. The road data and traffic data are used to form the road structural network, travel cost network, and Origin-Destination network that consisted of the multi-layer complex network; The eigenvector of each node was obtained with feature extraction that reflect the path choice and trip distance for nodes. The contiguity was guaranteed with the trip distance between each node and reference nodes, and the homogeneity was guaranteed with path choice of traffic between each node and all other nodes; Two-step division method was presented. This method fully exploits the diverse features of nodes to achieve accurate division, while simultaneously enhancing the processing speed for large-scale networks.

The subsequent sections of this paper are organized as follows: "Related work" (Sect. 2) presents a review of the literature. "Methodology" (Sect. 3) provides a detailed presentation of the TAZs Division model based on complex network theory, which consists of three modules. In "Case Study" (Sect. 4), the dataset for the Guangdong-Hong Kong-Macao Greater Bay Area is introduced, and the results and the comparison are conducted for verification. Finally, "Summary and Prospect " (Sect. 5) offers concluding remarks and outlines potential future research directions.

2 Related Work

Among several criteria applied toward TAZs division, these researches address the contiguity and/or the homogeneity constraints for obtaining well-defined TAZs. Contiguity is one of the most import criteria for this problem. Contiguity can be defined in terms of a network by equating each zone with a vertex and representing adjacency. Furthermore, homogeneity the represents the quality of a TAZ being clustered, maximizing homogeneity is the goal during the TAZs division process. Therefore, connectivity and homogeneity become the indexes to evaluate the effect of TAZs.

Some approaches have been explored in the realm of TAZs research. On the one hand, the TAZs facilitate traffic management and traffic safety. Li X. D. and Yang X.G. [2] introduced the concept of TAZs for urban road networks, aiming to simplify traffic control and management systems, enhance system reliability, and support system development. Their work involved analysing traffic correlation and similarity, laying the groundwork for defining TAZs mathematically. Moreover, Li R.M. and Gong X.Y. [3] proposed the concept of dynamic distributivity in urban road networks, conducting TAZs studies not only spatially but also temporally. For the traffic management, Lian J. [4] proposed a mobility pattern mining framework, facilitating trip-based TAZs computation while incorporating various auxiliary information. Weerasinghe O. and Bandara S. [5] presented a methodology for identifying Modified TAZs based on traffic flow distribution, subsequently redistributing macrolevel trip origins and destinations into Modified TAZs, considering both road networks and land-use characteristics to establish the relationship between trip production/attraction and land use. Additionally, for the traffic safety, Zeng Q. [6] developed a multivariate spatial model for analysing daytime and nighttime crash frequencies by injury severity within TAZs. Zhang C. [7] conducted a comprehensive analysis of zonal safety evaluation, introducing several criteria to assess crash risk at the zonal level and developing a negative binomial regression model to identify significant factors for unsafe zones.

On the other hand, the TAZs can be generated using different approaches, such as rule-based and data-driven division methods. Firstly, some rule-based approaches have been explored in follow studies. China Highway Engineering Consultants Corporation [8] and Yu H.J. [9] divided the TAZs with the administrative areas as the smallest units and merged the units with similar attributes. Ghadiri et al. [10] and Wang S.W. et al. [11] presented a TAZs method with regular geometric shapes, called regular TAZs to traffic zoning. Moreover, Assuncao et al. [12] summarize the neighbourhood structure by a minimum spanning tree (MST), and then partition the MST by successive removal of edges that link dissimilar regions. Martinez et al. [13] introduced a novel methodology for the design of Traffic Analysis Zones (TAZ) by utilizing a smoothed density surface of geocoded travel demand data. This approach aims to preserve data integrity by minimizing information loss during the transition from a continuous representation of trip origins and destinations to their discrete zonal forms. Wu S.M. et al. [14] analysed probability density and probability distribution of trip distance for calculating of traffic zone radius, and the calculating method and procedures of the traffic zone radius based on trip distance were put forward. Shirabe [15] established a spatial unit allocation model to solve the spatial aggregation division in a small range through the method of mixed integer linear programming. Secondly, Clustering methods with different attributes are frequently used in these researches (Wu N.N. [16], Xia D. et al. [17], Zhao L. & Li Y. [18], Yang B.Y. [19], Kim K. [20]). Furthermore, Chandra A. et al. [21] propose an enhanced spatial aggregation methodology for developing a freight Traffic Analysis Zone (TAZ) system. This approach employs a multi-objective optimization framework utilizing a genetic algorithm to achieve TAZ division. Thirdly, there are also some data-driven approaches for TAZs. Li Y. & Yan X. [22] proposed a two-level framework for traffic zone division using large-scale ride-hailing data from a big data perspective. The study area is partitioned into uniform grids through a traffic grid model to ensure consistent zone size. Xing X. et al. [23] and Dong H. et al. [24] propose comprehensive methods for traffic zone division based on mobile billing data, leveraging its advantages of high coverage, cost-effectiveness, and real-time updates. Yang B. et al. [25] develop a multi-source data-driven stepwise strategy to address the TAZ delineation problem. This approach establishes a zoning system where each zone exhibits homogeneous mobility patterns and consistent land use characteristics. Cai M. et al. [26] presents a technical framework of data-driven TAZ division, leveraging cellular signaling data and a clustering algorithm to address the TAZ division problem. Lastly, there are others or combined approaches. Sun L. et al. [27] presents the weighted C-space representation method based on the method of network science, and that is used to study the public transport network and establish the interaction intensity between bus networks, which is conducive to better exploring the characteristics of urban transport structure. Sun G. et al. [28] integrates hidden patterns within the data with user-defined rules. This method not only leverages traffic hotspots and travel patterns across regions but also accounts for practical constraints, such as the road network. Qi W.D. [29] presents the two regional transportation district traffic division and cell division of fuzzy clustering method and combined with the characteristics of land use in inter-city highway district.

From these researches, we conclude that the TAZs plays an important role on traffic management and planning, which helps to optimize the operation and service of the traffic system, improve traffic efficiency and safety. In our research, TAZs are utilized for traffic volume analysis and prediction during the feasibility study of a new highway project, which access the characteristics of the new highway and the influence on the parallel highway. However, a suitable and purpose-specific TAZs division method for the feasibility study of a new highway remains an unresolved issue. Moreover, traditional TAZs division method for highway is based on census data that is resource-intensive and covers only a fraction of the population and areas in a city. This study utilizes road data, trip data, and a two-step division algorithm to finish the TAZ division. Additionally, a TAZ division model based on complex network theory is proposed.

3 Methodology

3.1 Problem Formulation

The accurate definition and delineation of TAZs is the foundation of the research. To facilitate the model's expression and establish clear guidelines, the following definitions are introduced.

Definition 1: With consideration of the closed operation in the highway, the object for TAZs division is the set of nodes that encompass both entrances and exits of the highway. Each node is characterized by a list of node attributes. The mathematical formula is as follows:

$$v_i = [a_{i1}, a_{i2}, \ldots, a_{im}], v_i \in V \tag{1}$$

where, v_i is node i, a is node attribute, m is the number of node attribute, V is the set of nodes.

Definition 2: TAZs are comprised of multiple zones, each of which encompasses a specific subset of nodes. Additionally, each node is exclusively associated with only one TAZ, and the union of all TAZs constitutes the entirety of nodes within the study area. The mathematical formula is as follows:

$$Z = \{z_1, z_2, \ldots, z_i, \ldots, z_n\}$$

$$s.t. \begin{cases} z_i \cap z_j = \emptyset, i \neq j \\ z_1 \cup z_2 \cup \ldots \cup z_n = V \end{cases} \quad (2)$$

where, Z is the set of TAZ. z_i and z_j are the traffic analysis zone i and j.

Definition 3: The TAZs Division model has function of partitioning nodes with similar characteristics of nodes in their respective zone and the critical requirement of ensuring contiguity within each TAZ. The mathematical formula is as follows:

$$z = f(V) \quad (3)$$

where z is TAZ, f is the function of the TAZs Division model.

3.2 TAZs Division Model Based on Complex Network

The TAZs Division model comprises three modules: module A, B, and C. Module A is the multi-layer network that consists of the road structural network, travel cost network, and Origin-destination network. In module B, feature extraction is conducted based on Module A to extract both structural and clustering features. Module C is the two-step division method that achieve the coarse and fine division of nodes, respectively. The technology framework of the model is illustrated in Fig. 1

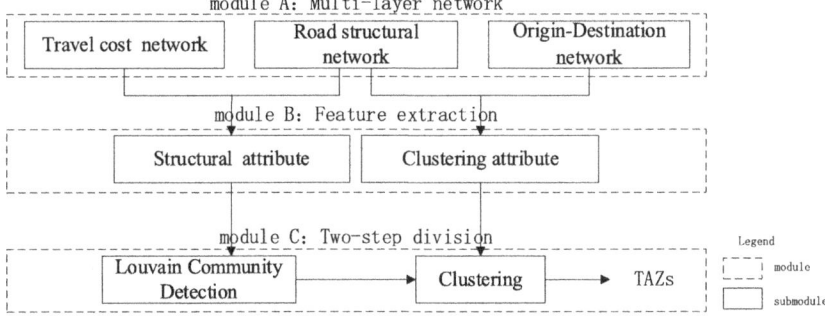

Fig. 1. TAZs Division model based on complex network

3.2.1 Module A: Multi-layer Network

The Module A encompasses three fundamental components: road structural network, travel cost network, and Origin-destination network. The road structural network serves as a directed graph $G = (V, E, A)$ representing the road structure for highway network. As in the prior research (Wen Z.Y. [30]), these elements consist of nodes and edges. Nodes represent the toll stations in the highway, denoted as $V = \{v_1, v_2, \ldots v_m\}$ that is a set of m nodes, The edges are represented as $E = \{e_1, e_2, \ldots e_n\}$ that consist of n directed road section. The adjacency matrix for nodes, denoted as $A \in \mathbb{R}^{n \times n}$, and a_{ij}

signifies the existence of a connection between two nodes: if the edge $e_{ij} = (v_i, v_j)$ exists in the network, then $a_{ij} = 1$, otherwise $a_{ij} = 0$. Additionally, in travel cost network, the weight of edge is the generalized travel cost, defined as $e_{ij} = \{v_i, v_j, w_{ij}\}$, where e_{ij} represents the edge from node v_i to node v_j, and w_{ij} represents the weight of e_{ij}. This weight is indicative of the generalized cost encompassing toll, operating cost and time value. Furthermore, the Origin-destination network incorporates the information pertaining to each trip, such as vehicle's plate number, type, origin, destination, departure and arrival time. All trips in the study area forms the Origin-destination network. Finally, by integrating these diverse components within the multi-layer network, this study establishes a comprehensive framework for analysing and modelling transportation systems effectively.

3.2.2 Module B: Feature Extraction

Metrics of feature extraction can be classified into two types of attributes for nodes: structural and clustering attributes, involving coarse and fine divisions, i.e., the two-step division. The structural attribute is determined by the derivative of the sum of minimum weighted path between nodes, and it is mathematically expressed as follows in Eq. (4). The higher the value is, the nodes are more closely connected in the travel cost network.

$$LA(s,t) = \frac{1}{d_{st}} = \frac{1}{\sum_{e_{ij} \in path(s,t)} w_{ij}}$$

$$s.t. \; s \in O, t \in D \tag{4}$$

where $LA(s, t)$ represents value of structural attribute, d_{st} is the sum of the minimum weighted path between origin s and destination t, and O, D are node sets of origin and destination, respectively. w_{ij} represents the trip cost of section e_{ij}.

On the other hand, the clustering attributes of each node comprise the trip distance and trip path choice, which are generated from the relationships of reference nodes and reference edges. In the feasibility study of new construction project, the reference nodes and reference edges are typically situated on new roads and their parallel sections, as illustrated in Fig. 2, the blue nodes and lines are reference nodes and reference edges, respectively, the dashed blue line is the new road, the solid blue lines represent its parallel sections. The trip distance attribute reflects node's position relative to the reference nodes, and the continuity of TAZs is ensured by similar trip distances; While the trip path choice attribute represents node's traffic assignment on the reference edges, adhering to the principles of Wardrop equilibrium (Wardrop [31]), this equilibrium theory posits that the network will reach a state of balance when users opt for the shortest routes, The homogeneity of the nodes in the TAZs is ensured by similar trip path choice. The expression for clustering attributes is given by Eq. (5):

$$d(i) = [d_i^1, d_i^2 \cdots, d_i^\alpha]$$

$$p(i) = [p_i^1, p_i^2 \cdots, p_i^\beta], \; s.t. \; \sum_{j=1}^{\beta} p_i^j = 1 \tag{5}$$

where $d(i)$ denotes the trip distance vector of node i relative to the reference nodes, $p(i)$ denotes the trip path choice of node i on the reference edges, α represents the total number of reference nodes, and β is the number of the reference edges.

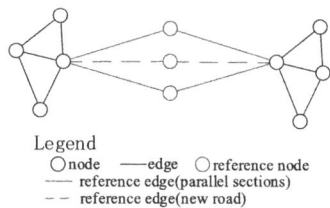

Legend
○ node —— edge ○ reference node
——— reference edge(parallel sections)
— — reference edge(new road)

Fig. 2. Reference nodes and edges

3.2.3 Module C: Two-Step Division Method

The two-step division method consists of the first step Louvain Community Detection algorithm with the structural attributes for coarse division, and the second step Clustering algorithm with the clustering attributes for fine division. In the Louvain Community Detection, community is considered a meso-unit in the complex network, characterized by tight internal connections and loose connections with other communities. This algorithm, renowned for its scalability and modularity maximization approach (Blondel et al. [32]), evaluates the change in modularity by moving node from its community to a neighbouring one. The modularity gain achieved through this moving is computed using Eq. (6) as below. The process is iteratively applied to all nodes until reaching the maximum in modularity.

$$\Delta Q = \frac{k_{i,in}}{m} - \gamma \frac{k_i^{out} \Sigma_{tot}^{in} + k_i^{in} \Sigma_{tot}^{out}}{m^2} \qquad (6)$$

where $k_{i,in}$ is the sum of $LA(s, t)$ in Eq. (4) from node i to other nodes in community C, m represents half of the sum of $LA(s, t)$ for all edges, k_i^{out}, k_i^{in} represent the outer and inner $LA(s, t)$ of node i, Σ_{tot}^{in}, Σ_{tot}^{out} represent the sum of in-going and out-going links incident to nodes in community C, γ is the resolution parameter, higher resolutions lead to more communities, while lower resolutions lead to fewer communities.

In the Clustering algorithm, the similarity measure is established using a modified K-means algorithm, i.e., weighted K-means algorithm. The coefficient λ is configured for different clustering attributes, with higher values indicating greater importance in the clustering process. The objective is to minimize the variance between nodes and the centroid of cluster. The centroid of each cluster represents the mean of nodes in the cluster, and the iteration continues until no further changes of TAZs. This formula is mathematically expressed in Eq. (7):

$$goal : \text{argmin} \sum_{k=1}^{K} \sum_{X_i \in C_k}^{N} d(X_i, V_k)$$

$$d(X_i, V_k) = \sum_{j=1}^{m} |x_{ij} - v_{kj}|^2 + \lambda \sum_{j=m+1}^{m+n} |x_{ij} - v_{kj}|^2 \qquad (7)$$

$$= d_1(X_i, V_k) + \lambda d_2(X_i, V_k)$$

where $d(X_i, V_k)$ is the variance between node i and the centroid of cluster k, x represent the values in Eq. (5) and v represent the values of the centroid of cluster, C_k is the node set in the cluster k, N represent the number of nodes in the cluster k, K is the number of clusters. d_1 and d_2 represent the variance of trip distance and trip path choice, respectively, and λ denotes the coefficient of the clustering attributes.

4 Case Study

4.1 Case Description

The validation of the TAZs Division model takes place with the highway network of the Guangdong-Hong Kong-Macao Greater Bay Area (GHM Greater Bay Area) in China. As illustrated in Fig. 3, the traffic between west and east sides of the Pearl River is allocated to the group of cross-river roads that consists of five distinct sections designated as A, B, C, D, and E. Notably, Section D is the new construction road presently under construction. Sections A, B, C, and E represent the parallel sections of Section D. The traffic data consist of the vehicle trip from July 5, 2021 to July 11, 2021 in the GHM Greater Bay Area.

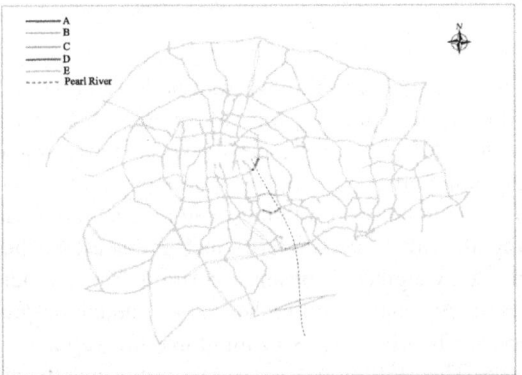

Fig. 3. Highway network in the GHM Greater Bay Area

4.2 Result

4.2.1 Network Establishment

Based on the dataset, we have constructed the multi-layer complex network of highway in the Greater Bay Area that comprises the road structural network, travel cost network and Origin-destination network. Figure 4 visually depicts the architecture of this network, and there are 589 nodes in total. The establishment of these multi-layer components provides the foundation for the analysis.

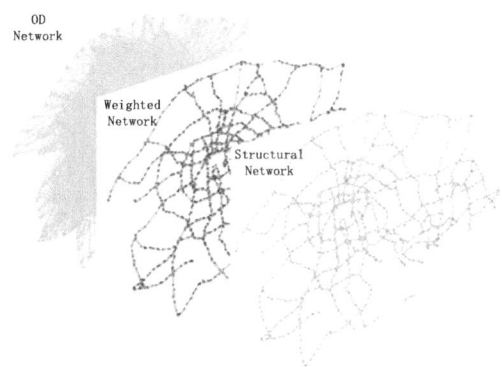

Fig. 4. Multi-layer complex network of highway in the GHM Greater Bay Area.

4.2.2 Result of Louvain Community Detection

As in the above Fig. 4, there are 589 nodes, and 589 X 588 directed edges can be formed, i.e., 346332 edges. The weight of each edge can be calculated with Eq. (4). With the Louvain Community Detection algorithm, these nodes have been partitioned into 8 communities when the modularity reaches the maximum value with $\gamma = 0.6$ in Eq. (7). Figure 5 illustrates the spatial distribution and arrangement of these communities, and adjacent nodes with same colour are in the same community. The identification and delineation of these communities are of great significance, as it sheds light on the underlying structure and connectivity patterns within the complex network, contributing valuable foundation to the second step.

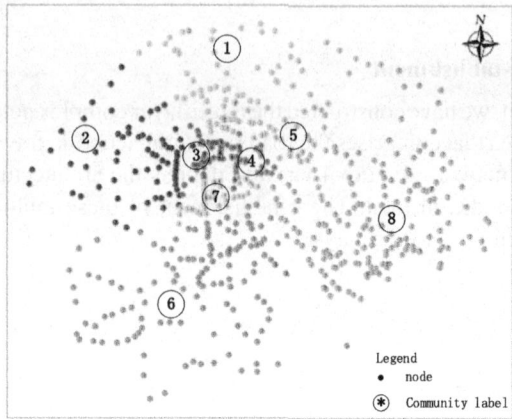

Fig. 5. The result of Louvain Community Detection

4.2.3 Result of Clustering

The fine division is conducted as the second step base on the community, and the algorithm of weighted K-means is adopted. For the reference nodes and reference edges, the parameters α, β in Eq. (5) are contingent upon the new road and parallel sections, i.e., the group of cross-river roads. In detail, the reference nodes are positioned at the centres of the five sections, while the reference edges correspond to the bidirectional sections, thus leading to $\alpha = 5$ and $\beta = 10$, corresponding to section A, B, C, D, and E in Fig. 3. Subsequently, utilizing Eq. (7), each community has been effectively partitioned into distinct TAZs for fine division. As a result, there are 37 TAZs in total. The geographical distribution and positions of these TAZs are visually presented in Fig. 6, and adjacent nodes with same color are in one TAZ.

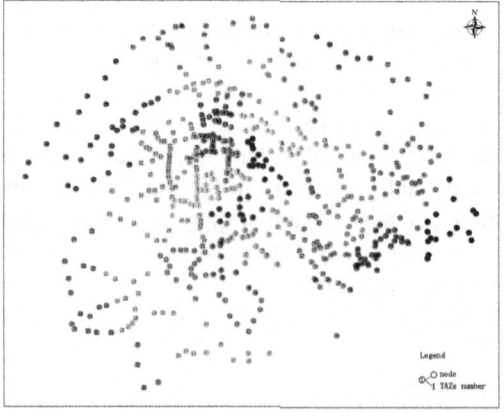

Fig. 6. The result of TAZs of the two-step division method

4.3 Validation

4.3.1 Evaluation Metrics

To validate the TAZ division results, this study evaluates the performances of the proposed methods using three key metrics as follow.

Caliński-Harabasz Score(CHS). The CHS was defined by Caliński & Harabasz[33], by calculating the CHS score through assessment of the inter and intra variance, we gauge the performance of these TAZs, wherein a higher score indicates a superior model performance. The CHS is defined by Eq. (8).

$$s = \frac{SS_B}{k-1} \div \frac{SS_{BW}}{N-k} \qquad (8)$$

where k is the total number of clusters, while N is the number of nodes within the network. SS_B, SS_{BW} represent the variance of intra-cluster and inter-cluster, respectively.

Contiguity. The contiguity is one key characteristic for obtaining well-defined TAZs, and the number of the discontinuity reflect the quality of the contiguity. A smaller value means better continuity.

Homogeneity. The homogeneity is quantized with the gini coefficient. The Eq. (9) provides the mathematical expression. Gini is the average of K clusters' Gini coefficient, and the lower the Gini coefficient, the more homogeneous the TAZs are.

$$Gini = \frac{1}{K} \sum_{k=1}^{K} \left(1 - \frac{1}{N} \sum_{j=1}^{N} \left(2 \sum_{i=1}^{j} w_i - w_j \right) \right) \qquad (9)$$

where, *Gini* is the value of Gini coefficient, w is the clustering attribute in Eq. (5).

4.3.2 Methods for Comparison

This study compares the outcomes of four TAZ division methods: (a) the proposed method, with parameters aligned as previously described; (b) the administrative area method, with results derived from the Shen-Zhong Link report [8]; (c) the proposed method, using trip distance as the clustering attribute for each node; and (d) the proposed method, with trip path choice as the clustering attribute for each node.

4.3.3 Analysis of the Result

The outcomes of the division process achieved by the four methods are presented in Fig. 7 (a–d), each corresponding to method (a–d). Nodes sharing the same color indicate their assignment to the same TAZ. For a comparative assessment, the evaluation metrics for each method are summarized in Table 1.

From the above Table 1, the obtained results present compelling evidence in favour of the TAZs Division model based on complex network, i.e., the method (a). In the Caliński-Harabasz Scores, the 138.02 of method (a) represents a significant improvement of 137.5% compared to the method (b); In the number of discontinuities, the discontinuities of method (a) represent the best of the four results; In the homogeneity, the Gini for

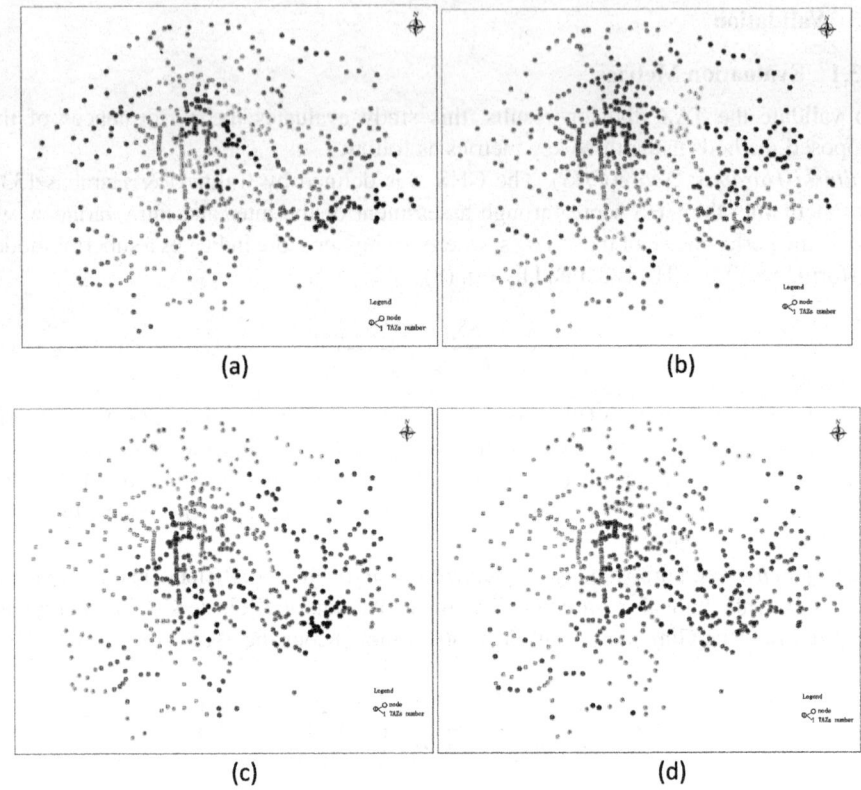

Fig. 7. TAZs result in the GHM Greater Bay Area

Table 1. Comparison results of metrics

Method	Number of TAZs	CHS	Number of discontinuities	Gini
a	37	138.02	4	0.235
b	29	58.11	6	0.260
c	37	120.54	6	0.263
d	37	23.79	18	0.348

method (a) is 0.235 that represents better homogeneity. In summary, the proposed method demonstrates superior performance in traffic zone division across all three evaluation metrics. This noteworthy enhancement attests to the model's superior characteristics. As a consequence, we can confidently assert that the TAZs Division model based on complex network exhibits a more reasonable and robust performance, thereby illustrating its value and applicability in the transportation analysis.

5 Summary and Prospect

In this study, we present and analyze the TAZs Division model based on complex network for new construction road's feasibility study and the case study of the highway network of the GHM Greater Bay Area in China. The model encompasses three core components: the multi-layer network, feature extraction, and two-step division method. By leveraging the multi-layer network, we obtain the structural and clustering attributes which are necessary for the subsequent feature extraction process. The two-step division method then effectively partitions the study area based on these attributes. Through the implementation of the model for the GHM Greater Bay Area that consists of 589 nodes, we successfully delineated 37 TAZs, resulting in the remarkable improvement in the evaluation metrics. This substantial improvement addresses the issue of dissimilar characteristics within TAZs, leading to enhanced homogeneity.

While the TAZs Division model base on complex network in this study has exhibited considerable advantages, there are still certain limitations to acknowledge. The current study data solely pertains to the study area, neglecting transit traffic flow, which has already formed an established trend in highway operation. To achieve a more accurate TAZs division, it becomes crucial to incorporate all traffic flow data, encompassing both study area and transit traffic. This comprehensive approach promises to further refine the model's performance and offers an exciting avenue for future research in tackling TAZs division challenges.

References

1. Lu, H.P.: Theory and Method in Transportation Planning. Tsinghua University Press, pp. 74–85 (2006)
2. Li, X.D., Yang, X.G.: Study on traffic zone division based on spatial clustering analysis. Comput. Eng. Appl. **45**(5), 19–22 (2009). https://doi.org/10.1109/CLEOE-EQEC.2009.5194697
3. Li, R.M., Gong, X.Y.: Traffic network time-space division of dynamic traffic assignment. J. Graduate School Chin. Acad. Sci. **23**(4), 520–526 (2006). https://doi.org/10.3969/j.issn.1002-1175.2006.04.014
4. Lian, J., Li, Y., Huang, S.L., Zhang, L.: Mining mobility patterns with trip-based traffic analysis zones: a deep feature embedding approach. 2019 IEEE Intelligent Transportation Systems Conference (ITSC), IEEE (2019). https://doi.org/10.1109/ITSC.2019.8917148.
5. Weerasinghe, O., Bandara, S.: Modified traffic analysis zones approach for the estimation of passenger flow distribution in urban areas. J. Urban Plan. Develop., **149**(1) (2023). https://doi.org/10.1061/(ASCE)UP.1943-5444.0000881
6. Zeng, Q., Wang, F.Z., Wang, Q.F., Pei, X., Yuan, Q.: Bayesian multivariate spatial modelling for crash frequencies by injury severity at daytime and nighttime in traffic analysis zones. Transp. Lett. **15**(6), 553–560 (2023). https://doi.org/10.1080/19427867.2022.2072459
7. Zhang, C., Yan, X., Ma, L., An, M.: Crash prediction and risk evaluation based on traffic analysis zones. Math. Probl. Eng. **2014**(3), 1–9 (2014). https://doi.org/10.1155/2014/987978
8. China Highway Engineering Consultants Corporation. The Feasibility Study of Shen Zhong Link. Report, p. 145 (2015)
9. Yu, H.J.: Research on some technology of traffic zone in transportation planning. Masters dissertation, XiDian University (2008)

10. Ghadiri, M., Rassafi, A.A., Mirbaha, B.: The effects of traffic zoning with regular geometric shapes on the precision of trip production models. J. Transp. Geogr., **78**(JUN), 150–159 (2019), https://doi.org/10.1016/j.jtrangeo.2019.05.018
11. Wang, S.W., Sun, L.S., Rong, J., Yang, Z.F.: Transit traffic analysis zone delineating method based on Thiessen Polygon. Sustainability, **6**(4), 1821–1832 (2014). https://doi.org/10.3390/su6041821
12. Assuncao, R.M., Neves, M.C., Camara, G., Freitas, C.D.C.: Efficient regionalization techniques for socio-economic geographical units using minimum spanning trees. Int. J. Geogr. Inf. Sci. **20**(7), 797–811 (2016)
13. Martinez, L.M., Viegas, J.M., Silva, E.A.: A traffic analysis zone definition: a new methodology and algorithm. Transportation **36**(5), 581–599 (2009). https://doi.org/10.1007/s11116-009-9214-z
14. Wu, S.M., Cheng, G.Z., Pei, Y.L.: Method of traffic zone division based on trip distance. Appl. Mech. Mater. **2013**(409), 1184–1187 (2013). https://doi.org/10.4028/www.scientific.net/amm.409-410.1184
15. Shirabe, T.: A model of contiguity for spatial unit allocation. Geogr. Anal. **37**(1), 2–16 (2005). https://doi.org/10.1111/j.1538-4632.2005.00605.x
16. Wu, N.N.: Urban environment-oriented traffic zoning based on spatial cluster analysis. J. Environ. Inf. **15**(2), 111–119 (2015). https://doi.org/10.3808/jei.201000171
17. Xia, D., Wang, B., Li, Y., Rong, Z., Zhang, Z.: An efficient mapreduce-based parallel clustering algorithm for distributed traffic subarea division. Discrete Dyn. Nat. Soc., (2015). https://doi.org/10.1155/2015/793010
18. Zhao, L., Li, Y.: Study on urban road network traffic district division based on clustering analysis. In 2018 Chinese Automation Congress (CAC) IEEE, pp. 3556–3560 (2018). https://doi.org/10.1109/CAC.2018.8623160
19. Yang, B.Y.: Research on traffic analysis zones division method based on multi-source data. HIT(Harbin Institute of Technology) (2020), https://doi.org/10.27061/d.cnki.ghgdu.2020.001332
20. Kim, K.: Identifying the structure of cities by clustering using a new similarity measure based on smart card data. IEEE Trans. Intell. Transp. Syst., **21**(5), 2002–2011 (2020), https://doi.org/10.1109/TITS.2019.2910548
21. Chandra, A., Sharath, M.N., Pani, A., Sahu, P.K.: A multi-objective genetic algorithm approach to design optimal zoning systems for freight transportation planning. J. Transp. Geogr., **92**(103037) (2021). https://doi.org/10.1016/j.jtrangeo.2021.103037
22. Li, Y., Yan, X.: Research on traffic zone partition method based on two-level partition theory. IOP Conf. Series: Mater. Sci. Eng. **688**(2), 022015 (2019). https://doi.org/10.1088/1757-899X/688/2/022015
23. Xing, X., Huang, W., Song, G., Xie, K.: Traffic zone division using mobile billing data. 2014 11th International Conference on Fuzzy Systems and Knowledge Discovery (FSKD) Xiamen, China, pp. 692–697 (2014). https://doi.org/10.1109/FSKD.2014.6980919
24. Dong, H., Wu, M., Ding, X., Chu, L., Jia, L., Qin, Y., Zhou, X.: Traffic zone division based on big data from mobile phone base stations. Transp. Res. Part C: Emerg. Technol., **58**(B), 278–291 (2015). https://doi.org/10.1016/j.trc.2015.06.007
25. Yang, B., Tian, Y., Wang, J., Hu, X., An, S.: How to improve urban transportation planning in big data era? A practice in the study of traffic analysis zone delineation. Transp. Policy **127**, 1–14 (2022). https://doi.org/10.1016/j.tranpol.2022.08.002
26. Cai, M., Hong, L., Xiong, C.: Data-driven traffic zone division in smart city: framework and technology. Sustain. Energy Technol. Assess. **2022**(52), 102251 (2022). https://doi.org/10.1016/j.seta.2022.102251

27. Sun, L., Lu, Y., Lee, D.H.: Understanding the structure of urban bus networks: The C-space representation approach. Cota Int. Conf. Transp. Prof. **2015**, 1557–1567 (2015). https://doi.org/10.1061/9780784479292.143
28. Sun, G., Chang, B., Zhu, L., et al.: TZVis: visual analysis of bicycle data for traffic zone division. J. Visualization **22**, 1193–1208 (2019). https://doi.org/10.1007/s12650-019-00600-6
29. Qi, W.D.: Study on the traffic district division and induced traffic forecase of intercity expressway. Chang'an University (2010)
30. Wen, Z., Weng, X., Zhang, P.: Evaluating the connectivity and imbalance contribution of new sections towards highway network: a complex network perspective. J. Adv. Transp., **2023**, 13 (2023), Article ID 6616512. https://doi.org/10.1155/2023/6616512
31. Wardrop, J.G.: Some theoretical aspects of road traffic research. Proc. Inst. Civ. Eng. **1**(3), 325–362 (1952). https://doi.org/10.1680/ipeds.1952.11259
32. Blondel, V.D., Guillaume, J.L., Lambiotte, R., Lefebvre, E.: Fast unfolding of communities in large networks. J. Stat. Mech: Theory Exp. **2008**(10), P10008 (2008). https://doi.org/10.1088/1742-5468/2008/10/P10008
33. Caliński, T., Harabasz, J.: A dendrite method for cluster analysis. Commun. Stat.—Theory Methods **3**(1), 1–27 (1974). https://doi.org/10.1080/03610927408827101

Forecasting Vehicle Mobility on Various Segments of the Urban Transport Network in the Fuzzy Paradigm

Ramin Rzayev[1(✉)] and Emil Ahmadov[1,2]

[1] Institute of Control Systems, Vahabzadeh Str. 68, AZ1141 Baku, Azerbaijan
raminrza@yahoo.com
[2] Intelligent Transport Management Centre of the Ministry of Internal Affairs of the Republic of Azerbaijan, Heydar Aliyev Ave. 137, Baku, Azerbaijan

Abstract. The mobility of vehicles on a selected segment of the transport network of an urban agglomeration is one of the four main parameters of transport flow. The article proposes an approach to estimating this parameter based on visual observation data. As an example, a section of Heydar Aliyev Avenue in Baku (Azerbaijan), characterized by the intensity of vehicle traffic, was selected. The information base of the study was made up of sensor readings from the Technical Vision System of the Intelligent Transport Management Centre of the Ministry of Internal Affairs of the Republic of Azerbaijan, which recorded vehicle speeds at the exit of this section every 10 s. The analysis and assessment of transport mobility was carried out in the paradigm of the fuzzy time series forecasting, reflecting the daily dynamics of changes in the interval of vehicles passing a selected segment of the urban transport network. The choice of the fuzzy paradigm is dictated by considerations relative to the weakly structured of averaged data from visual observation of vehicle speeds.

Keywords: Transport Network · Transport Flow · Transport Flow Parameter · Fuzzy Time Series · Forecasting

1 Introduction

The development of an intelligent transport management system in a large urban agglomeration is one of the solutions to extremely complex problems. However, the current level of development of info-communications and global navigation systems, Artificial Intelligence tools, including technical vision systems, and the effective use of active and passive sensors of various types and purposes [1], provide researchers with ample opportunities to solve this problem. In the context of this problem, one of the integral tasks is the construction of predictive models for projecting the parameters of traffic flow (TF) in a large agglomeration in the short term (for example, within one hour), as well as for identifying anomalies in traffic flow.

To date, a huge amount of knowledge has been accumulated on the topic of short-term forecasting of TF parameters, a review of which allows us to identify the following established approaches [2–4]: regression models; time series models; neural network models; support vector machine. At the same time, algorithms for solutions obtained using these methods and models, as a rule, work with data presented in the form of ordinary (crisp) numbers. However, in the process of monitoring TF, the results of observations obtained even with the use of the most advanced computer vision tools should generally be considered as weakly structured, i.e. those that are known to belong to a certain type. In this case, a more adequate representation of visual observation data of the TF is an interval of the form $z \in [z_{min}, z_{max}]$ or, even better, a verbal statement like $z =$ "HIGH", which can be described in the form of a suitable fuzzy set. The weak structure of visual traffic observation data predetermines a fuzzy approach to predicting TF parameters, one of which is the average travel time of transport along a selected segment of the transport network (TN). Over the past decades, the efforts of many scientists have obtained impressive results in the field of forecasting volatile time series using fuzzy methods of data analysis [5–9].

2 Problem Definition

According to [10], the TN of the urban agglomeration is mathematically interpreted in the form of a dynamic graph G, $g_i \in G$ oriented edges (arcs) of which symbolize real TN segments between nodes separating road sections. The orientation of the arc g_i is established by the direction of vehicle movement on the corresponding section, and the average time of its passage along this section is generally determined by a two-dimensional function $f: G \times T \to R$, which at a specific time $t \in T$ determines the value of this TF parameter in the form $p = f(g_i, t)$. From a formal point of view, the physical location of any segment $g_i \in G$ is determined by a two-valued function $x(g_i, \tau)$, $\tau \in [0, 1]$, with starting and end points $(N(g_i, 0), E(g_i, 0))$ and $(N(g_i, 1), E(g_i, 1))$, respectively.

The goal of this study is to construct an algorithm for predicting the mobility of vehicles on the selected segment $g_i \in G$. As an example, the TN segment in Baku (Azerbaijan) was selected, which reflects g section of Heydar Aliyev Avenue with the coordinates of the starting and end points (40°24′15″N, 49°53′05″E) and (40°23′55″N, 49°52′28″E), respectively (see Fig. 1).

Fig. 1. Section of Heydar Aliyev Avenue (Baku, Azerbaijan).

In this case, the initial data are:

- sensor readings of the computer vision system, which in real time (every 10 s during the day) records vehicle speeds (v) at the exit of the selected TN segment;
- segment length $S = 1063.68$ m;
- speed limit $v_0 = 70$ km/h (19.44 m/sec), established as a sign at the entrance to this road section.

To estimate transport mobility on segment g, i.e. time τ for vehicles to pass this segment, taking into account the values of v_0, v and S, the following kinematics equations are applied

$$\begin{cases} S = v_0\tau + a\tau^2/2, \\ v = v_0 + a\tau. \end{cases}$$

Excluding acceleration from consideration by transforming.

$$2(S - v_0\tau)/\tau^2 = (v - v_0)/\tau,$$

the following formula for calculating the average travel time of segment g is obtained

$$\tau = 2S/(v + v_0) \tag{1}$$

The application of formula (1) for vehicle speeds v, recorded by technical vision system sensors at the exit of segment g and averaged for every 10 s of observations, made it possible to form the time series ATTS $= \{\Delta t_k\}_{k=1}^{144}$ (Average Transit Time Series), reflecting the dynamics of changes in vehicle transit time intervals of this section for 144 steps of check measurements. This time series is presented in two ways: in the form of Table 1 and graphically in Fig. 2.

Table 1. Time series of vehicle mobility on segment g.

k	Time	v (m/sec)	Δt (sec)	k	Time	v (m/sec)	Δt (sec)
1	19.02.2024 00:09	20.14	53.74	139	19.02.2024 23:09	20.42	53.37
2	19.02.2024 00:19	20.33	53.48	140	19.02.2024 23:19	20.14	53.74
3	19.02.2024 00:29	20.58	53.15	141	19.02.2024 23:29	20.33	53.48
4	19.02.2024 00:39	20.08	53.82	142	19.02.2024 23:39	20.14	53.74
5	19.02.2024 00:49	20.56	53.18	143	19.02.2024 23:49	19.39	54.78
...	....	......	......	144	19.02.2024 23:59	18.97	55.38

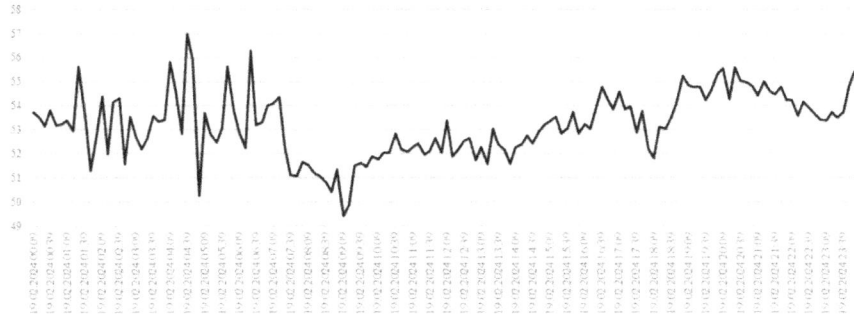

Fig. 2. Time series of vehicle mobility on segment g.

A significant feature of the historical data of the ATTS is that due to the imperfection of the recognition system and a number of other reasons, the readings of the vision system sensors cannot be considered absolutely accurate, that is, these readings *a priori* contain minor errors. This forces us to consider the historical data of the ATTS as weakly structured, which dictates the need to apply the mathematical apparatus of fuzzy logic to their processing. In addition, "immersion" of historical data in a fuzzy environment makes it possible to find patterns within the time series in terms of fuzzy sets and operate with them, which is impossible with ordinary numbers. Therefore, instead of ATTS, we will consider the corresponding fuzzy time series FATTS = $\{A_k\}_{k=1}^{144}$ (Fuzzy Average Transit Time Series), where A_k is an appropriate fuzzy set that describes the value of the time interval Δt_k of passage by vehicles of section g for the k-th moment of observation. The fuzzy set (FS) A_k is represented as the following tuple [11]

$$\{\mu_{A_k}(\Delta t_k)/\Delta t_k\}, \mu_{A_k}(\Delta t_k) \to [0,1], k = 1 \div 144, \qquad (2)$$

where $\mu_{A_k}(\Delta t_k)$ is the membership function that determines the degree of belonging of the interval value Δt_k to the fuzzy set A_k.

3 Fuzzification of Sensor Readings and FATTS Formation

The fuzzy set A_k, characterized by tuple (2), is an evaluative concept and it can be used as a qualitative criterion for assessing the values of travel time intervals based on speed sensor readings. The optimal number of such evaluation criteria is established step by step as follows [9].

Step 1. Sorting intervals Δt_k ($k = 1 \div 144$) into an increasing sequence $\{\Delta t_{p(i)}\}$, where p is a permutation that ranks the values of time intervals in ascending order, so that $\Delta t_{p(i+1)} \geq \Delta t_{p(i)}$.

Step 2. Calculation of the average value on the totality of all pairwise distances $d_i = |\Delta t_{p(i+1)} - \Delta t_{p(i)}|$ according to the formula

$$AD = \frac{1}{n-1} \sum_{i=1}^{n-1} \left| \Delta t_{p(i)} - \Delta t_{p(i+1)} \right|, \qquad (3)$$

and standard deviation according to the formula

$$\sigma_{AD} = \sqrt{\frac{1}{n-1} \sum_{i=1}^{n-1} (d_i - AD)^2}. \tag{4}$$

Step 3. Elimination of anomalies – sharply prominent values that must be reset. The values of pairwise distances that do not satisfy the condition

$$|AD - \sigma_{AD}| \le d_i \le |AD + \sigma_{AD}|. \tag{5}$$

are subject to reset.

Step 4. After re-calculating the average value AD over a set of pairwise distances remaining after the release of anomalous values, the corresponding number of qualitative assessment criteria (m) is calculated using the formula

$$m = [D_2 - D_2 - AD]/(2 \times AD), \tag{6}$$

where $D_{\min}$ and $D_{\max}$ are the minimum and maximum values of Δt_k, respectively; $D_1 = D_{\min} - AD$; $D_2 = D_{\max} + AD$.

Applying formulas (3) and (4) to the set $\{\Delta t_{p(i)}\}$, the average value $AD = 0.0530$ and the standard deviation $\sigma_{AD} = 0.0921$ was obtained, respectively. Rejecting d_i that do not satisfy condition (5) or, more specifically, the condition.

$$0.0391 = |0.0530 - 0.0921| \le d_i \le |0.0530 + 0.0921| = 0.1451,$$

Using formula (6), the final value of the average value for the totality of the remaining pairwise distances d_i was obtained: $AD = 0.0747$. In this case, the universe covering the ATTS data range is the segment $U = [D_1, D_2]$, where $D_1 = D_{\min} - AD = 49.51 - 0.0747 = 49.33$, $D_2 = D_{\max} + AD = 56.98 + 0.0,0747 = 57.06$. According to (6), the number of fuzzy subsets of the universe U, describing the qualitative criteria for assessing historical data of the ATTS, is determined as:

$$m = [57.07 - 49.33 - 0.0747]/(2 \times 0.0747) = 51.21 \approx 51.$$

To construct appropriate fuzzy sets, symmetric trapezoidal membership functions are used [8], which in the context of the problem under consideration are specified in the following form

$$\mu_{A_j}(x) = \begin{cases} 0, & x < a_{j1} \\ (x - a_{j1})/(a_{j2} - a_{j1}), & a_{j1} \le x \le a_{j2}, \\ 1, & a_{j2} \le x \le a_{j3}, \\ (a_{j4} - x)/(a_{j4} - a_{j3}), & a_{j3} \le x \le a_{j4}, \\ 0, & x > a_{j4}, \end{cases} \tag{7}$$

where $a_{j2} - a_{j1} = a_{j3} - a_{j2} = a_{j4} - a_{j3}$; $j = 1 \div 51$. Based on (7), the corresponding 51 symmetric trapezoidal membership functions were identified (see Fig. 3), the parameters of which are summarized in Table 3.

Forecasting Vehicle Mobility on Various Segments   167

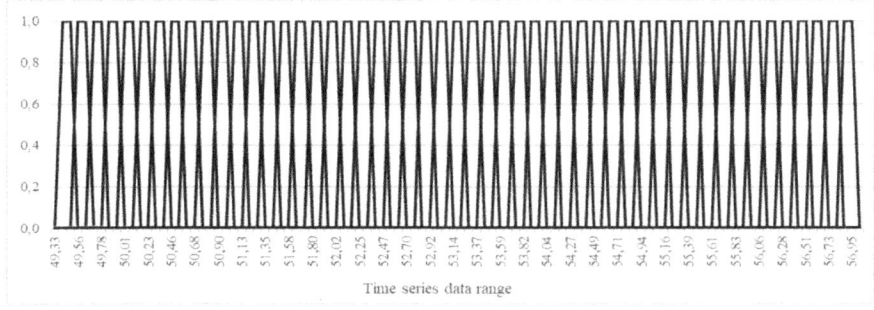

Fig. 3. Trapezoidal membership functions.

Fuzzification of historical data of the ATTS is carried out according to the principle [9]: Δt_k is described by the fuzzy set A_j ($j = 1 \div 51$) to which it belongs with the largest value of the membership function $\mu_{A_j}(\Delta t_k)$. If Δt_k falls into the segment $[a_{j2}, a_{j3}]$ (projection of the upper base of the j-th trapezoid onto the abscissa, see (7)), then its fuzzy analogue is A_j. Thus, the value $\Delta t_{33} = 52.46$ for 19.02.2024 05:29 (see Table 1) from the interval [52.40, 52.47] is described by the fuzzy set A_{21} (see Table 2).

Table 2. Table captions should be placed above the tables.

FS	Trapezoidal function parameters				FS	Trapezoidal function parameters			
	a_{j1}	a_{j2}	a_{j3}	a_{j4}		a_{j1}	a_{j2}	a_{j3}	a_{j4}
A_1	49.33	49.41	49.48	49.56	A_{27}	53.22	53.29	53.37	53.44
....	....	....	....	....	A_{28}	53.37	53.44	53.52	53.59
A_7	50.23	50.31	50.38	50.46	A_{29}	53.52	53.59	53.67	53.74
....	....	....	....	....	....	....	....	....	....
A_{24}	52.77	52.85	52.92	53.00	A_{49}	56.51	56.58	56.66	56.73
A_{25}	52.92	53.00	53.07	53.14	A_{50}	56.66	56.73	56.80	56.88
A_{26}	53.07	53.14	53.22	53.29	A_{51}	56.80	56.88	56.95	57.06

In other cases, additional calculations are required. In particular, for the value $\Delta t_{89} = 52.93$ (19.02.2024 14:49, see Fig. 2) we have: $\mu_{A24}(52.93) = 0.9178$ and $\mu_{A25}(52.93) = 0.0822$ (see Fig. 4). Therefore, A_{24} is chosen as the fuzzy analogue.

After fuzzifying historical data, the corresponding analogue of FATTS was generated and it is summarized in Table 3.

To build a predictive model, internal relationships are identified that reflect cause-effect relations between historical data. Depending on the number of premises in the fuzzy relation of the form "If..., then...", internal relationships are divided into groups of 1st, 2nd and higher orders. Internal relationships of the 1st order are grouped according to the principle: if the FS, for example, A_{14} is sequentially connected with A_1 and A_{17}, then the 1st order group is localized relative to it: $A_{14} \Rightarrow A_1, A_{17}$ (see Table 4, G8).

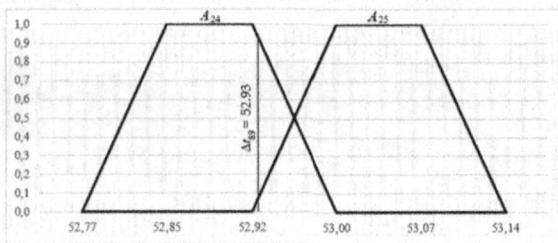

Fig. 4. Qualitative evaluation criteria in the form of fuzzy sets A_{24} and A_{25}.

Table 3. FATTS table view.

k	Time	v (m/sec)	Δt (sec)	FS	k	Time	v (m/sec)	Δt (sec)	FS
1	19.02.2024 00:09	20.14	53.74	A_{30}	139	19.02.2024 23:09	20.42	53.37	A_{26}
2	19.02.2024 00:19	20.33	53.48	A_{28}	140	19.02.2024 23:19	20.14	53.74	A_{30}
3	19.02.2024 00:29	20.58	53.15	A_{26}	141	19.02.2024 23:29	20.33	53.48	A_{28}
4	19.02.2024 00:39	20.08	53.82	A_{30}	142	19.02.2024 23:39	20.14	53.74	A_{30}
5	19.02.2024 00:49	20.56	53.18	A_{26}	143	19.02.2024 23:49	19.39	54.78	A_{37}
...	....	......	......	...	144	19.02.2024 23:59	18.97	55.38	A_{41}

The 1st order internal relationship between the quantities Δt_k and Δt_{k+1} can be interpreted as the fuzzy implication "If Δt_k is A_r, then Δt_{k+1} is Δt_{k+1}", where $k = 1 \div 144$; $r, p = 1 \div 51$. Thus, the internal relationship of the 1st order $A_{10} \Rightarrow A_7$ (G_5) between the values Δt_{52} (50.79) and Δt_{53} (50.42) is interpreted by the implication "If Δt_{52} is A_{10}, then Δt_{53} is A_7". If the internal relationship of the 1st order is presented as $A_k \Rightarrow A_{p1}, A_{p2}, ..., A_{ps}$, where $p_1, p_2, ..., p_s = 1 \div 51$, then in the form of the fuzzy implication it looks as "If Δt_k is A_r, then Δt_{k+1} is A_{p1} or A_{p2} or ... or A_{ps}".

In particular, the internal relationship of the 1st order $A_{12} \Rightarrow A_{10}, A_{12}, A_{16}$ (G_6, see Table 4) between the values Δt_{51} (51.06) and Δt_{52}(50.79), Δt_{46} (51.09) and Δt_{47}(51.06), Δt_{47} (51.06) and Δt_{48}(51.64), Δt_{50} (51.16) and Δt_{51}(51.06) is interpreted as "If Δt_k is A_{12}, then Δt_{k+1} is A_{10} or A_{12} or A_{16}". Accordingly, the internal relationship of the 2nd order, for example, $A_7, A_{14} \Rightarrow A_1$ is interpreted in the form of the fuzzy implication "If Δt_k is A_7 and Δt_k is A_{14}, then Δt_{k+1} is A_1", or, for example, the relationship $A_{20}, A_{21} \Rightarrow A_{18}, A_{23}$ is interpreted as "If Δt_k is A_{20} and Δt_k is A_{21}, then Δt_{k+1} is A_{18} or A_{23}".

Table 4. Groups of internal relationships of the 1st order.

Group	Fuzzy relation	Group	Fuzzy relation
G_1	$A_1 \Rightarrow A_4$	G_{27}	$A_{33} \Rightarrow A_{29}, A_{31}, A_{33}, A_{34}, A_{36}, A_{42}$
G_2	$A_4 \Rightarrow A_{15}$	G_{28}	$A_{34} \Rightarrow A_{15}, A_{18}, A_{20}, A_{38}$
G_3	$A_6 \Rightarrow A_{30}$	G_{29}	$A_{35} \Rightarrow A_{31}, A_{37}$
G_4	$A_7 \Rightarrow A_{14}$	G_{30}	$A_{36} \Rightarrow A_{24}, A_{35}, A_{40}$
G_5	$A_{10} \Rightarrow A_7$	G_{31}	$A_{37} \Rightarrow A_{33}, A_{34}, A_{37}, A_{41}$
G_6	$A_{12} \Rightarrow A_{10}, A_{12}, A_{16}$	G_{32}	$A_{38} \Rightarrow A_{36}, A_{37}$
G_7	$A_{13} \Rightarrow A_{22}$	G_{33}	$A_{39} \Rightarrow A_{38}$
G_8	$A_{14} \Rightarrow A_1, A_{17}$	G_{34}	$A_{40} \Rightarrow A_{37}, A_{42}$
G_9	$A_{15} \Rightarrow A_{12}, A_{14}, A_{15}, A_{20}, A_{25}, A_{28}$	G_{35}	$A_{42} \Rightarrow A_{29}, A_{30}, A_{33}, A_{39}$
G_{10}	$A_{16} \Rightarrow A_{15}, A_{18}, A_{20}$	G_{36}	$A_{44} \Rightarrow A_6, A_{36}$
G_{11}	$A_{17} \Rightarrow A_{16}, A_{19}, A_{26}$	G_{37}	$A_{47} \Rightarrow A_{26}$
...	....	G_{38}	$A_{51} \Rightarrow A_{44}$

4 FATTS Forecasting

Various rules are used to determine fuzzy predicts and defuzzify them [5, 7]. In the context of solving our problem, the essence of some of them is as follows. If historical data Δt_k is represented by the fuzzy set A_p, which within the totality of time series data forms only one internal relationship of the 1st order, for example, in the form of a fuzzy relation $A_p \Rightarrow A_s$ ($p, s = 1 \div 51$), then the forecast for the next $(k + 1)$-th period is considered as FS A_s. In the case of the group of relationships, for example, of the form $A_p \Rightarrow A_{s1}, A_{s2}, \ldots, A_{sj}$, ($s_1, s_2, \ldots, s_j = 1 \div 51$), then the union $A_{s1} \cup A_{s2} \cup \ldots \cup A_{sj}$ is the fuzzy forecast for the $(k + 1)$-th period.

To defuzzify fuzzy predicts, the following two rules are applied.

Rule 1. For the fuzzy relation $A_p \Rightarrow A_s$, where A_p is the fuzzy analogue of the value of the interval Δt_k at the k-th moment, the forecast in nominal terms for the next $(k + 1)$-th period is the abscissa of the middle of the upper base of the trapezoid, reflecting the FS A_s. In particular, this is confirmed by the rule of defuzzification (or point estimation) of the fuzzy set A, which is implemented by the formula [12]

$$F(A) = (1/\alpha_{\max}) \int_0^{\alpha_{\max}} M(A_\alpha) d\alpha, \tag{8}$$

where $A_\alpha = \{u | \mu_A(u) \geq \alpha, u \in U\}$ is the α-level sets ($\alpha \in [0, 1]$); $M(A_\alpha)$ is the cardinal number of the corresponding α-level set, calculated by the formula $M(A_\alpha) = (1/m) \sum_{k=1}^{m} u_k$, $u_k \in A_\alpha$. In particular, for the FS $A_7 = \{0/50.23, 1/50.31, 1/50.38, 0/50.46\}$ (see Table 2), which is the forecast in the conjunction $A_{10} \Rightarrow A_7$ (G_5, Table 5) for $0 < \alpha < 1$ accordingly we have:

$$\Delta \alpha = 1, A_{7,\alpha} = \{50.31, 50.38\}, M(A_{7,\alpha}) = (50.31 + 50.38)/2 = 50.345.$$

Then, according to (8), the forecast in nominal terms is calculated as

$$F(A_7) = (1/1) \int_0^1 M(A_{7\alpha})d\alpha \approx M(A_{7\alpha}) \cdot \Delta\alpha = 50.345 \cdot 1 = 50.345.$$

Rule 2. For the fuzzy relation $A_k \Rightarrow A_j, A_i, A_p$, where A_k is the fuzzy analogue of the value of the interval Δt_k at the k-th moment, the crisp forecast for the next $(k+1)$-th period is calculated as the arithmetic mean of the abscissa of the midpoints of the upper bases of trapezoids corresponding to the fuzzy sets A_j, A_i and A_p [7]. In particular, according to this rule, the predict for the check period (19.02.2024, 07:49), preceded by A_{12}, is calculated based on the internal 1st order relationship $A_{12} \Rightarrow A_{10}, A_{12}, A_{16}$ (see Table 5) as follows:

$$predict = [(50.75 + 50.83)/2 + (51.05 + 51.13)/2 + (51.65 + 51.73)/2]/3 = 51.19.$$

Thus, using Rules 1 and 2 for relationships of the 1st and 2nd orders, predicts were obtained, which are summarized in Table 6 and 7, respectively. The geometric interpretation of the predictive models is presented in Fig. 5. At the end of Tables 5 and 6, the values of statistical criteria for assessing the adequacy of models are presented [13]: MSE (Mean Squared Error), MAPE (Mean Absolute Percentage Error) and MPE (Mean Percentage Error), which are calculated by the corresponding formulas:

$$\text{MSE} = \frac{1}{m}\sum_{k=1}^{m}(F_k - A_k)^2, \text{MAPE} = \frac{1}{m}\sum_{k=1}^{m}\frac{|F_k - A_k|}{A_k} \times 100\%, \text{MPE} = \frac{1}{m}\sum_{k=1}^{m}\frac{F_k - A_k}{A_k} \times 100\%$$

where m is the length of the FATTS; A_k is the actual indicator Δt_k at the k-th moment of observation; F_k is predict of A_k. When interpreting these metrics, their characteristics should be taken into account. For example, MSE, being one of the most common metrics of forecasting errors, allows to evaluate the accuracy of the forecast in absolute units of measurement, while MPE and MAPE show the deviation in percentage terms. In particular, MAPE can be useful for comparing the forecast accuracy of different models when working with different ranges of data.

Table 5. 1st order predictive model for FATTS.

k	Time	Δt (sec)	FS	Fuzzy output of the model	Predict
1	19.02.2024 00:09	53.74	A_{30}		
2	19.02.2024 00:19	53.48	A_{28}	$A_{19} \cup A_{24} \cup A_{26} \cup A_{28} \cup A_{37}$	53.30
...	...	...	...	...	...
143	19.02.2024 23:49	54.78	A_{37}	$A_{19} \cup A_{24} \cup A_{26} \cup A_{28} \cup A_{37}$	53.30
144	19.02.2024 23:59	55.38	A_{41}	$A_{33} \cup A_{34} \cup A_{37} \cup A_{41}$	54.71
MSE1					0.7635
MAPE1					1.1609
MPE1					-0.0610

As can be seen from Table 6 and 7, the MSE indicators for the 1st and 2nd order prognostic models are $\text{MSE}_1 = 0.7635$ and $\text{MSE}_2 = 0.0934$, respectively. According to

Table 6. 2nd order predictive model for FATTS.

k	Time	Δt (sec)	FS	Fuzzy output of the model	Predict
1	19.02.2024 00:09	53.74	A_{30}		
2	19.02.2024 00:19	53.48	A_{28}		
3	19.02.2024 00:29	53.15	A_{26}	$A_{26} \cup A_{30}$	53.48
...	...	...	...	...	...
143	19.02.2024 23:49	54.78	A_{37}	A_{37}	54.83
144	19.02.2024 23:59	55.38	A_{41}	A_{41}	55.42
MSE2					0.0934
MAPE2					0.2316
MPE2					-0.0048

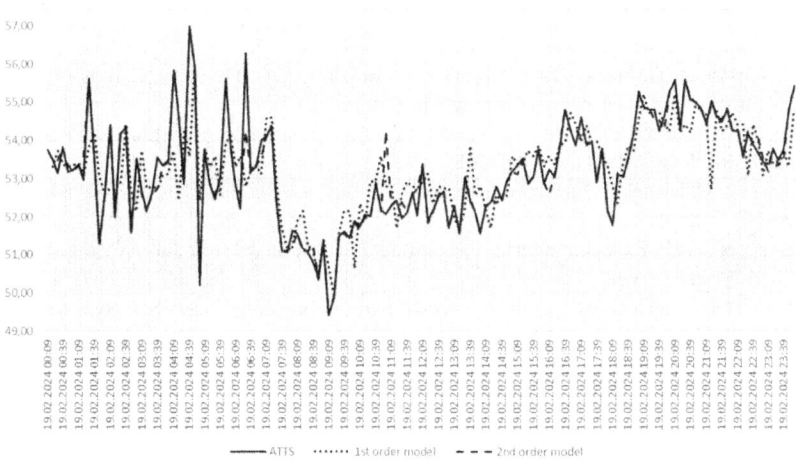

Fig. 5. Predictive models of 1st and 2nd orders.

the MAPE criterion, the magnitude of the forecast error is demonstrated as a percentage in comparison with the actual values of the FATTS: MAPE1 = 1.1609% and MAPE2 = 0.2316% According to the MPE indicator, which is a more informative criterion for assessing the adequacy of the forecasting model, acceptable "biases" of the 1st and 2nd order prognostic models $MPE_1 = -0.0610\%$ and $MPE_2 = -0.0048\%$ were obtained, as not exceeding the normative 5% threshold to the left of zero.

5 Conclusion

Time series forecasting in the fuzzy paradigm makes it possible to identify internal connections in the form of fuzzy relations and build predictive models of the 1st and higher orders on their basis. However, in the approach under consideration, where a sufficiently large number of qualitative assessment criteria are established in the form of appropriate fuzzy sets, the construction of higher-order models loses its meaning, because in this case, the evaluation criteria are identified with fuzzy sets reflecting nominal historical data of the time series.

The fuzzy approach proposed in the article is capable of supporting predictive and prescriptive analytical decisions by connecting previously collected data from sensors of the Technical Vision System that permanently monitors the transport system of a large urban agglomeration. The fuzzy time series forecasting algorithm makes it possible to reproduce models of changes in the dynamics of the average travel time interval of vehicles in all sections of the transport system and for various traffic scenarios.

References

1. Klein, L.A.: Traffic detector handbook. Federal Highway Administration, Turner-Fairbank Highway Research Center (2006)
2. Vlahogianni, E.I.: Short-term traffic forecasting: where we are and where we're going. Transp. Res.—Part C: Emerg. Technol., **43**(1), 3–19 (2014)
3. Bolshinsky, E.: Traffic flow forecast survey. In: Bolshinsky, E., Freidman, R. (eds.) Technion—Israel Institute of Technology, Technical Report (2012)
4. Faouzi, N.E.: Data fusion in intelligent transportation systems: progress and challenges. A Survey, Inf. Fusion **12**(1), 4–10 (2011)
5. Song, Q., Chissom, B.S.: Fuzzy time series and its models. Fuzzy Sets Syst. **54**, 269–277 (1993)
6. Kumar, N., et al.: Fuzzy time series forecasting of wheat production. Int. J. Comput. Sci. Eng. **2**(3), 635–640 (2010)
7. Chen, S.M.: Forecasting enrollments based on fuzzy time series. Fuzzy Sets Syst. **81**, 311–319 (1996)
8. Cheng, C.H., Chang, J.R., Yen, C.A.: Entropy-based and trapezoid fuzzification fuzzy time series approaches for forecasting IT project cost. Technol. Forecast. Soc. Chang. **73**, 524–542 (2006)
9. Ortiz-Arroyo, D., Poulsen, J.R.: A weighted fuzzy time series forecasting model. Indian J. Sci. Technol. **11**(27), 1–11 (2018)
10. Liu, X.: Dynamic graph shortest path algorithm. Lect. Notes Comput. Sci. **7418**, 296–307 (2012)
11. Zadeh, L.A.: The concept of a linguistic variable and its application to approximate reasoning. Inf. Sci. **8**(3), 199–249 (1975)
12. Andreichikov, A., Andreichikova, O.: Analysis, synthesis, planning decisions in the economy. Finance and Statistics, Moscow (2000). (in Russian)
13. Kim, S., Kim, H.: A new metric of absolute percentage error for intermittent demand forecasts. Int. J. Forecast. **32**(3), 669–679 (2016)

The Emerging ICT for Exposure Reduction in Urban Mobility: Transport Models for Risk Cycle

Francesco Russo[1(✉)], Antonio Comi[2], and Corrado Rindone[1]

[1] Dipartimento di Ingegneria dell'Informazione, Delle Infrastrutture e dell'Energia Sostenibile, Università Mediterranea di Reggio Calabria, Reggio Calabria, Italy
francesco.russo@unirc.it
[2] Department of Enterprise Engineering, University of Rome Tor Vergata, Rome, Italy

Abstract. The problem of risk in urban systems is becoming increasingly important. Inside the risk, the exposure component allows significant risk reductions to be achieved. Evacuation is the most important action to reduce exposure. The paper analyzes the risk cycle starting from the main elements: transport system models; training and exercises; emerging information and communication technologies (ICT). The method followed is to identify the main variables of risk reduction, the time deltas (i.e., time between event occurence and effect), expressed for the different scenarios: reality, planning, training and exercises. With these expressions, an advanced risk cycle model is formalized, which considers time deltas as the main reference. The formalization of a dynamic model with the use of emerging ICTs is then proposed.

The theoretical result is important because it allows planners to highlight the role of ICT in reducing evacuation times, both in planning and in exposure and training, and therefore in the reality of the occurrence of an event, introducing ICTs in the day-to-day modelling processes. The interpretation of formal results is important because it allows public decision-makers, technicians and planners, as well as researchers to place all models and operational results in a dynamic process.

Keywords: Risk · Exposure · Evacuation · Assignment · Transportation System · Information and Communication Technologies

1 Introduction

The question of risk in large urban realities due to anthropogenic or natural events has become central in the twenty-first century. The disasters in New York, Twin Towers 2003, and New Orleans, Hurricane Katrina 2005, have paid attention to the importance of knowing in depth all the components of risk.

Risk can be defined as the multiplication of three probabilities: the probability that an event occurs, defined as the occurrence, the probability that an element does not resist to the effects of the event, defined as vulnerability, and the probability relating to the possibility of removing people or things from the effects of the events, defined as exposure. This formalization, used in many fields, has its origin in the study of the risk related to the transport of hazardous materials ([1–3]).

Important lines of research have been activated over time regarding the occurrence and vulnerability. For the occurrence, it is enough to think of the forecasting systems for natural events, such as river floods, violent hurricanes, volcanic lava, tsunamis, or military systems for events with direct anthropogenic origin, such as attacks of various types. For vulnerability, reference can be made to all lines of research in the field of science and technique of civil construction. The problem of exposure, and its reduction, is the least one addressed by research.

The basic theme of exposure is given by the time that elapses between the occurrence of the event and the time in which the effects on the population and on things manifest themselves, here for convenience defined as *temporal delta*. Events and the time/temporal deltas connected to them are classified in the literature and it is evident that with the same event, the time delta is connected to a spatial delta, i.e. the distance between the occurrence of the event and the realization of the effect. The importance of exposure and its reduction immediately emerges as the time delta increases. The temporal deltas in the risk can be close to zero as in the earthquake or delayed as in the tsunami. The deltas in attacks have a wide distribution that derives from the type of attack.

In the context of exposure with time deltas sufficient for a significant reduction, only specific topics such as those of nuclear power plants, those of fires in buildings, and those for the transport of dangerous goods are studied in the literature ([3–7]). The main issue for the reduction of exposure is that of evacuation, in fact, at the limit, if it is possible to evacuate the entire population from the area subject to the effects of the event, the risk for the population becomes zero.

It should be noted that the US, after the serious natural and anthropogenic events of the beginning of the century, has invested significant resources and has developed impressive lines of research, constantly supporting the work of university and state departments aimed at studying risk reduction. It is sufficient to mention the role of the Federal Emergency Management Agency (FEMA), the importance it has had, and has [8], and the responsibilities it assumes with results that are constantly monitored, see Hurricane Katrina [9]. On the other hand, it seems that much attention has not been given to the large state apparatuses that are most confronted with the US, namely the EU and China, where research and implementations are not yet significant, at least in the knowledge of the authors.

The main theme is therefore that of the *transport system models* (TSM) to be used for risk conditions. The cultural delay that emerges is to use models and algorithms developed for ordinary traffic conditions ([10–14]), even for conditions that occur in conditions of risk with significant temporal delta. The works that make use of the TSM still consider static conditions for the System ([15–19]).

The interest therefore becomes that of having models that allow planners to study what happens in transport systems in emergency conditions, analyzing, in theory, what happens in the time delta, then planning and designing the measures so that an organized evacuation allows a drastic reduction in exposure. It is possible to recall some theoretical works with experimental comparisons carried out in Italy regarding the risk conditions and the role of transport systems where the following are analyzed: demand [20], supply [21], assignment and design [22], planning [23]. For the study of potential population accumulation nodes during an emergency, it is interesting to use the models developed for the analysis of centrality in transport networks ([24–26]). Some developments of models are for the evacuation of ships [27] and for ports [28]. The models were also used to study behavior modifications during Covid [29].

The use of specific models makes it possible to address the different phases into which risk cycles are divided at international level, i.e., the evolution of risk conditions and therefore the actions that can be carried out for each phase ([30, 31]). The first phase is called mitigation, in this phase the specific risks to be analyzed are identified and the main actions to be put in place are planned, before, during and after the occurrence of an event and therefore of the effects. The second phase is called preparedness, in these phases all the actions necessary to reduce risk are implemented with respect to the three components of occurrence, vulnerability and exposure. Actions are developed to increase, through exercise and teaching, the qualitative and quantitative capacity of managers and end users to respond to the emergency by putting into practice the planned behaviors. The third phase is the one called response, and is to be implemented when emergencies occur in reality and concerns all the activities to be carried out to respond to the emergency. The fourth phase is recovery, in this phase with varying degrees of importance we intervene to restore the functioning of the territorial system, bringing it back to ordinary conditions.

TSMs become crucial for the development of the first three phases because they make it possible to estimate the possible events on the transport networks, and therefore in relation to the time deltas the possibility of saving lives, or, in mathematical terms, to estimate evacuation times shorter than the time deltas related to the risk considered.

The availability of *emerging ICTs* that allow systems to be supported in ordinary conditions [32] provides the possibility of improving operation in extraordinary conditions determined by risk. This note refers to the new ICT technologies, the technologies of the ITS (intelligent transportation system) type are not considered, now widely absorbed and used in the analysis and design of transport systems, such as traffic light control. It refers to a set of technologies, developed in other fields, already mature and which can be systematically used in the transport sector on an urban scale. In this sense, they are defined as emerging ICTs, not because they are emerging in themselves, but because they are emerging with respect to the systematic use in the transport sector on an urban scale, and in the case under study, in the extraordinary conditions determined by risk.

On the basis of what has been seen, the first question that emerges is to define what are the basic factors for analyzing the exposure component of risk, considering the emerging ICTs. The next question is to verify how the basic factors can contribute to an advanced model of risk cycle, and therefore what is the overall role of ICTs in a model that considers all dynamic components. Figure 1 represents outline of the questions posed and the solutions identified.

To answer these questions, the paper, after the introduction in which the various topics are defined, is divided into two sections. Section 2 presents the three basic factors for the study of exposure: models of transport systems, training and exercises, ICT components. Section 3 presents the risk cycle first with reference to the transport sector in urban areas;

then the same section describes the progress of the cycle with the introduction of the ICT components. Finally, some conclusions are drawn regarding the results already obtained and the open lines of research.

The work is of interest both for technicians and planners, because it presents in a compact and structured way all the main issues related to risk and exposure reduction, it is of interest to researchers because it allows them to systematically compare the results achieved, providing important lines of research development.

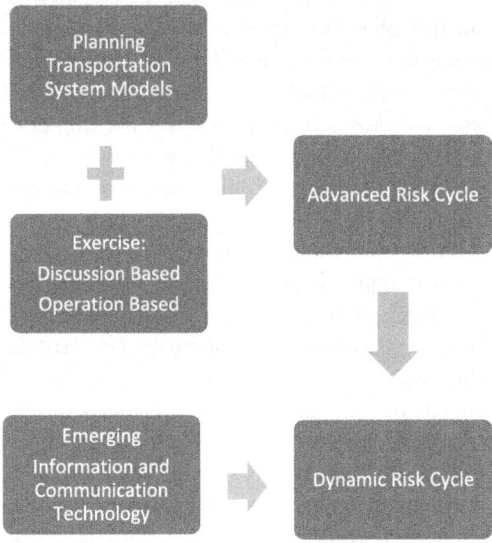

Fig. 1. Structure of paper

2 Setting of the Basic Factors

The role of ICTs is becoming increasingly important. The transport model system used in the planning process is first presented. The importance of training and exercise for the different risk conditions is summarized next. Finally, the emerging classes of ICT that are beginning to be used in transport systems under ordinary conditions are recalled in a synthetic way.

2.1 Transportation System Models

The Transportation System Model (TSM) is used to simulate both uncongested and congested system conditions.

The problem of the TSM develops on the basis of knowledge of demand and the structure of the network. Given these two factors as known, it is necessary to calculate the values of the link and path costs and of the link and path flows, congruent with each other and with demand. Figure 2 shows the two solution schemes. The first is related to the non-congested condition and it emerges that in a functional sequence, starting from the link costs, the link flows are obtained. The second is related to congested conditions and it emerges that congruence between flows and link costs is necessary, which cannot be obtained as a simple functional dependence. The reference equations for the TSM are reported in the literature ([10–14, 33]). In this note, reference is made to [34].

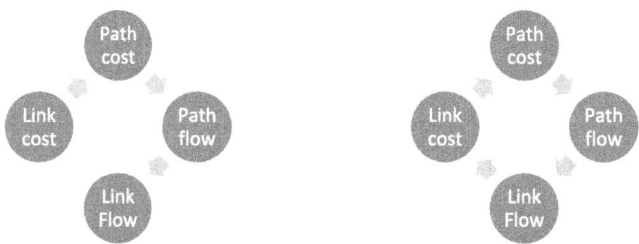

Fig. 2. Relationships between Link and Path Variables in Uncongested and Congested Network

The formulation of the problem of assignment and therefore of the calculation of time, in the case of non-congested time, can be traced back to the calculation for the link flow f of a simple function of the costs of link c and of the demand d:

$$f = f(c, d) \tag{1}$$

That is, from the knowledge of the link cost functions it is easy to calculate the link flows and therefore all the other significant variables of the system's performance.

The compact mathematical formulation with two equations allows planners to estimate the performance of the system in ordinary congested conditions, with a fixed-point model for which the functions of link costs and link flows, respectively:

$$c^* = c(f^*) \tag{2}$$

$$f^* = f(c^*, d) \tag{3}$$

The same structure can be used for extraordinary conditions, where again if the network were to be uncongested, the link costs and therefore the link flows would be immediately calculated, if instead it were to be congested it is again necessary to use appropriate algorithms that allow the solution to be obtained, in the event that it exists and is unique.

2.2 Training and Exercise in the Risk

According to USA approach ([35, 36]), training and exercises actions can be classified on: discussion-based, comprehending low-cost actions aimed at discussing evacuation plans and procedures; operation-based comprehending actions aimed at testing evacuation plans and procedures experimented in a real-world context. In particular, discussion-based are low cost actions ([37, 38]), including:

- seminars [39], or one-way informal discussions of existing plans and procedures to orient participants to plans and procedures; the main aim is to familiarize organizations and personnel with current or expected capabilities;
- workshops, or two-way informal discussions of existing plans and procedures designed to obtain feedbacks about evacuation plans and procedures; the main aim is to improve current or expected capabilities to face emergencies;
- table top exercises ([40, 41]), or discussion of an emergency scenario, designed for experimenting roles, procedures, responsibilities in the emergency situation; the main aim is to evaluate evacuation procedures and emergency protocol, experimented by the decision-makers and population developed along a hypothetical timeline of events;
- games, or simulations of emergency scenarios, usually experimented with combination of Virtual Reality and Serious Games, to simulate a realistic emergency scenario, in which the efficacy of defined rules and procedures for responding to real-life situations is evaluated.

Operation-based are more complex actions [42], including:

- drills, based on the reproduction of a single function (e.g., a health unit); the main aim is to evaluate results of the tested function;
- functional exercises, based on a simulation of the progress of a possible emergency scenario; the main aim is to replicate a system in an emergency condition and testing on real world what happen to resources and equipment;
- full-scale exercises, based on a complex simulation on an emergency scenario, developed in a real context, in a real-time; the main aim is to test more organization and more participants.

The level of preparedness can be measured by adopting a set of quantitative indicators. Training and exercises actions, indicated with $a^s{}_k$, contribute to increase the preparedness by discussing and experimenting planned procedures [43]. Potential effects, measured in terms of exposure reduction, can be estimated before and after a generic action. It is possible to introduce a vector of exposure indicators, specified and calculated before and after the action [44].

By fixing the scenario S, the effects of the action k can be measured by specifying the function $g_h(\cdot)$, for calculating the exposure indicator $e^s{}_h$:

$$e^s_h = g_h\left(a^s_k\right) \qquad (4)$$

where $e^s{}_h$ indicator, is estimated considering what happen between before and after the $a^s{}_k$ action.

2.3 Emerging ICTs

Different classes of ICTs can be used to support risk reduction in the various phases of the cycle mentioned above ([30, 31]). The technologies summarized here are defined as emerging, as mentioned above, not because they constitute a novelty in themselves, but because they are only very recently applied to transport systems, and some components are only as an experimental phase. The classes currently used in transportation analysis ([32, 45–49]), which include various specific technologies, are five.

Internet of Things (IoT), generally defined by a set of intelligent sensors that produce quantitative data and the Internet network that connects them to each other and to other systems; it is therefore not defined as an autonomous technology but rather as a System of technologies.

Blockchain (BC) or, otherwise called, internet of values given by a set of transactions locked in single chains related by an internet network; this technology allows the transition from a central organization of data, relating to the values exchanged, to a distributed organization that keeps all the synchronized data available to all the connected servers; the change is made on only one server and is carried over to the others; important BCs are those developed for container logistics on an international scale.

Big Data (BD), a set of data of orders of magnitude such that it cannot be managed with spreadsheets or traditional data processors, based on multidimensional matrices; They are defined by a large volume, a great speed of arrival and a great variety.

Artificial Intelligence (AI), a set of data and algorithms, which make it possible to identify alternatives, with respect to a defined problem and the available data, the alternatives can be ordered so that the first of them can be chosen autonomously by the algorithm itself.

Digital Twin (DT), defined as a system of models and algorithms that simulates a real system, i.e. providing it with the same inputs that can be given to the real system, generates the same outputs as the real system.

This classification does not consider the ICTs systems required for the autonomous vehicle of different levels. Not even the different levels with the related speed and performance capacity of the telematic connection networks are considered.

It is important to underline the ICTs mentioned are emerging, i.e. they have been used for a short time in cities, and therefore suffer from various technical limitations, while in well-defined environments, where they have been repeatedly used, they allow the best results to be obtained. A clear example is that of the use in ports where today the ICTs described constitute the main pillars of the Port Community System, and therefore allow to significantly increase the performance of the port itself, without particular investments in new infrastructures.

3 Risk Cycle Models

3.1 Advanced Risk Cycle in the Transportation System

In relation to what has been briefly said in the introduction, it is possible to reconstruct the risk cycle. It is interesting to reconstruct the first three phases, because in them it is possible to greatly reduce the risk and therefore the damage to people and/or property.

The main topic is that of the time delta, so it is useful to define some variables that allow you to identify the probability that exposure will be reduced.

The temporal delta can be specified with respect to different events for which assume different values. For clarity the temporal deltas are considered in relation to the time needed to evacuate every last person from the area subject to risk, and again for clarity things are completely neglected.

They are defined as follows:

Δ_{risk} the real time between the occurrence of the event and the effects on the people present in the risk area, the variable can be variously specified by inserting the condition to the safe place or to another;

Δ_{No_Plan} the evacuation time of the last person, in the absence of an evacuation plan;

Δ_{Plan} the evacuation time from the model of the last person, in the presence of an evacuation plan.

$\Delta_{No\_Exp\_Tea}$ the evacuation time of the last person, having not performed any training and exercise activities;

Δ_{Exp_Tea} the evacuation time of the last person, after performing various training and exercise activities.

Based on the definitions given, it follows

$$\Delta_{Plan} < \Delta_{No\_Plan} \tag{5}$$

and for a Δ_{risk} sufficiently high

$$\Delta_{Plan} < \Delta_{risk} < \Delta_{No\_Plan} \tag{6}$$

That is, the capacity of modelers and public decision-makers must be such as to organize an evacuation plan in which the time delta, planned, to evacuate the last person is less than the time delta of the risk in question. Figure 3 shows the different phases of the cycle.

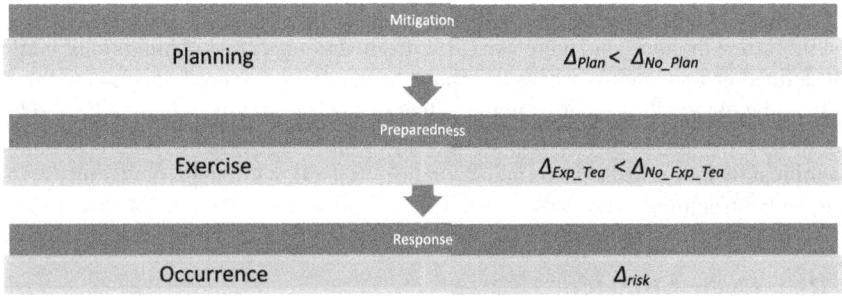

Fig. 3. Phases of the risk cycle, time deltas

Relations (5) and (6) always hold in the hypotheses of Δ_{risk} high enough.

Equation (6), however, only expresses the possibility that by making a good plan, you will have the effect on people as a model. That is, it implies a relationship between a real datum and a datum from a model. The plan could, for example, be done very well but not known by anyone. To get to the levels of Δ_{Plan} it is necessary to develop a system of activities such as those schematically seen above.

Going back to the definitions given, it follows

$$\Delta_{Exp\_Tea} < \Delta_{No\_Exp\_Tea} \tag{7}$$

Then, it can be assumed that not having done any training activity has the same time delta as not having planned:

$$\Delta_{No\_Plan} = \Delta_{No\_Exp\_Tea} \qquad (8)$$

The relationship (8) therefore represents the worst condition in which a city can find itself, i.e., the basic point. The point of arrival or the goal to be achieved is given by the relationship (6).

The importance of carrying out exercises and training is highlighted considering that it can be:

$$\Delta_{Exp\_Tea} < \Delta_{risk} \quad or \quad \Delta_{risk} > \Delta_{Exp\_Tea} \qquad (9)$$

Depending on whether or not Training and Exercises have succeeded in involving many people, according to the plan setting, until they approach the theoretical time delta of the plan, which constitutes the calculated probabilistic limit.

By combining the various equations in the end, it follows

$$\Delta_{Plan} < \Delta_{Exp\_Tea} < \Delta_{risk} < \Delta_{No\_Exp\_Tea} = \Delta_{No\_Plan} \qquad (10)$$

It is important to note the role that TSM models play, in the formulations given by (1) or (2) and (3) in the evaluation of Δ_{Plan}. If the plan is made using TSM models, the best results are obtained, and the Δ_{Plan} is indicated by Δ_{Plan_TSM}.

Similarly, if the exercises are done according to the FEMA framework [35], the results can be checked and Δ_{Exp_Tea} is indicated by $\Delta_{Exp\_Tea\_eg}$. Therefore, the relations expressed by (10) have to be rewritten as:

$$\Delta_{Plan\_TSM} \leq \Delta_{Exp\_Tea\_eg} < \Delta_{risk} < \Delta_{No\_Exp\_Tea} = \Delta_{No\_Plan} \qquad (11)$$

3.2 ICTs in the Dynamic Risk Cycle

Emerging ICTs are making it possible to develop the risk cycle considerably. The same structure defined in the previous paragraph, and summarized by the relation (11), can be redefined in an advanced way, being able to further reduce both the Δ_{Plan_TSM} and the $\Delta_{Exp\_Tea\_eg}$.

It is possible to introduce the knowledge provided to users by the real-time conditions of the system from the IoT, and the memory of the scenarios elaborated by the plan and deepened in the exercises through the BDs, both levels of knowledge can be combined in various ways using the resources of the AI, the presence of any transactions is solved by BC. This knowledge can be included in Eqs. (2) and (3).

It is important to highlight that emerging ICTs are playing a very important role in the development of the smart city, as indicated by the EU, and as verifiable in all international experiments ([50–52]). In fact, e-ICT is one of the three pillars of the smart city, together with mobility and energy in ordinary conditions. In the case of risk, the smart city requires the integration of e-ICT with mobility [53].

The emerging technologies mentioned can be fully exploited by switching to dynamic models, which allow the updating of both utilities and choices to be simulated.

In literature, classical formalization is made on the elaboration of individual knowledge from one day to the next [34] or with scheduled models [54]. In the case of the risk with Δ_{risk} large enough to allow an evacuation, the time Δ_{risk} can be divided into periods, each equal to the time it takes to process the information.

In the presence of risk, it is necessary to separately identify what contributions can be made in the user's updating process, with respect to the two evolutions in period y and at time t.

For certain attributes, it is feasible to estimate the value that was observed (tested) during evacuation drills in the past or the value that the user acquired during risk-planning training sessions. These values can and should be used as a basis of experiential knowledge of temporal processes with respect to y. $X[y-1]$, $X[y-2]$ do not indicate the values of the days, or periods "physically" preceding day y, but the previous ones in which the exercises or training were carried out.

Thus, generally, the path costs, on day y at time t, may be stated as depending on the path attributes X, and can be written as follows:

$$C[t, y] = \psi(X[t], \ldots, X[y-1], X[y-2], \ldots) \tag{12}$$

Learning mechanisms according to the classical approach, for the ordinary conditions on the inter-period process can be evaluated as:

$$X_{hk}^{fo}[y] = \gamma \cdot X_{hk}^{exp}[y-1] + (1-\gamma) \cdot X_{hk}^{fo}[y-1] \tag{13}$$

where

- $X_{hk}^{fo}[y]$ is the value of h-th of path k predicted/calculated on day y;
- $X_{hk}^{exp}[y-1]$ is the value of h-th of path k encountered/tested on day y-1;
- γ ($\in]0, 1]$) is the encountered/tested weight.

The Eq. 13 can be transferred to the extraordinary conditions, exploiting the knowledge coming from knowledge obtained in previous evacuation drill as well as from the data used in designing the evacuation plan EP (i.e., data collected during evacuation drills), or using e-ICT, stored in BD:

$$X_{hk}^{BD,fo}[y] = \gamma \cdot X_{hk}^{BD,exp}[y-1] + (1-\gamma) \cdot X_{hk}^{BD,fo}[y-1] \tag{14}$$

The path used (experienced) on day y-1 (e.g., evacuation drill) and the path k' not used but taken into consideration in the day y-1, $K_{ov}[y-1]$, can both be calculated by Eq. 14 coherently using *EP* stored in *BD*.

The other fundamental process to consider is the one that evolves in time t. Also, for this process, it is necessary to highlight what happens in extraordinary conditions.

In these conditions, there are two types of information related to: the event (exogenous to TSM) and the traffic (endogenous to TSM).

Given the extraordinary conditions, user learning (for both endogenous and exogenous fields) can only be achieved through information arriving from outside. Therefore, the update occurs at each time t of day y. The single tool through which the update takes place can be summarized in the IoTs mentioned above. IoTs can range from those for the detection of exogenous parameters to those for the detection of endogenous parameters.

In this last case, the real-time network status can be obtained through real-time vehicle sensors (e.g., IoT).

Finally, it is possible to merge the past (*BD*) and real-time (*IoT*) values of the *h*-th past attribute at time *t* of the current day *y*, as follows:

$$X_{hk}^{fo}[t,y] = \xi \cdot X_{hk}^{BD,fo}[t,y] + (1-\xi) \cdot X_{hk}^{IoT}[t,y] \quad (15)$$

where

- $X_{hk}^{IoT}[t,y]$ is the value of *h*-th attribute X_{hk} realized at *t* of the current day *y*; this information is made available via the *Internet of Things* and allows to show how the network performance is evolving at the moment; for instance, it can show the travel time (X_{hk}) that other vehicles are testing out at day *y* in order to travel at time *t* on the same path *k*; it should be noted that this information is updated for each time *t* across the entire network;
- $X_{hk}^{BD,fo}[t,y]$ is the value of attribute X_{hk}, provided by EP at time *t* of day *y*;
- $\xi (\in \,]0, 1])$ is the weight assigned to the value without real-time information provided by *BD* at time *t* of day *t*; the value of ξ is considered fixed but it can also be viewed more broadly as variable with *t*, moving to *0* for the link where the vehicle is travelling (i.e., user is experimenting with real-time value).

Based on what has been seen in this paragraph, the values of the time deltas from the plane with TSM Δ_{Plan_TSM} can be further reduced. Indicating with $\Delta_{Plan\_TSM\_ICT}$ the temporal delta from the plane, evaluated with dynamic models that make use of ICT, it has:

$$\Delta_{Plan\_TSM\_ICT} \leq \Delta_{Plan\_TSM} \quad (16)$$

In the same way proposed for the mitigation phase and therefore for the planning phase, the dynamic model can be extended to the entire exercise and training process using DTs, in this case the $\Delta_{Exp\_Tea\_eg}$ can be further improved by the systematic use of DTs and by the inclusion in DTs of models of the type expressed by (14), arriving at a delta that can be expressed as $\Delta_{Exp\_Tea\_eg\_ICT}$, and it has:

$$\Delta_{Exp\_Tea\_eg\_ICT} \leq \Delta_{Exp\_Tea\_eg} \quad (17)$$

With these two results, (16) and (17) obtained with the use of emerging ICT it is possible to rewrite (11) in a more advanced way:

$$\Delta_{Plan\_TSM\_ICT} \leq \Delta_{Exp\_Tea\_eg\_ICT} \leq \Delta_{Plan\_TSM} \leq \Delta_{Exp\_Tea\_eg} < \Delta_{risk} < \Delta_{No\_Exp\_Tea} = \Delta_{No\_Plan}$$

The latter relationship is interesting because it provides a direction of development both for the preparation of the plans and for the training and exercise phases, which are often underestimated.

4 Conclusions

The paper analyzed the issue of the risk cycle, with particular reference to the reduction of exposure in urban transport systems. The paper introduces the general theme of the use of emerging ICTs in the context of transport systems and highlights the central role they have in the risk cycle, allowing to reduce the time deltas of the plan and operation. At the end this translates into the saving of human lives.

On the one hand, the work is part of the new lines of research developed internationally for risk reduction, and on the other hand it allows the development of multiple experiments such as: Innovative exercises both discussions based and operation based, specifications and calibrations of models for the updating of costs and choices, development of dynamic assignment models. All these experiments have emerging ICTs as a central element.

The work presented can be of great interest to researchers because it proposes a set of necessary developments, and of considerable use for technicians and politicians who have the responsibility for decisions in risk conditions.

Funding. This study was carried out within the research project "RISK: Recovery Increasing by Social Knowledge"-2022B4TT2M (CUP C53D23004800006), PIANO NAZIONALE DI RIPRESA E RESILIENZA (PNRR) Missione 4 "Istruzione e Ricerca"—Componente C2, Investimento 1.1 "Fondo per il Programma Nazionale di Ricerca e Progetti di Rilevante Interesse Nazionale (PRIN)", D.D. n.104 del 2 febbraio 2022, ERC SH7 "Human Mobility, Environment, and Space". This paper reflects only the authors' views and opinions, neither the European Union nor the European Commission nor the Italian Ministry for Universities and Research can be considered responsible for them.

References

1. American Institute of Chemical Engineers ed: Guidelines for chemical transportation safety, security and risk management. Wiley, Hoboken, N.J (2008)
2. Kletz, T.: Major hazard aspects of the transport of dangerous substances. J. Loss Prevent. Process Indus., **5**, 254 (1992). https://doi.org/10.1016/0950-4230(92)80055-D
3. Russo, F., Rindone, C.: Risk Assessment in Urban Area by Transport of Dangerous Goods. In: Ksibi, M., Sousa, A., Hentati, O., Chenchouni, H., Lopes Velho, J., Negm, A., Rodrigo-Comino, J., Hadji, R., Chakraborty, S., and Ghorbal, A. (eds.) Recent Advances in Environmental Science from the Euro-Mediterranean and Surrounding Regions (4th Edition). pp. 873–875. Springer Nature Switzerland, Cham (2024). https://doi.org/10.1007/978-3-031-51904-8_190
4. Goldblatt, R.: Development of Evacuation Time Estimates for the Davis Nuclear Power Station (1993). https://www.nrc.gov/docs/ML0502/ML050250240.pdf
5. Goldblatt, R., Reuben, B., Weinisch, K.: Evacuation Planning, Human Factors, and Traffic Engineering Developing Systems for Training and Effective Response, https://onlinepubs.trb.org/onlinepubs/trnews/trnews238evacplanning.pdf, (2005)
6. University of Maryland—Fire and Rescue Institute: Latest announcements, http://www.mfri.org, (2024)
7. Milazzo, M.F., Lisi, R., Maschio, G., Antonioni, G., Spadoni, G.: A study of land transport of dangerous substances in Eastern Sicily. J. Loss Prev. Process Ind. **23**, 393–403 (2010). https://doi.org/10.1016/j.jlp.2010.01.007
8. Rose, A., et al.: Benefit-cost analysis of FEMA hazard mitigation grants. Nat. Hazards Rev. **8**, 97–111 (2007). https://doi.org/10.1061/(ASCE)1527-6988(2007)8:4(97)
9. Roberts, P.S.: Policy Review. FEMA after Katrina. 137, (2006)
10. Ben-Akiva, M.E., Lerman, S.R.: Discrete choice analysis: theory and application to travel demand. MIT Press, Cambridge, Mass (1985)

11. Sheffi, Y.: Urban transportation networks. Prentice Hall, Englewood Cliff, NJ (1985)
12. Cascetta, E.: Transportation Systems Engineering: Theory and Methods. Kluwer Academic, Dordrecht; Boston, MA (2001)
13. Ben-Akiva, M., Walker, J., Bernardino, A.T., Gopinath, D.A., Morikawa, T., Polydoropoulou, A.: Integration of choice and latent variable models. Perpetual motion: Travel behaviour research opportunities and application challenges. 431–470 (2002)
14. Train, K.: Discrete Choice Methods with Simulation. Cambridge Univ. Press, Cambridge (2003)
15. Sheffi, Y., Mahmassani, H., Powell, W.B.: A transportation network evacuation model. Transp. Res. Part A **16A**(3), 209–218 (1982)
16. Sheffi, Y.: Urban Transportation Networks: Equilibrium Analysis With Mathematical Programming Methods. Prentice-Hall, Englewood Cliffs, N.J (1985)
17. Van Zuilekom, K., Van Maarseveen, M., Van Der Doef, M.: A decision support system for preventive evacuation of people. In: Van Oosterom, P., Zlatanova, S., and Fendel, E.M. (eds.) Geo-information for Disaster Management. pp. 229–253. Springer Berlin Heidelberg, Berlin, Heidelberg (2005). https://doi.org/10.1007/3-540-27468-5_16
18. Henke, I., Troiani, G., Pagliara, F.: An analysis of the vulnerability of road networks in response to disruption events through accessibility indicators specification. Transp. Plan. Technol. **47**, 628–655 (2024). https://doi.org/10.1080/03081060.2024.2329650
19. Tai, C.-A., Lee, Y.-L., Yau, J.-T.: A study of evacuation behavior during earthquakes. Int. J. SDP. **9**, 874–884 (2014). https://doi.org/10.2495/SDP-V9-N6-874-884
20. Russo, F., Chilà, G.: Safety of users in road evacuation: modelling and DSS for demand. Presented at the SUSTAINABLE DEVELOPMENT 2009 , Cyprus April 21 (2009). https://doi.org/10.2495/SDP090431
21. Musolino, G.: Methods for risk reduction: modelling users' updating utilities in urban transport networks. Sustainability. **16**, 2468 (2024). https://doi.org/10.3390/su16062468
22. Vitetta, A.: Network design problem for risk reduction in transport system: a models specification. Int. J. TDI. **6**, 283–297 (2022). https://doi.org/10.2495/TDI-V6-N3-283-297
23. Russo, F., Rindone, C.: Planning in road evacuation: classification of exogenous activities. WIT Trans. Built Environ. **116**, 639–651 (2011). https://doi.org/10.2495/UT110541
24. Rindone, C., Russo, A.: A Network Analysis for HSR Services in the South of Italy. In: Gervasi, O., Murgante, B., Garau, C., Taniar, D., C. Rocha, A.M.A., and Faginas Lago, M.N. (eds.) Computational Science and Its Applications—ICCSA 2024 Workshops. pp. 217–232. Springer Nature Switzerland, Cham (2024). https://doi.org/10.1007/978-3-031-65318-6_15
25. Tesoriere, G., Russo, A., De Cet, G., Vianello, C., Campisi, T.: The centrality of Italian airports before and after the COVID-19 period: what happened? European Transport/Trasporti Europei. 1–16 (2023). https://doi.org/10.48295/ET.2023.93.2
26. Russo, A., Campisi, T., Bouhouras, E., Basbas, S., Tesoriere, G.: Sustainable Maritime Passenger Transport: A Network Analysis Approach on a National Basis. In: Gervasi, O., Murgante, B., Rocha, A.M.A.C., Garau, C., Scorza, F., Karaca, Y., and Torre, C.M. (eds.) Computational Science and Its Applications—ICCSA 2023 Workshops. pp. 195–207. Springer Nature Switzerland, Cham (2023). https://doi.org/10.1007/978-3-031-37120-2_13
27. Wang, L., Zhou, P., Gu, J., Li, Y.: Numerical simulation of passenger evacuation process for a cruise ship considering inclination and rolling. JMSE. **12**, 336 (2024). https://doi.org/10.3390/jmse12020336
28. Polydoropoulou, A., Bouhouras, E., Karakikes, I., Papaioannou, G.: Enhancing Climate Resilience in Maritime Ports: A Decision Support System Approach. In: Gervasi, O., Murgante, B., Garau, C., Taniar, D., C. Rocha, A.M.A., and Faginas Lago, M.N. (eds.) Computational Science and Its Applications—ICCSA 2024 Workshops. pp. 241–252. Springer Nature Switzerland, Cham (2024). https://doi.org/10.1007/978-3-031-65329-2_16

29. Saha, T., Lee, K., Hyun, K.K., Cassidy, J., Jang, S.: Understanding travel behaviors and mobility challenges faced by older adults during the COVID-19 pandemic. JAL. **4**, 177–187 (2024). https://doi.org/10.3390/jal4030012
30. Oloruntoba, R., Sridharan, R., Davison, G.: A proposed framework of key activities and processes in the preparedness and recovery phases of disaster management. Disasters **42**, 541–570 (2018). https://doi.org/10.1111/disa.12268
31. Schmitt, T., Eisenberg, J., Rao, R.R.: Improving disaster management: the role of IT in mitigation, preparedness, response, and recovery. National Academies Press, Washington, D.C. (2007). https://doi.org/10.17226/11824
32. Schroten, A., Van Grinsven, A., Tol, E., Leestemaker, L., Schackmann, P.P., Vonk-Noordegraaf, D., Van Meijeren, J., Kalisvaart, S.: Research for TRAN Committee—The impact of emerging technologies on the transport system. European Parliament, Policy Department for Structural and Cohesion Policies, Brussels (2020)
33. Ben-Akiva, M., Bierlaire, M., Koutsopoulos, H.N., Mishalani, R.: Real Time Simulation of Traffic Demand-Supply Interactions within DynaMIT. In: Gendreau, M. and Marcotte, P. (eds.) Transportation and Network Analysis: Current Trends. pp. 19–36. Springer US, Boston, MA (2002). https://doi.org/10.1007/978-1-4757-6871-8_2
34. Cantarella, G.E., Watling, D., de Luca, S., Di Pace, R.: Dynamics and stochasticity in transportation systems: tools for transportation network modelling. Elsevier, Amsterdam, Netherlands (2019)
35. US Federal Emergency Management Agency. FEMA: Homeland Security Exercise and Evaluation Program, https://www.fema.gov/emergency-managers/national-preparedness/exercises/hseep, last accessed 2024/08/30
36. Russo, F., Rindone, C.: Planned and implemented actions by exercises. In: Gervasi, O., Murgante, B., Garau, C., Taniar, D., C. Rocha, A.M.A., and Faginas Lago, M.N. (eds.) Computational Science and Its Applications—ICCSA 2024 Workshops. pp. 28–40. Springer Nature Switzerland, Cham (2024). https://doi.org/10.1007/978-3-031-65308-7_3
37. Borell, J., Eriksson, K.: Learning effectiveness of discussion-based crisis management exercises. Int. J. Disas. Risk Reduct. **5**, 28–37 (2013). https://doi.org/10.1016/j.ijdrr.2013.05.001
38. Rindone, C., Barabino, B.: Risk reduction by urban evacuation: a review of mobility solutions for users with special needs. In: Proceedings of EMCEI conference 2023. , Morocco
39. Rindone, C., Moschella, M.: Experimentation on Risk Reduction by Training: A Framework for Seminar Activity. In: Gervasi, O., Murgante, B., Garau, C., Taniar, D., C. Rocha, A.M.A., and Faginas Lago, M.N. (eds.) Computational Science and Its Applications—ICCSA 2024 Workshops. pp. 129–143. Springer Nature Switzerland, Cham (2024). https://doi.org/10.1007/978-3-031-65308-7_10
40. Rindone, C., Moschella, M.: Training and exercises for disaster risk reduction: survey design of a table top activity for mobility in evacuation condition. In: Proceedings of CrossMed Conference (2024)
41. Russo, F., Rindone, C.: Experimentation for risk reduction by training and exercises: the role of table top activities. In: Proceedings of EMCEI conference 2024. , Morocco (2024)
42. Roud, E., Gausdal, A.H., Asgary, A., Carlström, E.: Outcome of collaborative emergency exercises: Differences between full-scale and tabletop exercises. Contingencies Crisis Mgmt. **29**, 170–184 (2021). https://doi.org/10.1111/1468-5973.12339
43. Skryabina, E.A., Betts, N., Reedy, G., Riley, P., Amlôt, R.: The role of emergency preparedness exercises in the response to a mass casualty terrorist incident: a mixed methods study. Int. J. Disas. Risk Reduct. **46**, 101503 (2020). https://doi.org/10.1016/j.ijdrr.2020.101503
44. Russo, F., Rindone, C.: Methods for risk reduction: training and exercises to pursue the planned evacuation. Sustainability. **16**, 1474 (2024). https://doi.org/10.3390/su16041474

45. Comi, A., Russo, F.: Emerging information and communication technologies: the challenges for the dynamic freight management in city logistics. Front. Future Transp. **3**, 887307 (2022). https://doi.org/10.3389/ffutr.2022.887307
46. Russo, F., Comi, A.: Sustainable urban delivery: the learning process of path costs enhanced by information and communication technologies. Sustainability. **13**, 13103 (2021). https://doi.org/10.3390/su132313103
47. Haghshenas, S.S., Guido, G., Haghshenas, S.S., Astarita, V.: The Role of Artificial Intelligence in Managing Emergencies and Crises within Smart Cities. In: 2023 International Conference on Information and Communication Technologies for Disaster Management (ICT-DM). pp. 1–5. IEEE, Cosenza, Italy (2023). https://doi.org/10.1109/ICT-DM58371.2023.10286925
48. Astarita, V., Haghshenas, S.S., Haghshenas, S.S., Guido, G., Martino, G.: Bibliometric-Based Literature Review on Artificial Intelligence in Disaster Response. In: Proceedings of AIIT 4TH international conference. Greening the way forward: sustainable transport infrastructure and systems (2024)
49. Astarita, V., Guido, G., Haghshenas, S.S., Haghshenas, S.S.: Risk reduction in transportation systems: the role of digital twins according to a bibliometric-based literature review. Sustainability. **16**, 3212 (2024). https://doi.org/10.3390/su16083212
50. European Commission. EC: Communication from the commission smart cities and communities European Innovation Partnership, 2012, https://ec.europa.eu/transparency/documents-register/detail?ref=C(2012)4701&lang=en, (2012)
51. Pintor, L., Uras, M., Colistra, G., Atzori, L.: Monitoring People's Mobility in the Cities: A Review of Advanced Technologies. In: Menozzi, R. (ed.) Information and Communications Technologies for Smart Cities and Societies. pp. 25–42. Springer Nature Switzerland, Cham (2024). https://doi.org/10.1007/978-3-031-39446-1_3
52. Karamanlis, I., Nikiforiadis, A., Botzoris, G., Kokkalis, A., Basbas, S.: Towards sustainable transportation: the role of black spot analysis in improving road safety. Sustainability. **15**, 14478 (2023). https://doi.org/10.3390/su151914478
53. Russo, F., Rindone, C.: European Smart cities towards disaster risk reduction. In: Proceedings of CrossMed Conference (2024)
54. Cascetta, E., Biggiero, L., , A., Russo, F.: A system of within-day dynamic demand and assignment models for scheduled inter-city services. Presented at the Proceedings of Seminar D&E on Transportation Planning Methods at the 24th PTRC Summer Annual Meeting , England (1996)

Concept of a Virtual Test Field for Inland Waterway Transport

Jason Sutanto[✉], Christian Hürten, Maximilian Jarofka, Frédéric E. Kracht, and Dieter Schramm

University of Duisburg Essen, Lotharstraße 1, 47057 Duisburg, Germany
jason.sutanto@uni-due.de

Abstract. In recent years, the shipping industry has increasingly turned to automation to improve operational and economic efficiency, safety, and sustainability. However, the high complexity and safety requirements of automated technical systems present challenges in system design and testing, resulting in time-consuming processes and higher costs when testing with real systems. Therefore, an alternative testing environment is needed to implement automation functions within the shipping industry in a cost-effective manner. The joint project "Data bases, infrastructures and technologies for virtual testing of automation functions – VERA" aims to implement a virtual test field for the development and evaluation of automation functionalities as well as simulation of traffic scenarios in the inland waterway transport sector. This paper discusses the features of the virtual test field as well as the modular system architecture. Key modules of the virtual test field include a physically accurate motion model for inland vessels, a traffic module that generates waterway traffic based on AIS data, and a scenario management system that handles simulation parameter variation and scenario settings.

Keywords: smart shipping · virtual test field · inland waterway transport · traffic simulation

1 Introduction

The shift towards more environmentally friendly and smarter mobility through digital transformation is not limited to the automotive industry, but also impacts the shipping sector as a vital component of the supply chain. Although automation in the shipping industry and particularly in inland navigation is still in its early stages, it is constantly improving. Various levels of automation are being developed simultaneously, ranging from ship assistance systems to teleoperated vessels and highly automated self-driving vessel. In view of the major challenges in inland navigation, e.g. the lack of qualified nautical personnel, the highly competitive pressure, and the resulting tasks in the context of automation, there is a considerable need for the development of future-oriented solutions. [1].

The development and testing of the technologies and functionalities mandatory for automation requires a suitable research infrastructure that is tailored or can be adapted to different technology readiness levels (TRLs) depending on the development progress. Test fields are one such infrastructure in which complex technical systems can be investigated under realistic conditions. In addition to real test fields on real waterways, this also includes virtual test fields as versatile and universally applicable development and test platforms.

Due to the intrinsic characteristics of the inland waterway transport sector, it is not viable to procure such a test field at feasible cost. Rather, such real test fields are more suitable for a later phase of development. Moreover, since it is part of the public transport system, its operation should not be hindered for economic and safety reasons.

Although a virtual test field could not fully replace a physical test field, it offers an efficient platform to test and simulate prototypical models under custom-defined operating conditions. The advantages of virtual testing include the possibility to define and simulate risk-free testing of safety-critical cases or rare-occurring traffic scenarios that may occur very seldomly under normal operating conditions, as well as parallel computing which can potentially speed up the development and testing phases. This is especially beneficial in the inland shipping sector, where the volume of traffic is much lower than that of the automotive area [2]. The more limited availability of real-world data in inland shipping compared to the automotive sector can slow down development processes, particularly when using data-driven methods such as machine learning. Virtual testing hereby serves as a tool to generate new data to train and test models without the difficulty and danger in conducting a real-life maneuver and the need to set up specific traffic scenes, which may hinder the normal operation of other vessels. Another advantage is the ability to conduct and replay deterministic simulations. Given the same initial conditions, the outcome of a simulation can be made reproducible. This consistency allows for more precise analysis and prediction of system behavior, allowing for easier and faster development.

Furthermore, the development and testing of automated vessels or systems is currently hindered by the lack of a valid legal framework. Virtual test fields offer an opportunity to begin developments and tests immediately, without the need for a legal framework. This is particularly beneficial due to the time pressure faced during the development phase. Additionally, the knowledge gained from these virtual test fields can greatly contribute to the acceleration of the development of a legal framework for automated systems. A comparable project which can benefit from the test field is the project "methods for the safe design of automation and remote monitoring in inland navigation – *SAFEBin*" which is funded by the Federal Ministry for Digital and Transport Affairs of Germany. This project aims to develop the basic principles and legal framework for the risk assessment of systems for automated ship navigation in German inland waterways [3]. This synergy between the virtual test field and the legal framework can have significant benefits for further development projects and demonstrations. Once the project is completed, the results will be made available to various users in the research, industry, and administration sectors.

In the *VERA* research project, a virtual test field to develop and test automation functionalities specifically for inland waterway transport is to be developed. The *VeLABi* (test and control center for autonomous inland vessels) simulator infrastructure [4] is to be used during development as a generic Human-Machine-Interface (HMI) system, which includes a 360° projection screen and a ship bridge with common operating elements and designed according to EN 1864:2008, as well as a versatile research platform for further projects concerning simulation in the shipping sector [5].

The simulation environment will encompass real-life waterways in the German federal state of North Rhine-Westphalia and surrounding areas, mainly focusing on parts of the Rhine River with its varying current flow across the length and width of the river. Two canals without significant current flow, namely the Rhine-Herne Canal and the Dortmund-Ems Canal, are also of particular interest. The Dortmund-Ems Canal is designated as a physical test field for other projects. The creation of digital twins of automated systems as well as Software- and Hardware-in-the-Loop (SiL and HiL) tests of such systems are to be made possible through suitable interfaces. Functionalities of the virtual test field are to be designed modularly to allow not only full-scale tests, but also partially and fully isolated tests of subsystems and functions within the context of automated ship navigation, guidance, and control. This includes tasks such as environment perception, object recognition, localization, positioning, situation overview, path and trajectory planning, steering and control, monitoring, as well as communication.

This paper describes the main features to be implemented as well as a proposed system architecture of the virtual test field planned to be created as part of the *VERA* research project.

2 State of the Art

Simulation-based testing using HiL has been a crucial part of the development process in the automotive industry since the late 20th century [6]. The exponential increase in computing power has brought practical use to Digital Twin (DT) technology, where individual components are integrated into a complete virtual system that is able to mimic the state of its physical counterpart via real-time data exchange [7].

The even more far-reaching logical continuation is the embedding of DTs in a complete simulation environment, known as virtual test fields. Systems within a virtual test field should be able to communicate and interact with one another as well as with the environment realistically. While it is still not very widespread in use, virtual test fields are growing in popularity in recent years as vehicle automation and connected driving become research topics of interest.

Consequently, the German Federal Ministry for Digital and Transport has expressed its willingness to fund the development of test fields in e.g. road and highway networks, ports, and inland waterways [8]. Several projects are underway in different parts of Germany with the goal of creating a virtual test field. *CAPTN Förde Areal* aims to build a research vessel for use in the Kieler Förde and to create a physical and virtual test field of the area [9]. The creation of a digital test field encompassing the Spree-Oder waterway in Berlin and Brandenburg is the objective of the project *DigitalSOW* [10]. Similarly, *HANNAH* focuses on the Schlei for its area of interest [11]. A complete list of funded research projects by the Federal Republic of Germany with the aim of creating physical and virtual test fields for road networks, waterway networks, and ports are catalogued respectively in [12, 13], and [14].

3 Concept

3.1 Objectives

The *VeLABi* infrastructure contains a rudimentary prototypical virtual test field that was designed according to the specifications of previous research projects *AutoBin* [1], *FernBin* [15], and *ELLA* [16]. The goal of *VERA* is therefore to further develop the available infrastructure into a multifunctional and modular virtual test field. Essential features to be developed include:

- Geometric and hydraulic modelling of specific inland waterways, such that the simulation of relevant navigation scenarios within the area is possible.
- Population of the terrain with static and dynamic objects, which are instanced and catalogued.
- Illustration of surrounding (intermodal) traffic through traffic simulations.
- Simulation of maneuvering and propulsion dynamics of inland vessels on rivers and canals.
- Consideration of relevant external influences on the vessel's movement, such as the numerical calculation of localized wind loads based on real flow data as well as the water current.
- Realistic implementation of virtual visual (e.g. camera, LiDAR, RADAR) and nautical (e.g. GNSS and AIS receiver) sensors, including measurement uncertainties and error models to enable evaluation of failure scenarios.
- Functionalities for simulation configuration, management, and replication.

The modular implementation of all (sub-)components is of central importance to enable a high degree of flexibility in the composition of test scenarios. Furthermore, a higher-level monitoring and management system for the execution and evaluation of test scenarios is to be realized through a module that generates various parameters and events according to predefined distribution functions. In addition to these developments, the project will also demonstrate the first exemplary applications of assistance systems:

- Extension of the ECDIS (Electronic Chart Display and Information System) with processed static and simulation-based dynamic information on the waterway and surrounding traffic.
- Testing the calculation of risk metrics to determine the hazard levels as a preliminary study for the development of a real-time capable risk assessment.

3.2 General Construct

Figure 1 depicts a schematic diagram of the general system architecture of the virtual test field. The most apparent component is the HMI. The HMI consists of input devices such as the ship control levers and radio telecommunication as well as feedback information shown on the bridge displays. While the *VeLABi* infrastructure is one such HMI to be used, it is not limited to a single interface. Multiple HMIs controlled by multiple operators can be implemented in the case that multiple vessels are to be controlled or a takeover mechanic between two or more vessels such as the teleoperation system in the

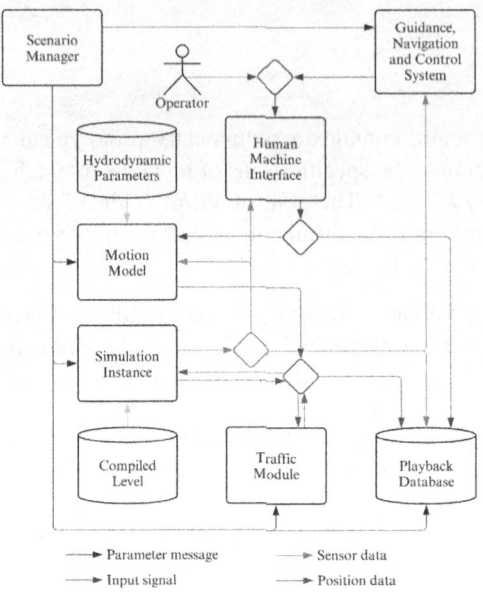

Fig. 1. Overview of the system architecture

project "Remote-controlled, coordinated driving in inland navigation – *FernBin*" [15] should be tested.

Aside from the operator, input signals may also be received from the GNC (guidance, navigation, and control) system. The GNC system uses signals from a GNSS signal generator and contains real-time capable hardware for trajectory planning and control. Radio signals including correction data (e.g. simulating errors during RTK mode) are generated with a real-time computer. Realistic uncertainties in the GNSS data are moreover considered. Statistical uncertainties are implemented and signal errors as well as failure modes are emulated depending on the signal reception in the simulation environment (i.e. bridges and other construction elements that may degrade signal connection). [17].

3.3 Motion Model

The input signals (e.g. engine order telegraph and rudder state) are then passed on to the vessel motion model(s), shown schematically in Fig. 2. The motion model calculates the vessel's position and speed based on the input signals as well as given hydrodynamic parameters obtained from knowledge or experimental results. Therefore, the input signals are passed to the actuator models which are then processed and further passed to the machine models. Engine speed and rudder angle are inputs to the propeller and rudder models which output the forces and torques. These internal forces and torques are combined with external forces and torques calculated from various aerodynamic and hydrodynamic effect models.

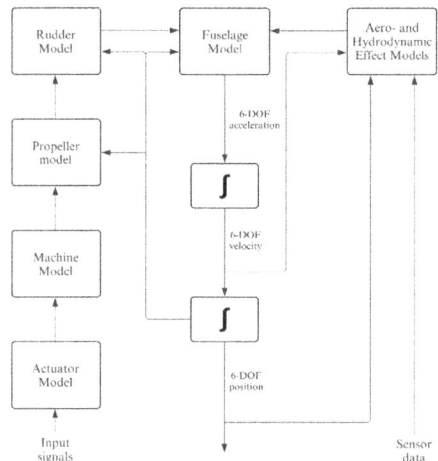

Fig. 2. Overview of the motion model

The aero- and hydrodynamic models rely on the integrated acceleration values from the fuselage model. A list of considered models and their required data for calculation is as follows:

- Wind model: 3D wind vector data.
- Flow model: 3D current flow data.
- Motion model for a depth-limited waterway: water depth or riverbed topology.
- Motion model for a laterally restricted waterway: waterway width or riverbank topology.
- Motion model for ship-to-ship interactions (e.g. close encounters, overtaking): vessel geometry and movement parameters.
- Sea state model: sea states

Due to the modular design of the simulation environment, different motion models can be used depending on the requirements of the simulation runs. For simulating traveling with almost constant speed, the Abkowitz-type model integrating shallow water effects proposed in [18] is implemented.

For maneuvering at slow and varying speeds, such as in a harbor, a modular model is implemented as described in [19]. This motion model is based on the following differential equations:

$$m(\dot{u} - vr) = X$$
$$m(\dot{v} + ur) = Y$$
$$I_z \dot{r} = N$$

The external forces X and Y and the torque N are calculated as a function of the planar velocities u and v, the turning rate r, their derivatives in time and the control variables rudder angle δ and propeller speed n.

$$X = f_X(u, \dot{u}, v, \dot{v}, r, \dot{r}, \delta_1..., n_1...)$$
$$Y = f_Y(u, \dot{u}, v, \dot{v}, r, \dot{r}, \delta_1..., n_1...)$$
$$N = f_N(u, \dot{u}, v, \dot{v}, r, \dot{r}, \delta_1..., n_1...)$$

Given the modular structure of the model, various external influences can be integrated into this model, such as the effects of wind or the use of a bow thruster.

The output of the fuselage model is the acceleration of the vessel in all three rotational and translational axes. The integration of these values produces the vessel's calculated velocity and position which are then passed on to the simulation instance(s) and the traffic module.

3.4 Simulation Environment

The simulation instance is executed in the UNREAL ENGINE 5 (UE5) game engine [20] as a compiled level. Multiple simulation instances can be active simultaneously, such as the case when running the test field in the *VeLABi* simulator, as shown in Fig. 3.

Fig. 3. The virtual test field simulated in *VeLABi*

The simulation instance is schematically represented in Fig. 4. Game engines such as UE5 and UNITY are a popular software of choice to be used in simulators [21, 22] due to their flexibility to fit in most use cases, real-time rendering capabilities, built-in physics simulations, and a high-level frontend interface for development. Previously, the UNITY game engine was used as the software for simulation and visualization in completed projects such as *AutoBin* and *FernBin*. The choice to migrate to UE5 was based on the more complex requirements of the project that would not be feasible to be realized in UNITY.

After evaluating the features present in UNITY and UE5, it was apparent that UE5 was better suited for creating large-scale worlds due to its 64-bit IEEE-754 floating-point number implementation (compared to UNITY's 32-bit implementation), thereby massively reducing the order of magnitude of floating-point precision error. The need for a shifting coordinate system (e.g. Floating Origin [23]) in UE5 for the scale of the project is therefore completely eliminated. UE5 also comes with built-in features that optimizes the performance of large worlds, such as Nanite [24], a geometry system allowing for dynamic generation of Level of Detail (LOD) meshes, and World Partition System [25], a level management system for dynamic loading of parts of a world. Moreover, the UE5 engine code is open source [26] and it uses C + + as its scripting language, allowing for greater flexibility in adding custom features and lower-level control such as memory management.

The generation of the level environment is done both automatically and manually. External data sources such as GIS (Geographic Information System) and water survey data are to be collected and an automated procedure to generate navigation areas is to be developed. After the area topology is created, it is to be populated with buildings and vegetation. Photogrammetry data (if available) may also be used to further add detail to the scenery. Objects in the area are then manually modified using classical 3D mesh modelling methods where necessary to better resemble its real-life counterpart, particularly in places of interest such as ports. Navigation-relevant objects, such as landmarks and nautical traffic signs, are furthermore catalogued.

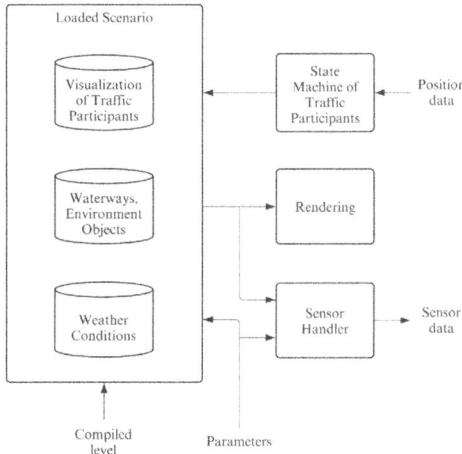

Fig. 4. Overview of the simulation instance

Each simulation instance in UE5 receives position data from the motion model and outputs sensor data measured from the generated 3D mesh data. The state of each traffic participant is determined using the position data and then visually rendered into the scene. The compiled level features a 3D virtual environment complete with waterways and other objects, accompanied by a dynamic weather system that has a direct impact on other systems. Sensors and cameras utilize the 3D mesh data to render images of the scene and pass on the data to other subsystems of the virtual test field.

Each virtual sensor is its own individual module and is developed as a general implementation rather than being built into e.g. a vessel class. This modularity enables easy swapping or expansion of sensors as needed. It also allows for parallel computing on different instances to optimize resource utilization and enhance system scalability. The virtual sensors encompass visual sensors (RGB camera, LiDAR, RADAR), motion and environmental sensors (compass, speed sensor, depth sensor, IMU), and communication-related sensors (GNSS receiver, AIS signal generator).

Data from the simulation instances are augmented accordingly with error data to mimic the behavior of their real-life counterparts. These include noise behavior (which can be weather-and-daytime-dependent) and uncertainties for the sensors, as well as packet losses and latencies for communication-related data such as mobile radio and GNSS correction data. Malfunctions, faults, failures of individual data, and single or entire sensor packages can be simulated and reproducibly toggled on or off. The sensors are integrated to the scenario manager for reproducible and random control of their behavior.

3.5 Traffic Simulation and Scenario Management

The use of microscopic traffic simulation (traffic module) is crucial for accurately representing realistic traffic. This involves calculating simulation runs based on recorded AIS data for specific traffic situations. Defined traffic scenarios are processed and made available for playback in the simulator. By adjusting input parameters such as weather, daytime, water levels, traffic volume, and fleet composition, a wide range of randomized scenarios can be created in the virtual test field. This allows for testing automation systems and generating realistic and dynamic traffic situations using statistics. Custom scenarios are also possible to be defined, such that historical traffic scenarios can be simulatively reproduced. An extension for SUMO (Simulation of Urban Mobility), a tool used for traffic simulations [27], is to be developed to include inland navigation and focus on intermodal transport chains. Figure 5 depicts a basic traffic simulation of the inland waterway on the Rhine River in SUMO.

Fig. 5. Inland waterway traffic simulation in SUMO

The scenario manager is an essential function of the virtual test field, allowing users to create and manage simulations. A test scenario is defined by various factors such as the shipping area, traffic participants, vessel properties, and environmental conditions. The parameters for each of these factors can be defined with variation ranges and distribution functions. Pseudo-random events, which can involve variations in environmental conditions or malfunctions/failures, can be generated based on distribution functions. Within a scenario, multiple test runs can thus be implemented, with all parameters and any data flow (input, sensor, and position data) documented and saved in a playback database. This ensures that any simulated scenario is deterministic and thus can be reproduced and replayed, allowing easier evaluation during the development and testing phases.

3.6 Networking

ROS2 (Robot Operating System 2) [28] has proven in past projects to be a suitable framework for distribution of generated data. Parameter messages from the scenario manager are transmitted via custom ROS2 messages to other subsystems of the virtual test field. The communication in ROS2 is based on the FastDDS middleware, an open-source DDS (Data Distribution Service) implementation by eProsima for real-time systems [29].

By using FastDDS, the need to implement complex network programming is eliminated. Communication between endpoints i.e. nodes is established via a publish-subscribe pattern. Endpoints that send data (i.e. publishers) transmit data to topics, whereas endpoints that receive data (i.e. subscribers) listen only to topics of interest. With the use of DDS APIs, systems communicating with one another do not need to be in the same programming language, e.g. a C + + simulation instance can communicate with a motion model written in Python. Because of the modular structure of the architecture, subsystems can easily be added, replaced, or removed even during HiL testing just by using suitable network interfaces in ROS2 or plain FastDDS.

The modularization principle also applies to the network infrastructure. By establishing a secure VPN connection, the subsystems of the virtual test field do not need to be run on the same network. This is especially beneficial as not all hardware for the test field may be present at the same location. For instance, the *VeLABi* simulator is located at the DST institute in Duisburg, a copy of the ship bridge used in *VeLABi* as a secondary HMI is at the the Chair of Mechatronics of the University of Duisburg-Essen, and the GNC system is operated by the Institute of Automatic Control at the RWTH Aachen University.

4 Conclusion

This paper discusses the system architecture to be implemented in the *VERA* research project which aims to build a virtual test field encompassing waterways in the German federal state of North Rhine-Westphalia and the surrounding area. The virtual test field serves as an advanced development tool to test automation functionalities and simulate traffic scenarios specifically for inland vessels and inland waterway transport.

The *VeLABi* infrastructure is designated as the fundamental starting point for the test field implementation and as the generic HMI to be used. Core features of the test field include realistic modeling of the designated area, microscopic traffic simulation, implementations of ship external and internal forces and torques, virtual sensors, and a simulation scenario management system.

Acknowledgement. The research is funded by the Federal Ministry for Digital and Transport BMDV and conducted within the scope of the joint research project "Data bases, infrastructures and technologies for virtual testing of automation functions – VERA" (Original in German: *Datengrundlagen, Infrastrukturen und Technologien für Virtuelle Erprobungen von Automatisierungsfunktionen*), part of the "Funding guideline for the research and development of digital test fields on federal waterways (DTW II)" with the funding number 45DTW2V06B.

References

1. Bakshande, F., et al.: The AutoBin project - Key concepts, status, and intended outcomes. In: Autonomous Inland and Short Sea Shipping Conference - AISS2020. Duisburg (2020)
2. Freight transport statistics - modal split. https://ec.europa.eu/eurostat/statistics-explained/index.php?title=Freight_transport_statistics_-_modal_split. Accessed 03 Apr 2024
3. SAFEBin – SmartShipping. https://www.smartshipping.info/safebin/. Accessed 10 Apr 2024
4. Kracht,F., et al.: VeLABi – Research and control center for autonomous inland vessels. Automatisierungstechnik **70**(5), 411–419. (2022)
5. Inland navigation vessels - Wheelhouse - Ergonomic and safety requirements, EN 1864:2008
6. Brayanov, N., et al.: Review of hardware-in-the-loop -a hundred years progress in the pseudo-real testing. Electrotechnica & Electronica **54**, 70–84 (2019)
7. Singh, M. et al.: Digital twin: origin to future. Appl. Syst. Innov. **4**(2), 36 (2021)
8. Förderdatenbank - Förderorganisationen - Bundesministerium für Digitales und Verkehr [Funding database - Funding organizations - Federal Ministry for Digital and Transport] (in German). https://www.foerderdatenbank.de/FDB/Content/DE/Foerdergeber/B/bmdv-bundesministerium_fuer_digitales_und_verkehr.html. Accessed 04 Apr 2024
9. Förde Areal – CAPTN. https://captn.sh/foerde-areal/. Accessed 04 Apr 2024
10. DigitalSOW. https://www.digitalsow.de/en/digitalsow.html. Accessed 04 Apr 2024

11. HANNAH - Digitales Testfeld Schlei [HANNAH - Digital test field Schlei] (in German). https://www.binsmart.de/dms/hannah/. Accessed 04 Apr 2024
12. Testfeldmonitoring - Startseite [Test field monitoring - Homepage] (in German). https://www.testfeldmonitor.de/Testfeldmonitoring/DE/Home/home_node.html. Accessed 04 Apr 2024
13. Digitale Testfelder an Bundeswasserstraßen - Liste der Projekte [Digital test fields on federal waterways - list of projects] (in German). https://digitale-testfelder-wasserstrassen.bund.de/projekte/liste. Accessed 04 Apr 2024
14. Automatisiertes und autonomes Fahren [Automated and autonomous driving] (in German). https://www.digitest-hafen.de/erprobungsfelder/automatisiertes-und-autonomes-fahren/. Accessed 04 Apr 2024
15. Weber, T., et al.: Concept of a teleoperation system for inland shipping vessels. In: 2022 IEEE 25th International Conference on Intelligent Transportation Systems (ITSC), pp. 349–354. IEEE, Bilbao (2022)
16. ELLA – Entwicklungsplattform im Modellmaßstab für Manöver-Automatisierung [ELLA - Model-scale development platform for maneuver automation] (in German). https://www.dst-org.de/ella/. Accessed 02 Apr 2024
17. Koschorrek, P. et al.: Towards semi-autonomous operation of an over-actuated river ferry. Automatisierungstechnik **70**(5), 433–443 (2022)
18. Yang, Y., el Moctar, O.: A mathematical model for ships maneuvering in deep and shallow waters. Ocean Eng. **295** (2024)
19. Henn, R., Schweig, S.: Predictor-corrector approach to path control for inland vessels. In: AISS 2024 - Autonomous Inland and Short Sea Shipping Conference, Duisburg (2024) (under review)
20. Unreal Engine 5. https://www.unrealengine.com/en-US/unreal-engine-5. Accessed 06 Apr 2024
21. Michalík, D., et al.: Developing an unreal engine 4-based vehicle driving simulator applicable in driver behavior analysis—a technical perspective. Safety **7**(2) (2021)
22. Wang, Z., et al.: Digital twin simulation of connected and automated vehicles with the unity game engine. In: 2021 IEEE 1st International Conference on Digital Twins and Parallel Intelligence (DTPI), pp. 1–4. IEEE, Beijing (2021)
23. Thome, C.: Using a floating origin to improve fidelity and performance of large, distributed virtual worlds. In: 2005 International Conference on Cyberworlds (CW 2005), p. 270. IEEE, Singapore (2005)
24. Nanite Feature Documentation. https://eoshelp.epicgames.com/s/article/Nanite-Feature-Documentation?language=en_US. Accessed 04 Apr 2024
25. World Partition Feature Documentation. https://eoshelp.epicgames.com/s/article/World-Partition-Feature-Documentation?language=en_US. Accessed 04 Apr 2024
26. Contributing to Unreal Engine | Epic Developer Community. https://dev.epicgames.com/documentation/en-us/unreal-engine/contributing-to-the-unreal-engine?application_version=5.3. Accessed 04 Apr 2024
27. Eclipse SUMO - Simulation of Urban MObility. https://eclipse.dev/sumo/. Accessed 06 Apr 2024
28. ROS: Home. https://www.ros.org/. Accessed 06 Apr 2024
29. eProsima Fast DDS - ROS 2 Documentation: Humble documentation. https://docs.ros.org/en/humble/Installation/DDS-Implementations/Working-with-eProsima-Fast-DDS.html. Accessed 04 Apr 2024

Critical Characterization of Three-Phase Traffic Flow in Severe Condition

Bo Song, YongSheng Qian(✉), JunWei Zeng, and Xu Wei

School of Traffic and Transportation, Lanzhou Jiaotong University, Gansu, Lanzhou 730070, China
qianyongsheng@mail.lzjtu.cn

Abstract. To ensure driving safety in adverse weather conditions, this paper examines traffic flow during foggy days as a case study. Using cellular automata, it explores the critical features of three-phase traffic flow under such conditions. Initially, the impact of moderate fog on driving behavior and vehicle motion is analyzed. Next, the effect of brake lights on drivers is incorporated to enhance the MCD (Modified Comfortable Driving) model. A new model, MCD-S (Modified Comfortable Driving for Severe weather) is proposed. Finally, a comparative analysis of the spatio-temporal diagrams, fundamental diagrams, and speed volatility between the MCD and MCD-S models is presented. Results indicate that moderate fog decreases the density at the critical phase transition, thereby reducing traffic flow efficiency and increasing congestion. It significantly affects synchronous flow, transition from orderly to wide moving jams in the first-order phase transition. Additionally, moderate fog conditions lead to more stable vehicle following and reduced aggressive driving behavior.

Keywords: Traffic Engineering · Three-Phase Traffic Flow · Cellular Automata · Severe Weather · Critical Characteristics · MCD model

1 Introduction

1.1 A Subsection Sample

The complex traffic environment is an important reason for the frequent occurrence of traffic accidents. Among them, the impact of slippery road surface and visual obstacles due to severe weather is the most serious [1].

Many scholars have begun to focus on the effects of severe weather on traffic flow. By collecting the number of road accidents and other analyses, the impact of severe weather on road traffic flow speed, flow is studied. The Department of Transport of the United Kingdom conducted a survey for the first time during 1949–1950 to estimate the effect of weather on traffic flow [2]. Michelle L. Angel et al. [3] collected meteorological data and traffic sensor data in Florida. The results showed that the average travel speed decreased during rainfall.

Cellular automata are widely used to study traffic flow. The abstract description of traffic elements in a discrete manner provides a unique advantage for conducting traffic studies. Alex Donkers et al. [4] simulated traffic flow under special weather conditions by building a traffic flow model for special weather scenarios. It was concluded that bad weather affects the driving behavior and habits of drivers, leading to an increase in the energy consumption of cars. Liu et al. [5] established a microscopic model of following and lane changing traffic flow under foggy conditions based on cellular automata. The effects of foggy conditions on highway traffic flow were revealed through simulation. Li et al. [6] proposed a CA model under different degrees of rainfall, and the optimal speed limit value under rainfall conditions was proposed by variable speed limit control strategy. Gao et al. [7] calibrated the heeling model based on the measured data NGSIM, which made the traffic flow simulation results closer to the actual situation.

At present, there are many studies on traffic flow under severe weather conditions by scholars at home and abroad, but there are fewer researches on phase transiton mechanisms and observations of synchronized flow phases. In the three-phase traffic flow theory, the traffic flow is divided into free state and congested state. Among them, the free flow state has less interaction between vehicles (Free Flow, abbreviated as F), when vehicles can travel at maximum speed. Congested traffic flow can be further subdivided into Synchronizes Flow (S) and Wide Moving Jams (J). So the core of three-phase traffic flow theory is Synchronizes Flow [8]. Among them, the MCD (Modified Comfortable Driving model) model established by Jiang et al. improved the model on the basis of the CD model [9], because of its introduction of the vehicle's brake lights, so compared with other three-phase traffic flow models in simulating the microstructural characteristics of the traffic flow state to get more realistic results.

This paper propose a new model with cellular automata to better understand the effect of severe weather on the critical characteristics of three-phase traffic flow, even in the simple case where lane changing behavior is not considered, it can occur that severe weather makes the traffic flow state change.

2 Simulation Scenarios and Modeling Ideas

A periodic boundary (Fig. 1) is employed to investigate the critical features of three-phase traffic flow in adverse weather conditions. The periodic boundary not only has relatively simple traffic flow characteristics, but also can better reflect the basic behavior and laws of traffic flow, so the scenario is more accurate for theoretical analysis and modeling.

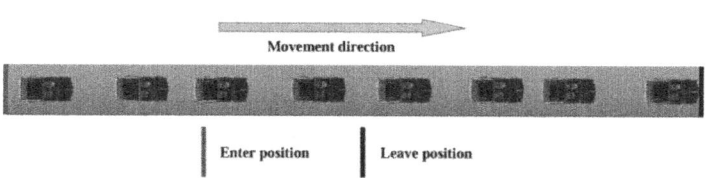

Fig. 1. Schematic diagram of a single-lane scenario

3 Model

The MCD model can accurately simulate various phenomena in traffic flow. In this section, for the microscopic effects of bad weather on driving behavior, the MCD model is improved to further obtain a CA model for three-phase traffic flow under severe weather conditions.

3.1 MCD Model

In the severe weather, for manual vehicle drivers, judging the distance of the vehicle can't rely on the outline of the car to judge. At this time, brake lights become the main factor for drivers to judge the distance. Among them, the MCD model proposed by Jiang Rui et al. [9]. Considers the influence of brake lights. Compared with other models, the rapid acceleration and deceleration of the vehicle is controlled, and it will be more applicable on the road in severe weather conditions. Due to space constraints, the MCD model will not be elaborated upon in this paper.

3.2 MCD-S Model

The following rules of the MCD model are all for the case when the road conditions are normal. When there is severe weather on the road, the driver's habits will change accordingly. Taking foggy days as an example, drivers' driving habits usually have the following changes: 1. Enhancement of vision. Severe weather may make the line of sight become blurred; 2. Increase the frequency of braking. Drivers need to leave more reaction time before braking to minimize accidents due to skidding; 3. Exercise slowly. Higher speeds increase the probability of emergencies.

In summary, this paper proposes a new model MCD-S (Modified Comfortable Driving of Severe weather). The new model adds a process of "multiple braking" on the basis of MCD. Under the influence of severe weather, when a vehicle traveling smoothly at a high speed for a long time encounters the deceleration of the vehicle in front of it, in order to prevent the vehicle from skidding and to maintain the desired speed, it will lightly brake and only let the brake light come on to remind the vehicle behind it to control the following distance. The difference between the new model and the MCD model is the update of the brake light status, which is changed to

$$
\begin{aligned}
&\text{if } (d_n \leq d_{sight}) \text{ and } b_{n+1}(t) = 1) \text{ then } b_n(t+1) = 1; \\
&\text{if } (v_n(t+1) < v_n(t)) \text{ then } b_n(t+1) = 1; \\
&\text{if } (v_n(t+1) > v_n(t)) \text{ then } b_n(t+1) = 0; \\
&\text{if } (v_n(t+1) = v_n(t)) \text{ then } b_n(t+1) = b_n(t);
\end{aligned}
\quad (1)
$$

d_{sight} indicates the extent to which the target vehicle is affected by the front brake lights, which can also be interpreted as the distance at which the vehicle in front can be noticed. Table 1 illustrates the importance of the relevant parameters in the follow model.

Table 1. List of variables in vehicle-following rule

parameters	Meaning
d_n	Distance between vehicle and front vehicle
v_n	Speed of target vehicle
b_n	Target vehicle brake light status
d_{sight}	Distance at affected by the front brake lights

3.3 Model Parameter Calibration

Our speed limit and distance standards are shown in Table 2 [10].

Under sunny conditions, the visibility on the highway can reach several kilometers, but usually, the actual driving is more concerned about the situation in front of a few hundred meters to one kilometer in order to be able to react in time. In this paper, the visibility under sunny conditions is taken as 1000 m, i.e. 400cells, and in this paper, the d_{sight} is taken as 100–200 m in medium fog for example.

Table 2. Definition of visibility in China's highway traffic management regulations

Definition	Increased visibility
No fog	>1000
Light fog	500–1000
Light fog	200–500
Medium Fog	100–200
Foggy	50–100
Dense Fog	<50

4 Simulation

The simulation scenario consists of a three-lane highway measuring 5000 m in length, with each cell measuring 2.5 m. Each vehicle occupies 2 cells, resulting in a total of 2000 cells along the highway. The simulation involves 20 independent runs, with each run lasting 3000 time steps. The results from the first 2000 steps may vary due to chance, so only the outcomes from the final 1000 steps are analyzed.

4.1 Fundamental Diagram Analysis

The fundamental diagram serves as the primary framework for differentiating the various states of three-phase traffic flow, effectively illustrating the macroscopic characteristics of traffic under varying road densities. Figure 4 presents the fundamental diagrams for both the MCD model and MCD-S, along with the corresponding flow and velocity under different conditions, using a parameter value of 10 while keeping the other parameters consistent with those detailed in Sect. 2.

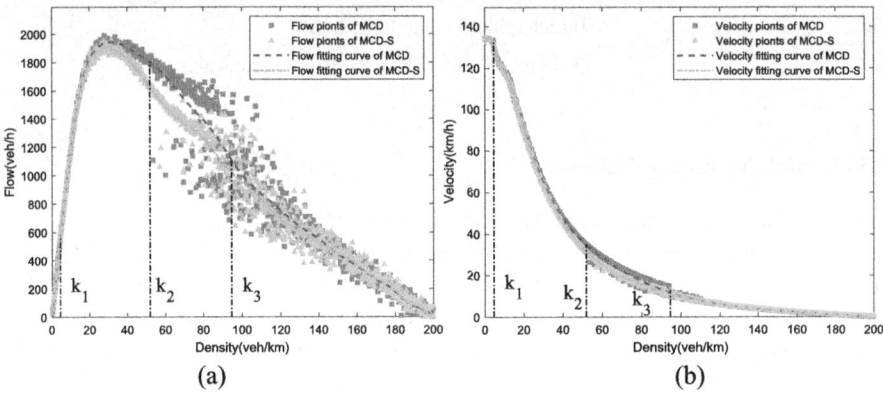

Fig. 2. The fundamental diagram of MCD model and MCD-S model

As illustrated in Fig. 2(a) and (b), the fundamental diagram of the MCD model consists of two branches. The upper branch arises from an initial uniform distribution of vehicle positions, while the lower branch originates from a state of congestion. When the density is extremely low ($k < k_1$), vehicles maintain the highest speed and the traffic flow state is synchronized flow. When the density exceeds k_1, vehicle speeds begin to decrease and light synchronized flow occurs. When the density is within the range of $k_1 < k < k_2$, the road flow and vehicle speeds intersect at the point where the two branches meet, marking a transition from light synchronous flow to heavy synchronous flow. Once the density surpasses k_2, the traffic flow experiences a first-order phase transition from free flow to synchronous flow. Conversely, if the system begins in a congested state, the traffic will exhibit wide moving jams.

The fundamental diagram of the MCD-S model follows a trend similar to that of the MCD model. When the density falls within the range of $k_1 < k < k_2$, drivers exhibit increased caution due to the effects of foggy weather. As a result, road flow diminishes, causing synchronous flow to encroach upon free flow at lower densities. Consequently, the traffic flow experiences a first-order phase transition from free flow to synchronous flow at these reduced densities. When the density exceeds k_2, the synchronous flow phase leads to a notable decrease in both road flow and the average speed of vehicles. When the density does not reach k_3, the synchronized flow undergoes a first-order phase transition towards the wide moving jams. When $k > k_3$, the traffic flow state is not significantly different from the normal condition.

Severe weather can diminish the density required for critical phase transitions in traffic flow, including the first-order transitions from free flow to synchronous flow and from synchronous flow to wide moving jams. Among the phases in three-phase traffic flow, the synchronous flow phase is particularly susceptible to the effects of foggy weather, demonstrating that it is one of the most unstable phases in traffic dynamics.

4.2 Spatio-Temporal Diagrams Analysis

As shown in Fig. 3, the spatio-temporal diagrams of lane of the MCD model at different phases of traffic flow are demonstrated at densities of 4veh/km, 30veh/km, 65veh/km and 85veh/km, respectively, with the time intervals taken from the last 1500 time steps of the model's stable operation, and the colors in the spatio-temporal maps represent the vehicle speeds.

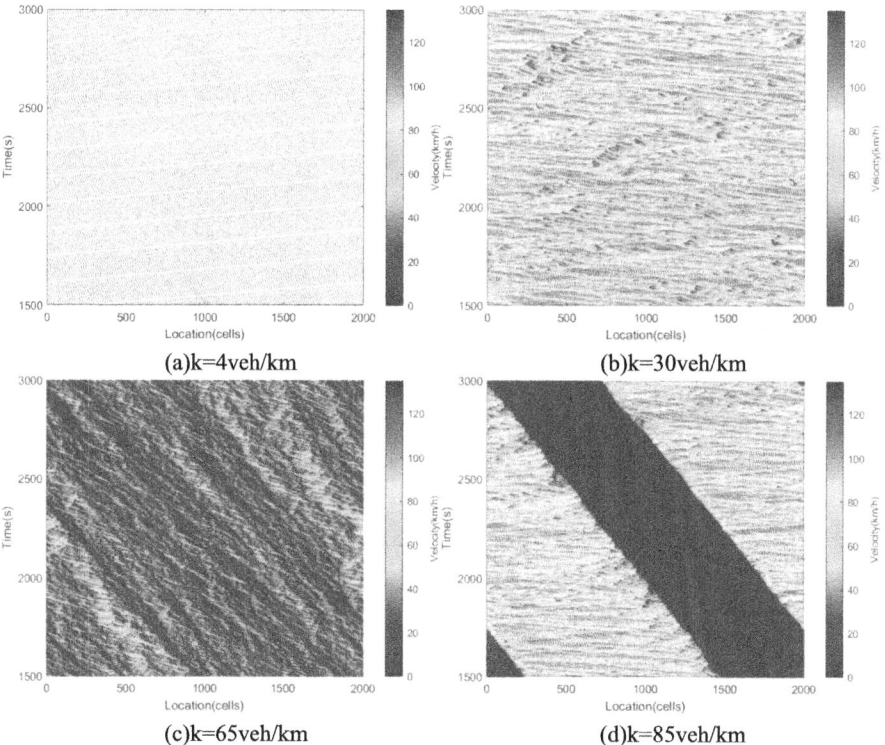

Fig. 3. Spatio-temporal diagrams of MCD model

When the density is at 4veh/km (Fig. 3(a)), there is minimal interaction between vehicles, allowing for high speeds to be maintained. As the density increases to 30veh/km (Fig. 3(b)), a low-speed region emerges in the spatio-temporal diagram, with light synchronous flow gradually encroaching upon the free-flow region. At densities of 65veh/km and 85veh/km, the traffic flow undergoes a complete transition from free flow to synchronous flow (Fig. 3(c)), or it may evolve into wide moving jams, where free flow and light synchronous flow coexist (Fig. 3(d)). When the density becomes sufficiently high, any initial distribution will lead to the coexistence of wide moving jams, free flow, and light synchronized flow.

As shown in Fig. 4, the spatio-temporal diagrams of the MCD-S model under foggy conditions at different phases of traffic flow are demonstrated at the same density.

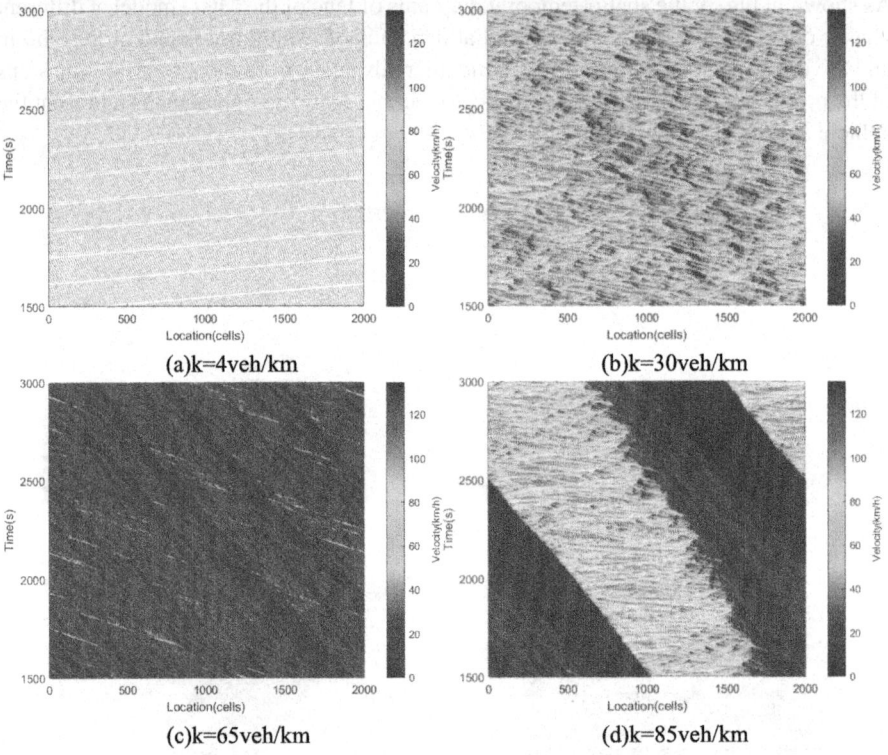

Fig. 4. Spatio-temporal diagrams of MCD-S model

When the density is 4veh/km (Fig. 4(a)), the foggy day has no effect on the free flow. At a density of 30veh/km (Fig. 4(b)), more low-speed regions appeared compared to the MCD model, and the foggy day had an effect on the light synchronous phase. When the density is 65veh/km and 85veh/km (Fig. 4(c)), the synchronous flow velocity is lower compared to the MCD model. When the simulation is started from congestion, the traffic flow is wide moving jams, free flow and light synchronous flow coexist (Fig. 6(4)), which is less different from MCD.

In summary, foggy weather influences drivers' behavior at the same maximum vehicle speed. The impact of fog is most significant during the synchronous flow phase of traffic, where it reduces the average speed of vehicles in synchronous flow or triggers a first-order phase transition from synchronous flow to wide moving jams. However, fog has minimal effects on both the free flow and wide moving jam states.

4.3 Speed Volatility Analysis

As shown in Fig. 5, the MCD model and MCD-S model vehicle speed fluctuations are analyzed at densities of 4veh/km, 30veh/km, and 85veh/km, respectively, so as to respond to the driver's behavior. The time interval is taken as the last 600 steps of the stable operation of the model, and the colors in the spatio-temporal plots represent the vehicle speeds.

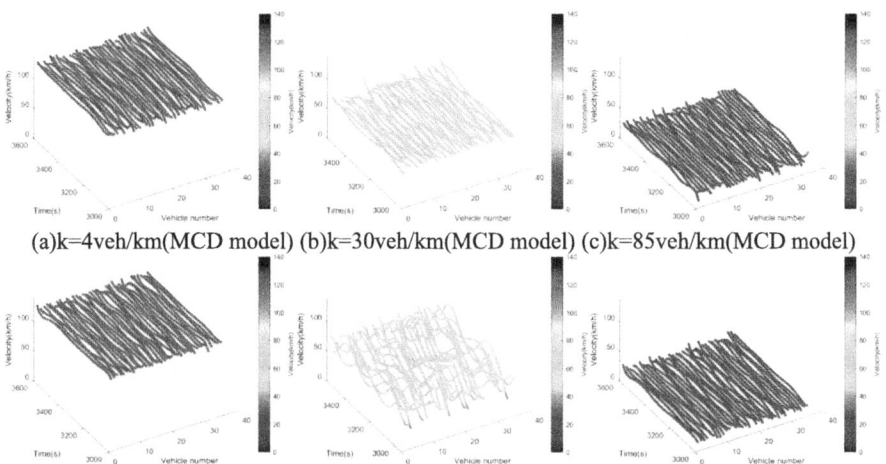

(a)k=4veh/km(MCD model) (b)k=30veh/km(MCD model) (c)k=85veh/km(MCD model)

(d)k=4veh/km(MCD-S model) (e)k=30veh/km(MCD-S model) (f)k=85veh/km(MCD-S model)

Fig. 5. The speed volatility of different traffic flow phase

Figure 5(a) and (d) illustrate that when traffic density is low and the flow state is free flow, the speeds of vehicles in both models remain stable throughout the driving process, with the free flow phase largely unaffected by foggy weather. In contrast, during the synchronous flow phase, as shown in Fig. 5(b) and (e), vehicle speed fluctuations are significantly larger under normal conditions. This is because drivers frequently accelerate and decelerate to maximize their speed, benefiting from better visibility. However, in foggy conditions, drivers tend to moderate their driving behavior, leading to reduced speed fluctuations and prompting a transition from synchronous flow to wide moving jams. This further highlights that synchronous flow is the phase most impacted by foggy weather. As depicted in Fig. 5(c) and (f), when traffic is in a state of wide moving jams, vehicle speeds, already diminished due to road density, are no longer significantly influenced by foggy conditions.

5 Conclusion

This study utilizes the three-phase traffic flow theory to develop a cellular automata model for traffic flow under severe weather conditions. It aims to capture the driving behavior and vehicle motion characteristics affected by foggy weather by analyzing traffic flow spatio-temporal diagrams, fundamental diagrams, and speed fluctuations. The key conclusions are as follows:

1. Foggy weather can reduce the density of traffic flow critical phase transition, including the first-order phase transition from free flow to synchronous flow or from synchronous flow to wide moving jams.
2. Among the phases in the three-phase traffic flow, the synchronous flow phase is most seriously affected by the foggy weather, which also proves that the synchronous flow is the most unstable phase in the traffic flow.
3. Foggy conditions can reduce the average speed of synchronous flow vehicles and make the synchronous flow to the wide moving jams the first-order phase transition.
4. Under foggy conditions, drivers will drive more cautiously and reduce frequent acceleration and deceleration behaviors in order to reduce the occurrence of traffic accidents, which makes the speed fluctuation of vehicles reduced.

In addition, different degrees and types of bad weather will also have an impact on the driving behavior of drivers, which will lead to the phase transition of traffic flow. The impact of different degrees of bad weather on drivers' driving behavior was not analyzed in detail in the study, and further research can be done on this part in the future.

Acknowledgement. This research received funding from several sources, including the National Natural Science Foundation of China (Grants No. 72361017, 52362047, and 71861024), the Major Research Plan of Gansu Province (Grant No. 21YF5GA052), the 2021 Gansu Higher Education Industry Support Plan (Grant No. 2021CYZC-60), the Natural Science Foundation of Gansu Province (Grant No. 18JR3RA119), the Excellent Doctoral Program of Gansu Province (Grant No. 23JRRA906), and the Double-First Class Major Research Programs funded by the Educational Department of Gansu Province (Grant No. GSSYLXM-04).

References

1. Liu, L.L., Weng, J.C., Rong, J.: Traffic flow characteristic of city expressway under snow weather. J. Transp. Inf. Saf. **30**(1), 10–14 (2012)
2. Tanner, J.C.: Effect of weather on traffic flow.Nature **169**(4290), 107–107 (1952)
3. Angel, M.L., Sando, T., Chimba, D.: Effects of rain on traffic operations on Florida freeways. Transp. Res. Rec.: J. Transp. Res. Board **2440**(1), 51–59 (2014)
4. Donkers, A., Yang, D., Viktorović, M.: Influence of driving style, infrastructure, weather and traffic on electric vehicle performance. Transp. Res. Part D: Transp. Environ. **88**, 102–569 (2020)
5. Liu, Z.H., et al.: Modeling and simulation of car following in fog based on cellular automata. J. Syst. Simul. **33**(10), 2399–2410 (2021)
6. Li, S.L., et al.: Variable speed-limit control of expressway in rainy days based on cellular automaton. J. Highw. Transp. Res. Dev.**40**(4), 194–200+208 (2023)
7. Gao, Y.F., et al.: An Improved car-following model for mixed passenger and freight traffic on expressway based on NGSIM data. J. Highw. Transp. Res. Dev. **40**(9), 187–196 (2023)
8. Kerner, B.S., Klenov, S.L., Wolf, D.E.: Cellular automata approach to three-phase traffic theory. J. Phys. A: Math. Gen. **35**(47), 9971–10013 (2002)
9. Jiang, R., Wu, Q.S.: Cellular automata models for synchronized traffic flow[J/OL]. J. Phys. A: Math. Gen. **36**(2), 381–390 (2003)
10. Wang, S.Y.: The analysis on dynamic features of road traffic flow in poor visibility based on the cellular automata model. Hefei University of Technology (2011)

Design and Implementation of Traffic Congestion Relief Strategies Based on Multi-objective Optimization Algorithms

Jun Zhao(✉)

College of Artificial Intelligence, Jiangsu Food and Pharmaceutical Science College, Huai'an, China
hyzhaojun@126.com

Abstract. This study chose Nanjing as a case study, comprehensively analysed the spatio-temporal distribution characteristics of traffic flow, and constructed an optimisation model based on the theory of vehicle queues. The study profoundly explored signal control strategies and developed an adaptive traffic signal control system. This system integrates real-time data acquisition, dynamic data processing, and intelligent decision-making support modules, thereby forming a highly automated and intelligent traffic management platform. The study verified the effectiveness of the optimisation strategy through simulation experiments utilising microscopic traffic simulation software VISSIM and SUMO. The results indicated that the average delay was decreased by 25%, and the emissions of CO and NOx were reduced by 15% and 20% respectively. The study put forward a multi-objective optimisation model based on mixed integer linear programming (MILP), which combines multi-objective optimisation algorithms with the requirements of modern traffic management and aims to achieve the dual goals of minimising traffic delay and reducing exhaust emissions. This comprehensive strategy holds significant practical significance and application value for addressing urban traffic congestion problems.

Keywords: Urban traffic congestion · Multi-objective optimization algorithm · Mixed integer linear programming (MILP) · Signal control strategy · Vehicle queuing theory

1 Introduction

Driven by globalization and regional integration, the urbanization process is accelerating, followed by the rapid expansion of traffic demand. Urban traffic congestion has increasingly become a key bottleneck restricting the sustainable development of cities. Urban traffic congestion not only causes huge time cost loss and environmental pollution, but also has a negative impact on the quality of life of residents and the economic vitality of cities. In light of these considerations, the effective identification and alleviation of urban traffic congestion has emerged as a pivotal and challenging issue within the domains of urban planning and transportation engineering. In order to effectively alleviate this problem, it is necessary to consider the dynamic characteristics of traffic systems and multi-objective optimization requirements, and develop more advanced traffic signal control methods. In this study, Mixed Integer Linear Programming (MILP) was used as an optimization tool to construct an optimization model of signal control strategy with the dual objective of minimizing vehicle delay and exhaust emissions.

In the process of constructing the model, we fully consider the spatial-temporal distribution characteristics of traffic flow in Nanjing and collect and analyze real-time traffic flow data in key areas such as Nanjing Yangtze River Bridge. Through the comprehensive consideration of multi-dimensional data such as vehicle arrival time, vehicle type and flow fluctuation, an optimization model based on vehicle queuing theory is established to realize the optimal planning of traffic signal timing.

2 Theoretical Basis and Literature Review

The design of strategies to ease traffic congestion is a complex, interdisciplinary, and systematic engineering task that requires the input of experts from a variety of fields. This section of this study will explore the latest theoretical and empirical research findings in the fields of traffic flow theory, optimization algorithms, intelligent transportation systems, and microscopic traffic simulation techniques.

The theoretical basis of urban traffic congestion relief strategy is rooted in traffic engineering, traffic flow theory, and network optimization theory in operations research. In the multi-objective optimization design of traffic signal control strategy, the Mixed Integer Linear programming (MILP) model is widely used because of its effectiveness in solving complex network optimization problems. MILP models can handle integer and linear constraints in signal timing to minimize traffic delays and environmental impacts (Cools et al., 2013; Liu & Liu, 2010) [1, 2]. The development of an adaptive traffic signal control system integrates advanced technologies such as real-time data acquisition, dynamic data processing, and intelligent decision support. Research progress in this area has shown that the adaptability and efficiency of traffic signal control systems can be significantly improved by using machine learning algorithms and data-driven models (Li & Ding, 2015; Van Lint et al., 2012) [3, 4]. Microscopic traffic simulation software, including VISSIM and SUMO, offers robust analytical instruments for the assessment and enhancement of traffic strategies. Such software is capable of simulating the intricate interactions of traffic flows at the microscopic level, as well as assessing the impact of disparate traffic management strategies on traffic flow conditions and environmental consequences (Behrisch et al., 2011; Wang et al., 2013) [5, 6]. Aiming at the sporadic traffic congestion problem of expressway, this paper proposed a prediction method based on improved cellular transmission model and a guidance strategy based on the definition of influence scope. Research results show that the proposed guidance strategy has a considerate mitigating effect on the occasional congestion of expressways (Li Xu, 2019) [7]. This study presents a method for multi-objective optimization of traffic signal control using a genetic algorithm-based fuzzy control model. The model employs average delay and number of stops as optimization objectives, utilizing a stochastic weighting approach (Qu Xin, 2017) [8]. A multi-objective optimization method of urban traffic signal control based on cellular transmission model is proposed, and multi-objective genetic algorithm is used to optimize the solution to realize the coordinated control of traffic signals (Li Zhenlong, 2016) [9].

The findings of this research demonstrate the advancements made in the domain of traffic engineering, particularly in the development of strategies for congestion mitigation and optimization of signal control. The integration of technologies such as Big Data, Artificial Intelligence, and the Internet of Things has significantly contributed to the refinement of urban traffic management, offering a more scientific and intelligent approach to this field. However, data-driven limitations (research is often constrained by the quality and completeness of data, which affects the accuracy and generalization ability of the model) and algorithmic adaptations [10–13] (existing multi-objective optimization algorithms [12–15] may be inadequate for adapting to the highly dynamic and nonlinear characteristics of the transportation system, especially when dealing with large-scale and high-dimensional problems) present significant challenges. Technological implementation challenges (the conversion of the theoretical model to practical applications faces technical complexity) also require further investigation. Furthermore, additional research is required to address the following challenges: cost-benefit trade-offs, compatibility with existing infrastructure, user behavior [16, 17] (in particular, the uncertainty of traffic participant behavior and the influence of socio-economic factors on traffic flow), and societal factors, as well as environmental impact assessment.

This study reviews the theoretical basis and key technologies of traffic congestion relief strategy design, analyzes the application of multi-objective optimization algorithm in traffic signal control strategy design [18], discusses the core technology of adaptive traffic signal control system, and the important role of microscopic traffic simulation software in strategy evaluation. These theoretical and technical results provide a solid foundation and advanced methodological guidance for the design and implementation of traffic congestion relief strategies in Nanjing.

3 Cause and Characteristic Analysis of Traffic Congestion

Taking the traffic of Nanjing as an example, the causes and characteristics of traffic congestion are explained. As cities continue to experience rapid growth, it is increasingly important to conduct a thorough examination of the factors contributing to traffic congestion in order to develop effective traffic management strategies. This study uses a combination of quantitative analysis and case study to deeply explore the causes of traffic congestion in Nanjing and uses advanced data analysis techniques to quantitatively identify the characteristics of congestion.

3.1 Macroscopic Cause Analysis

From the macro level, the causes of traffic congestion in Nanjing can be analyzed from three aspects: urban spatial structure, transportation infrastructure supply, and transportation policy. The unreasonable layout of urban spatial structure, such as the centralized urban layout, leads to the concentrated phenomenon of traffic flow. For example, a traffic flow simulation study for the city of Nanjing showed that the traffic flow in Xinjiekou business district was more than 1.5 times higher than usual during morning and evening peak hours on weekdays. Inadequate supply of transport infrastructure, such as inadequate road capacity and public transport service level, is also an important cause of traffic congestion. An evaluation of transportation infrastructure based on the city of Nanjing shows that a 10% increase in the coverage of subway lines can reduce the traffic congestion index of the surrounding area by about 5%.

3.2 Micro-cause Analysis

At the micro level, individual travel behavior and traffic flow dynamics have a direct impact on the formation of traffic congestion. The irregularity of driving behavior, such as frequent lane changes and sudden braking, can lead to unstable traffic flow. An empirical study in Nanjing showed that the irregularity of driving behavior is positively related to the severity of traffic congestion. In addition, the difference in vehicle performance is also a factor contributing to the instability of traffic flow. An analysis of the impact of vehicle performance in Nanjing shows that discrepancies in vehicle acceleration performance can lead to uneven distribution of traffic flow, which in turn affects the stability of traffic flow.

3.3 Feature Recognition Method

In order to quantitatively analyze the characteristics of traffic congestion, clustering analysis, time series analysis, and GIS spatial analysis were used in this study. Through clustering analysis of traffic flow data in Nanjing, traffic congestion patterns during morning and evening rush hours were identified. Time series analysis methods, such as ARIMA model, are applied to Nanjing traffic flow data to predict the dynamic change trend of traffic flow. In addition, GIS technology is used to visually analyze the spatial distribution of traffic congestion in Nanjing, revealing congestion hotspots and key road sections.

3.4 Characteristic Case Analysis

Through a case study, this study specifically shows the recognition results of traffic congestion features in Nanjing. For example, analysis of traffic flow data from the Yangtze River Bridge in Nanjing found that the morning peak on weekdays is usually between 7:00 and 9:00, while the evening peak is between 17:00 and 19:00. In addition, by analyzing the traffic incidents on the main arterial roads in Nanjing, it is found that traffic accidents and road construction are the main causes of local traffic congestion.

Through the above cause analysis and feature identification, this study provides a scientific basis for the intelligent discrimination and guidance strategy of traffic congestion in Nanjing. The research results will help traffic management departments to better understand the internal mechanism of traffic congestion and formulate more accurate and effective management measures.

4 Traffic Congestion Guidance Strategy

4.1 Design of Traffic Congestion Relief Strategy

Taking Nanjing City as an example to design the traffic congestion alleviation strategy of Nanjing city, this study uses a multi-objective optimization algorithm based on data-driven, especially for the signal control strategy of the city. We adopt an optimization model based on vehicle queuing theory with the main objective of reducing vehicle delays and improving the throughput at intersections. At the same time, considering the environmental characteristics of Nanjing, we pay special attention to reducing the idle time of vehicles to reduce exhaust emissions. Using a mixed integer linear programming (MILP) model, we solve for the optimal signal timing while satisfying the traffic rules and physical constraints.

When optimizing the signal control strategy in Xinjiekou area of Nanjing city, we collected traffic flow data for three consecutive months, including vehicle arrival time, vehicle type, and flow fluctuation. After optimization by the MILP model, the traffic efficiency of this intersection is improved by about 22%, and at the same time, the NOx emissions are reduced by about 18%.

The data collection process is primarily conducted by the Nanjing Municipal Transportation Bureau, utilizing traffic monitoring cameras, magnetic induction sensors, and GPS tracking devices to gather traffic flow data. This information is then integrated with meteorological data such as temperature, humidity, precipitation, and wind speed to optimize traffic signal control. The challenges faced in this endeavor include issues related to data quality, equipment coverage, real-time data processing capabilities, and variable weather conditions. To overcome these limitations, it is essential to establish a robust data quality monitoring system, optimize the layout of equipment, employ efficient computational techniques and algorithms, and integrate meteorological modeling into the overall framework.

The collection sample table is shown in Table 1.

Table 1. Table of traffic flow data

Serial number	Time period	Vehicle arrival time	Vehicle type	Flow count	Flow fluctuation	Signal period (seconds)	Vehicle delay (seconds)	NOx emissions (ppb)
1	07:00–07:10	07:00:15 07:00:30	car truck	60	increase	120	25	20
2	07:10–07:20	07:11:00 07:11:15	car bus	65	stability	120	20	22
3	07:20–07:30	07:22:00 07:22:30	car motorcycle	50	decrease	120	15	19

(*continued*)

Table 1. (*continued*)

Serial number	Time period	Vehicle arrival time	Vehicle type	Flow count	Flow fluctuation	Signal period (seconds)	Vehicle delay (seconds)	NOx emissions (ppb)
4	07:30–07:40	07:33:00 07:33:45	truck bus	45	increase	120	30	21
5	07:40–07:50	07:41:30 07:42:00	Car truck	70	stability	120	22	23
–	–	–	–	–	–	–	–	–

Note:
Vehicle arrival time: Two example time points within each 10 min are listed here.
Vehicle type: The main types of vehicles passing through the intersection during the time period are listed.
Flow count: Represents the total number of vehicles passing through the intersection in 10 min.
Traffic fluctuation: describes the trend of traffic in a time period.
Signal period: The signal period is assumed to be fixed at 120 s, which is 2 min.
Vehicle delay: represents the average sojourn time of a vehicle at an intersection.
NOx emissions: Represents the average NOx emissions monitored during the time period.

4.2 Development of an Adaptive Traffic Signal Control System

In the traffic congestion relief stage, we developed an adaptive traffic signal control system, which can analyze the traffic flow data of Nanjing in real time and dynamically adjust the signal timing. The system implementation framework includes a real-time data acquisition module, a traffic state evaluation module, a signal control strategy generation module and an execution module. In addition, we particularly consider the impact of special events on traffic flow in Nanjing, such as large public events or road construction. The system structure is shown in Fig. 1.

The structure and workflow of the adaptive traffic signal control system are explained as follows.

(1) Data acquisition layer: the basic layer of the system, responsible for collecting real-time traffic flow data of key traffic nodes in Nanjing. This layer includes a variety of data acquisition devices, such as traffic cameras, induction coils, and traffic volume detectors, which jointly monitor vehicle arrival times, types, and flow fluctuations.
(2) Data transmission layer: The collected data is transmitted to the central processing system through wired and wireless networks. This layer ensures the real-time and accuracy of the data, and provides raw input for subsequent processing.
(3) Data processing and analysis layer: receives data from the transport layer, and performs preprocessing, analysis and pattern recognition. The data preprocessing unit is responsible for cleaning and formatting the data, the data analysis unit applies statistical methods and algorithms to understand the characteristics of the traffic flow, and the pattern recognition unit identifies the laws and anomalies in the traffic flow.
(4) Decision support layer: Based on the processing and analysis results, the optimization algorithm unit uses multi-objective optimization strategies, such as MILP, to generate signal control strategies. The signal control strategy generation unit is

responsible for formulating specific signal timing schemes to reduce vehicle delays and exhaust emissions.
(5) Signal control layer: The strategy formulated by the decision support layer is implemented, and the timing of traffic lights is dynamically adjusted through the signal light control unit and the timing adjustment unit to optimize the traffic flow.

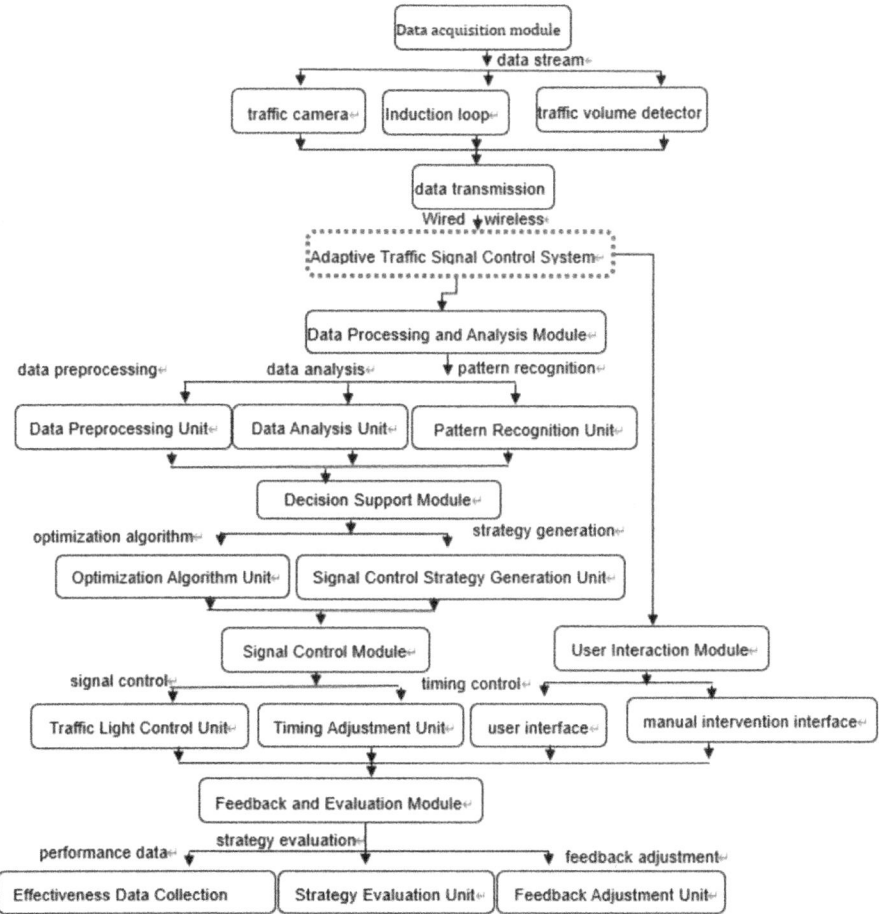

Fig. 1. Structure diagram of the adaptive traffic signal control system

(6) User interaction layer: It provides a user interface for traffic managers to monitor the system status and carry out necessary manual control through the manual intervention interface, increasing the flexibility and responsiveness of the system.
(7) Feedback and evaluation layer: collects the effect data after the system execution, evaluate the effect of the optimization strategy through the strategy evaluation unit, and adjust the control strategy according to the evaluation results by the feedback adjustment unit, forming a closed-loop optimization system.

The workflow of the whole system is a continuous cycle process, from data acquisition to signal control, and then to evaluation and feedback, to ensure that the traffic signal control system can adaptively respond to the complex traffic flow changes and special events in Nanjing, improve the intelligent level of traffic management, and realize the dual optimization of traffic efficiency and environmental quality.

4.3 Implementation of Traffic Congestion Mitigation Strategy

The traffic signal system in the Yangtze River Bridge area in Nanjing is selected for the implementation of the adaptive control strategy, and we collect real-time traffic data through traffic cameras and induction coils. The implementation results show that the average travel time of vehicles during peak hours is reduced by about 30%, and the volatility of traffic flow is effectively controlled.

Based on the hypothetical traffic simulation data, the following data and tables are a detailed analysis of the traffic simulation results for the morning peak hour in the Yangtze River Bridge area in Nanjing.

Simulation conditions:

- Simulation period: morning peak hour, 7:00–9:00.
- Simulation days: 5 working days.

The data table of the simulation results is shown in Table 2.

Table 2. Data table of simulation results

Metric	Percent optimization (mean)	before optimization (mean)	Percentage improvement
Travel Time (seconds)	60	45	25%
Vehicle delay (seconds)	25	15	40%
Queue Length (meters)	150	100	33.33%
NOx emission (g/h)	120	96	20%

Detailed analysis:
Travel time and delay analysis:

- Travel time: After optimizing the signal control strategy, the average travel time of vehicles through the simulated road segment is reduced from 60 s to 45 s, indicating a smoother flow of vehicles.
- Vehicle delay: The average delay is reduced from 25 s to 15 s, a reduction of 10 s, indicating a reduction in sojourn time and an increase in traffic efficiency.

Queue length analysis:

- Queue length: At signaled-intersection, the average length of vehicle queue is reduced by 1/3 from 150 m to 100 m, indicating that the signal adjustment effectively alleviates the congestion situation.

Environmental Impact Assessment:

- NOx emissions: By reducing vehicle delays and idle time, NOx emissions are reduced from 120 g/h to 96 g/h, a reduction of 20%, which has a positive effect on improving air quality.

Conclusion and Recommendations:

- Optimization of signal control strategies significantly improves the efficiency of traffic flow and reduces delays and emissions.
- Continuous monitoring of traffic conditions is necessary to adjust the signal control strategy in time to cope with changes in traffic flow.
- For special events, such as road construction, traffic guidelines should be planned and issued in advance to reduce the negative impact on traffic flow.

4.4 Evaluation of Traffic Congestion Guidance Strategies

In the evaluation stage, we used the microscopic traffic simulation software VISSIM and SUMO to conduct simulation experiments to evaluate the effectiveness of the congestion relief strategy in Nanjing. The evaluation metrics include the average vehicle delay, number of stops, travel time, and fuel consumption. Environmental assessment models such as Vehicle Emission Modeling (VEM) were also used to estimate the potential impact of the strategy implementation on air quality in Nanjing.

In evaluating the traffic signal control strategy in Nanjing city, we simulated the traffic flow state under different traffic signal control schemes. The evaluation results show that the optimized signal control strategy has significant effects on both vehicle delay reduction and environmental pollution reduction, with a 25% reduction in average delay and a 15% and 20% reduction in CO and NOx emissions, respectively.

Vehicle delays and number of stops:

- Average delay: After optimizing the signal control strategy, the average delay time of the vehicle was reduced from 120 s to 90 s, a 25% reduction.
- Number of stops: The average number of stops per vehicle decreased from 2.5 to 1.9, a 24% decrease.

Travel time and fuel consumption:

- Travel time: The average travel time for vehicles was reduced from 45 min to 33.75 min, a 25% reduction.
- Fuel consumption: Average fuel consumption was reduced from 3.0 L to 2.25 L, also a 25% reduction.

Environmental Impact Assessment:
CO emissions: The assessment showed that CO emissions were reduced from 1500 g to 1275 g, a 15% reduction.

- NOx emissions: NOx emissions were reduced from 2000 g to 1600 g, a 20% reduction.

  Conclusions and Suggestions:
- Optimization of the traffic signal control strategy effectively reduced vehicle delays and stops and improved roadway efficiency.
- Reduced travel time and fuel consumption have a positive impact on reducing operating costs and improving travel experience.
- The reduced emissions significantly contribute to the improvement of air quality in Nanjing and are consistent with environmental goals.
- Continued monitoring of traffic conditions and environmental indicators is recommended to further optimize traffic management efforts.

5 Conclusion

This study employs deep learning technology to develop a system capable of dynamically adjusting traffic signal timing, thereby optimizing traffic flow and reducing congestion. The research also proposes a multi-objective optimization strategy for traffic signal control, effectively managing and alleviating traffic congestion. Empirical studies demonstrate that these methods enhance the accuracy of traffic assessments and improve flow optimization efficiency, holding significant implications for the advancement of intelligent transportation systems and the modernization of urban traffic management.

References

1. Cools, M., Moons, E., De Schutter, B.: A review of urban traffic control models and algorithms. IEEE Trans. Intell. Transp. Syst. **14**(3), 1274–1284 (2013)
2. Liu, H.X., Liu, X.Y.: A multi-objective optimization model for traffic signal timing based on MILP. In: 2010 International Conference on Intelligent Computation Technology and Automation (ICICTA), pp. 62–66. IEEE (2010)
3. Li, Z., Ding, Y.: Adaptive traffic signal control using machine learning: a survey. IEEE Trans. Intell. Transp. Syst. **16**(1), 1–15 (2015)
4. Van Lint, J.W.C., Hoogendoorn, S.P., Van der Zijpp, N.J.: Impacts of traffic management measures on urban traffic flow: a machine learning approach. Transp. Res. Part C: Emerg. Technol. **22**, 52–68 (2012)
5. Behrisch, M., Bieker, L., Erdmann, J., Krajzewicz, D.: SUMO - Simulation of Urban MObility: an overview. In Proceedings of the 2011 International Conference on Advances in System Simulation, pp. 40–45 (2011)
6. Wang, Y., Zhang, Y., Guan, Y.: A review of traffic simulation models and their applications in traffic operations management. IEEE Trans. Intell. Transp. Syst. **14**(3), 1300–1311 (2013)
7. Li, X.: Research on dynamic definition of impact range of sporadic traffic congestion on expressway and guidance strategy. People's Public Security University of China (2019)
8. Qu, X.: Research on environment-friendly multi-objective optimization control method of traffic signal. Jilin University (2017)
9. Li, Z., et al.: Multi-objective optimization of intersection signal control based on genetic algorithm. Comput. Appl. **36**(S2), 82–84+88 (2016)
10. Quan, H., Zhao, J., Zhang, Q.: A Two-stage spectrum sharing scheme for the internet of vehicles driven by deep reinforcement learning. Radio Commun. Technol., 1–9 (2024)

11. Liu, Y., et al.: YOLO-T: an object detection model for complex traffic scenes. Shandong Sci., 1–12 (2024)
12. Kong, H.: Research and implementation of cooperative map matching algorithm based on the internet of vehicles. Wuhan University of Technology (2020). https://doi.org/10.27381/d.cnki.gwlgu.2020.000730
13. Tan, L.: Research on data processing and game control algorithms in traffic signal control systems. North China University of Technology (2010)
14. Xue, S.: Research on intelligent computing models and algorithms for complex information networks. Shanghai University (2018)
15. Yin, X.: Research on user mobile behavior pattern mining based on the Apriori algorithm. Beijing University of Posts and Telecommunications (2016)
16. Liming, S., Gangqi, L., Meining, L.: Research on active intelligent optimization method for traffic signals based on real-time image processing algorithm. Electr. Des. Eng. **32**(02), 171–175 (2024). https://doi.org/10.14022/j.issn1674-6236.2024.02.037
17. Wei, S., et al.: Method of group vehicle guidance and cooperative operation under vehicle-road cooperative environment. J. Transp. Eng. **22**(03), 68–78 (2022). https://doi.org/10.19818/j.cnki.1671-1637.2022.03.005
18. Yang, H.: Research on multi-objective optimization of traffic detector layout. Tianjin University (2020). https://doi.org/10.27356/d.cnki.gtjdu.2020.003695

Wide-Area Ship Movement Prediction Using Random Forests

Tanja Vähämäki[✉][iD], Farshad Farahnakian[iD], Paavo Nevalainen[iD], and Jukka Heikkonen[iD]

Department of Computing, University of Turku, 20014 Turku, Finland
{tmvaha,farshad.farahnakian,ptneva,jukhei}@utu.fi
https://www.utu.fi/en/university/faculty-of-technology/computing

Abstract. Maritime situational awareness requires real-time traffic prediction over a large area based on the Automatic Identification System (AIS). The second requirement is allowing input from all the traffic. We propose Random Forests (RF) for ship movement prediction and demonstrate how it can be adapted to varying zone shapes and anomaly detection tasks. We also apply it to the clustering of vessels to regularly and irregularly moving ships. Our research area is the Baltic Sea and the recording period of data is 26 July 2022 ... 12 August 2022. Results from the class of regularly behaving ships (499 ships out of 634) show 0.2 ... 2.1 km mean absolute error (MAE) over 15 min ... 2 h which reaches the same accuracy as many published cases with more expensive computational models. The prediction for all supported time intervals can be updated every 10 min, which makes the implementation practical for large-scale situational awareness systems.

Keywords: maritime surveillance system · abnormal vessel behavior · AIS data · Random Forest · time series prediction

1 Introduction

There is increasing attention to maritime traffic, pinpointing abnormalities and managing the traffic. Modern prediction solutions need to give a good predictive overview, be able to adapt to multiple categories of vessels and additional information like expressed destination.

Technically speaking, the current Automatic Identification System (AIS) allows long-term prediction, since basically all the essential data can be captured. A main obstacle is the efficient formatting and access of historical data. Although the state of the global financial and geopolitical system and weather conditions fluctuate year by year, a large history information helps especially with rarely occurring events. The volume of AIS messages is high, and this becomes more acute the larger areas are recorded and analyzed. Mere problem of transferring the AIS data to a central hub in real time becomes a problem, if the prediction concerns large areas (e.g. Northern Atlantic, English Channel and baltic Sea). There might be only few possible ways to implement continuous prediction: e.g. by discretizing the vessel trajectory data, having seasonal models, or having online stream data models which forget data in a controlled way. We experiment with a static Random Forest (RF) method under an assumption that the results help to estimate the potential of future Stream RF (SRF) implementations.

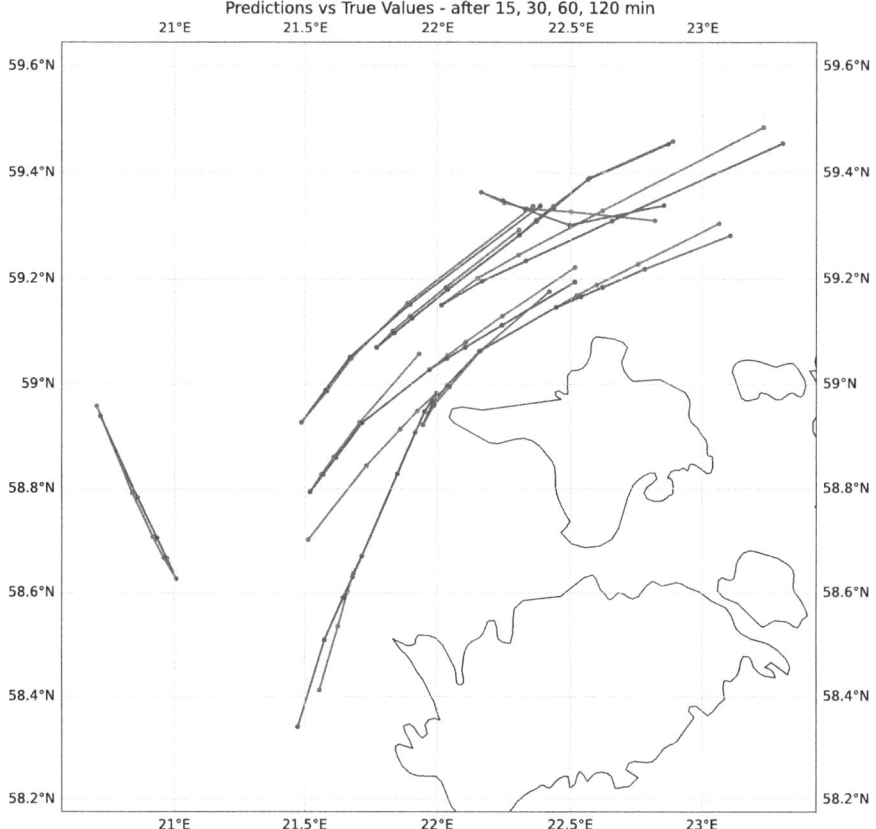

Fig. 1. Predicted (red lines) and actual (blue lines) trajectories of ships at 15, 30, 60, and 120 min to the future (red dots). (Color figure online)

In this study, we apply a RF method to predict ship movements over the Baltic Sea. A detail of the Baltic Sea with size 110 km × 130 km and some observed vessel trajectories (blue) and predictions (red) of the RF model are illustrated in Fig. 1. The time interval in question is 120 min and red dots represent predictions at 15 min, 30 min, 60 min and 120 min. As the main result, we present the prediction to 2 h in the future in Sect. 3, including a numerical comparison to a state-of-the-art method.

We also outline a method of classifying ships into regularly behaving ships and ships which are difficult to predict. We call these two categories 'regular' and 'irregular'. Irregularity is more as a pronouncement of the need of having separate models for these ships involved with military, cost guard, fishing or maritime infrastructure activities.

The feature importance is analyzed and an outline of an algorithm is included to give end-users a reliability measure of prediction. The algorithm is proven to be real-time for 15,30,60 and 120 min predictions, since it finishes the necessary computations in 10 min.

We intend to enhance the scope and versatility of predictive analysis of maritime navigation over a broad spectrum of maritime environments. A novelty is giving up the time prediction and focusing only on the route prediction, since this kind of prediction still can serve some special needs like optimizing the icebreaker movements.

The rest of this paper is organized as follows. Section 1.1 discusses some of the most important related works. AIS data and its vectorization process, and the formulation of computation is presented in Sect. 2. The quantitative results are in Sect. 3 while some qualitative results and possibilities of the method are in Sect. 4. The final conclusions are in 5.

1.1 Related Work

AIS data [27] has been the focus of a lot of research. The ship heading, position, speed and destination are sent to either a satellite receiver or vessel traffic services with area-specific radio receivers. Individual ships tend to send the AIS messages regularly, but there is a wide variation in the frequency of sending the message. A review [30] in concerns AIS data mining, which has 22 references, and ship behavior analyses, which has 37 references from the year 2019.

Extremely short-term prediction includes fluid dynamics and can be considered governing a 1–10 s time period, and the environment is considered invariant [32]. Short-term prediction concerns obstacle avoidance and course changes over 2 – 5 min [32], and a long-term prediction over 15 min – 2 h is about course planning, arrival time estimations and keeping up a maritime situational awareness [9]. Our work falls to this third category, as shown in Fig. 1. There, blue lines represent the observed history, and red lines are the predicted movement at 15 min, 30 min, 60 min and 120 min to the future (red dots). The short time period 15 min have been covered mainly to support quick anomaly alerts.

A probabilistic model of rendezvous in the future is presented in [23]. The model is able to give a trust interval. Our method relies on observed differences between decision trees, which brings only a subjective measure, but is computationally advantageous. Also, our approach omits the velocity data, which is essential both in randezvous and arrival time prediction [32].

The relation between environment and ship orientation, e.g. heading over ground (hog), is studied in [33]. The paper focuses on grounding prevention and detection, improving the original accuracy of the AIS signals.

The maritime industry has seen a surge in the development and application of predictive models to enhance navigational safety and operational efficiency. The Automatic Identification System (AIS) has emerged as a crucial source of data for ship movement prediction. Utilizing AIS data, researchers have classified various methods for predicting ship trajectories, emphasizing the role of machine learning (ML) in advancing maritime safety and logistics. Models such as Random Forests (RF), k-Nearest Neighbors (k-NN), and deep learning have been adapted to predict vessel locations, demonstrating the versatility and effectiveness of ML in handling complex, dynamic maritime data [12,24].

A review of existing studies reveals a diverse application of ML algorithms for ship movement prediction using AIS data across different maritime regions. For instance, in the Singapore Strait region, researchers have developed models to assess ship collision risks, highlighting the congested nature of global shipping routes [11]. The Gulf of Finland has served as another testing ground, where the k-NN method was employed for trajectory prediction, supporting emission control and route optimization efforts [24]. These studies underscore the global relevance of AIS-based ML models for maritime traffic management and safety.

In [8], the authors employed K-means method to cluster and group data from ships. Then, they used this clustered data to help an artificial neural network to learn how to predict where ships will go next. They found out that their method could correctly estimate the ships' movements about 70% of the time. In another study [6], they suggested a method utilizing a k-NN classifier to forecast maritime paths. This technique underwent evaluation using authentic data derived from AIS transmissions gathered in the vicinity of Malta. Testing outcomes indicate that their approach achieves a precision rate of 0.794, a recall rate of 0.785, and an overall accuracy of 0.931. In [20], they used a different ML technique called a support vector machine to predict which way ships would go after they left a specific area called the Nakanose traffic route. They looked at data from right before the ship left, 5 min before, and 10 min before, and trained their prediction system with this information. They found their proposed prediction system could predict the ship's movements with accuracy of 87.9%, 80.3%, and 77%, depending on when the data time interval. However, they also noticed a problem with their method because most of the data they used to teach the program came from just one type of situation.

RF [4] is a non-parametric prediction method, and adding one parameter as a limit of decision tree deviation, it becomes a parametric anomaly detection method. Detection of abnormal behavior of vessels usually utilizes parametric, non-parametric and clustering methods [7].

RF has been used in the context of maritime traffic prediction. A short-term speed prediction [25] considered only one ship, and although the same approach could work for larger categories of ships, speed prediction might not reach the set real-time constraints of our study. The trajectory similarity measure of [31] is used to compare within approx. 500000 trajectories destined to approx. 10000 ports. The similarity is used to predict the ship destination. We assume, that this approach should also use the AIS destination field with necessary reliability issues [29] being handled somehow. The study of [31] seems to be the only one with larger data than our case.

There are several approaches to keep the data and the RF state relevant over seasonal changes. One approach is a branch of methods called the Stream RF (SRF). Streaming Random Forests [1], Online Random Forests [13], and Hoeffding Trees [19] are invented for streaming data, where the history is not necessarily available in its entirety in a numerical form. All these methods are restricted to see each data entry only once, while they try to guarantee an accuracy comparable to basic RF, the method of this study. Online RF uses time tags for each data to maintain the active set of decision trees. Hoeffding Tree minimizes the amount of relevant data needed for a tree branching decision.

The asymptotic behaviour of the batch RF and the Multi-label Hoeffding Trees was studied in [21]. Based on this study, it seems plausible that some of these stream data RF versions could substitute the basic RF presented in this work, if properly parameterized and controlled. Section 4.3 has a outline of how to implement a stream data model in our case.

Isolation Forest [16], unlike our experiments, addresses the abnormalities specifically by not attempting to model the normal body of the data. Instead it seeks for minority subsets with feature values different from the main data. Isolation Forest tends to work well in its specific task when the feasibility of the prediction problem itself is already demonstrated by a Machine Learning method. Feature importance is a side-product of the RF [4]. It comes in many varieties. One is the Gini importance, which is built into the basic RF but has some weakness when compared to results of standard wrapper-based feature selection. Permutation based feature importance is based on re-arranging all values of a feature and seeing the prediction degradation while keeping the trained model constant. Shapley Additive exPlanations (SHAP) [17] has a game-theoretic approach to adjusting the feature importance reported by RF.

Several large AIS databases are reported in [35]. The largest scales are spatially global, concerning $10....15 \times 10^3$ ships and having timescales up to 8 years, Our scale of study is 650 ships over 2 weeks. It may seem modest when compared with largest maritime traffic databases, but the focus is more about the applicability, when compared to possible server resources per area. Our approach allows reading in, processing and forming a spatially large model with a very few tuning parameters.

One source [14] compares multi-origin-destination situation in an approx. 200 km × 200 km area as one of the examples. This case resembles our case, yet is smaller and with less traffic. There, deep learning based methods were better than machine learning based methods. Comparison between different methods over different cases is hard, though, since deviation errors (location or time error over a length of travel or duration of time) are seldom expressed numerically in the current research.

An excellent study using attention mechanism and Long Short-Term Memory [18] also reviews all modern Machine Learning methods used to ship movement prediction. The work includes a comment about the computational efficiency and prediction accuracy stating that these two are almost always exclusive.

One of the first real-time, online services for ship prediction is [36], which provides a prediction of one ship at a time using artificial neural network (ANN), where the computation happens in between the order and graphical response. The response time is comparable to our 0.11 sec per ship, even our proposal concerns a large area and multiple of ships. The current research concerns are about mending several data sources in a multistep procedure [26] and having a concurrent platform [28] for the maritime situational prediction as an online service.

In the current research, there is a separation to route prediction, arrival time and rendesvouz time prediction, and to a coupled time-location prediction. This makes it hard to compare various approaches and that is why we provide an indirect time accuracy by geometrical argumentation in Sect. 3. Also, a possible pre-screening of the data is not often made clear, and that led us to automate the division to regular and non-regular ships, see Section 2.3.

2 Materials and Methods

The AIS data collected covers 145×10^3 km^2 of the Baltic Sea, which itself is approx. 400000 km^2. The reason of partial coverage is the maximum reliable transmission distance of 150 km from the nearest AIS relay station. The data provider is the Finnish Transport Agency and their AIS entry capture rate reduces to zero at the latitude 57° N leaving the Southern Baltic Sea unobserved. The period is two weeks 26 July - 12 August 2022 from the Baltic Sea. The data was split into two weeks of train and three days of test data including 1054 and 634 vessels correspondingly. A ship trajectory consists of latitude and longitude pairs filtered to AIS entries with 5 min time interval. Figure 2 shows the train data with each vessel trajectory having a different color. The cut-off latitude at 57° N is clearly visible. The common seafaring routes are revealed as bundles of trajectories.

2.1 Data Preparation

The AIS data, sampled once every 5 min from the Baltic Sea between 26 July and 12 August 2022, was first divided into two weeks of training and three days of test data. This resulted the data sets containing 452 MB of data with 3,225,307 recordings and 110 MB of data with 778,078 recordings correspondingly. Some less significant entry fields like the name, call sign, estimated time of arrival, or ship type were omitted. The field destination was excluded because it brings in a complex text matching sub-problem, with the same port written in multiple different ways, in addition to a large amount of missing values. There is a study about the estimated validity and utility of the destination field [30], and this

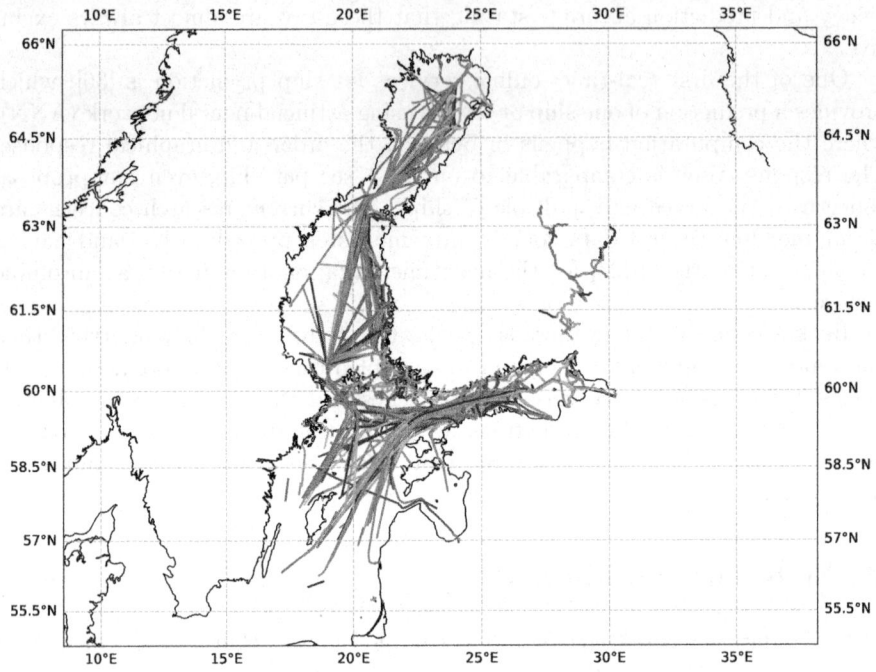

Fig. 2. Trajectories of 1145 vessels captured during 26 July 2022 - 9 August 2022 over the Baltic Sea. The coverage is not complete at the latitudes 55.5–57°. Some inland waterways are included, too.

should be a target of more study in the future. All the other fields in the AIS data were selected as attributes for an intrinsic feature selection of the training process in Random Forest models. They are shown in Table 1.

Table 1. Static AIS attributes specific to a vessel, dynamic attributes sampled once every five minutes from the AIS stream, and predicted labels.

Static Entry Attributes	mmsi, draught, length, width
Dynamic Entry Attributes	sog*, cog*, heading*, lon*, lat*, hours, minutes, weekdays, journey id
Labels	lon, lat +5, +15, +30, +60 and +120 min later

The AIS data can be divided to static (specific to ship mmsi) and dynamic (time-dependent) entry attributes. The static attributes used in this study are: *mmsi, draught, length, width*. The dynamic attributes chosen to this study are: *sog** (speed over ground), *cog** (course over ground), *heading**, *lon** (longitude), *lat** (latitude), *hours, weekdays*. The *hour* attribute had *minutes* included by

$hour := hour + minute/60$. Any change in *destination* field caused a new *journey_id* to be generated for each mmsi.

The dynamic features marked by an asterisk (*) had variants from past observations at 0, -5, -15, -30, -60, -120 min. That means these 5 features (*) were represented by a vector of length $5 \times 6 = 30$. Using *lat* value at -15 min as an example, the naming of these trace features follows the format *lat_15*. Labels used for teaching were positions (*lon, lat*) with variations $+5$, $+15$, $+30$, $+60$ and $+120$ min later (*lon_5* etc.). They were represented by a vector of length 5.

Invalid AIS entries were omitted by excluding recordings with missing values. Also, journeys of ships with discontinuity points of over 10 min in the time series when the ship was moving (sog over 4 knots) were omitted. Additionally, recordings with missing values in the past observations or as labels of the future positions of the ships were omitted. The final sizes of the training and test feature data sets became 91 MB and 41 MB, corresponding to 1,590,206 and 325,750 recordings, respectively. The number of recordings omitted from the original AIS data sets in the above phases is shown in Table 2. As one can see, about half of the recordings were omitted because of interruptions in sending AIS messages during the journeys of the ships.

Table 2. Number of recordings in the original AIS data, omitted recordings and the resulting feature data sets.

Recordings	Reason	Training data (%)	Test data (%)
AIS data		3 225 307	778 078
Omitted	Missing values	$-25\ 395$ (1)	$-5\ 502$ (1)
	Removed journeys	$-1\ 481\ 329$ (46)	$-434\ 210$ (56)
	Missing past obs.	$-67\ 576$ (2)	$-6\ 465$ (1)
	Missing future pos.	$-60\ 801$ (2)	$-6\ 151$ (1)
	All	$-1\ 635\ 101$ (51)	$-452\ 328$ (58)
Feature data		1 590 206	325 750

2.2 Random Forest in Trajectory Prediction

RF [4] is an added level above the decision trees, which in turn, consists of a collection of decision rules, alternative branches and responses. These elements are also called as decision, chance and end nodes, respectively. Each decision tree can either be maximized by the information gain of each branch it has, or by trying to force as homogeneous splits as possible. RF is an ensemble of decision trees, and bootstraps them [3] so that the training data is first randomly divided into subsamples and then features are randomly selected for a chosen number of trees. The idea is to reduce the correlation between the decision trees helping them to predict labels more independently. In RF, multiple decision trees are trained simultaneously and each of them predicts labels for the data instances. The prediction of the RF regressor is an average value of all the decision trees' predictions. Independent decision trees increase the stability of the model by reducing the variance of the results. The RF model is more unlikely to predict a label incorrectly than a single decision tree.

In our solution, two RF regressors were trained to predict the ship's next positional coordinates, one for longitudes and the other for latitudes over time intervals of 5, 15, 30, 60, and 120 min. The models were trained with the training and tested with the test feature data set. Z-score standardization was used to prevent the features with a larger scale from dominating. The prediction errors were calculated as Haversine distances between the predicted and true positional pairs of coordinates after 5, 15, 30, 60, and 120 min for all the pairs of coordinates in the ships' trajectories. The accuracy of the RF models was measured with the mean absolute errors (MAE) calculated for the ships from the above values.

The hyperparametes of the RF regressors were set as default values of the Scikit-learn library detailed in Table 3. The training time of the RF longitude and latitude regressors was 8227 and 7905 CPU seconds, totally about 4.5 h. The total time to forecast the next longitude and latitude coordinates (after 5, 15, 30, 60, and 120 min) was 0.06 and 0.05 CPU seconds correspondingly. The implementation was done with the configuration shown in Table 4.

Table 3. Hyperparameters used when training the RF regressors

Hyperparameter	Explanation	Value
n_estimators	No. of trees	100
max_depth	Max. depth of trees	None
min_samples_split	Min. samples per split	2
min_samples_leaf	Min. samples per leaf	1
max_samples	Max. samples per tree	None
max_features	Max. leafs considered	1
max_leaf_nodes	Max. leaf nodes	None
bootstrap	Bootstrap sampling	True
criterion	Criterion for splitting	squared_error
min_impurity_decrease	Min. impurity decrease	0
min_weight_fraction_leaf	Min. weight. fraction per leaf	0
ccp_alpha	Cost-complexity pruning	0
n_jobs	No. of parallel jobs	None
oob_score	Compute out-of-bag score	False
verbose	Verbosity of output	0
worm_start	Reuse of previous solution	False
Random_state	Random seed	42

Table 4. System configuration details

Model	HP Pavilion Laptop 14-ce0xxx
Processor	Intel(R) Core(TM) i5-8250U CPU @ 1.60 GHz 1.80 GHz
RAM	8 GB
Graphics card	Intel(R) UHD Graphics 620

2.3 Ship Irregularity

The RF regressors' predictions, average values of all the decision trees, can be understood as the most probable next coordinates for the ships. However, they are point estimates and the predictions of the individual decision trees may also illustrate probable optional trajectories for the ships. We interpret the predictions of the decision trees as alternative next coordinates that the decision trees have learnt based on the train data. The reason for alternative predictions of the decision trees may be that the ships in the train data have selected several different routes, e.g., in certain navigational forks, or the ships have not exhibited consistent movements in that area.

In cases where neither the RF regressor nor any of the decision trees have predicted the coordinates accurately enough (difference not within the set threshold value: 1,5 km for 30 min, 3 km for 60 min, and 6 km for 120 min forecasts), the positional coordinates are signaled as irregular, indicating a potentially abnormal behaviour of the ship in that specific spot.

We chose to use the predictions for 120 min to cluster the ships as regular or irregular. Each ship was given a degree of irregularity (DOI) which is a percentage of coordinates signaled as irregular over all the coordinates the ship's trajectory using 120 min forecasts. If the DOI value exceeded the set threshold value (1%) the ship was clustered as irregular indicating potentially abnormal behaviour of the ship.

This can be formulated in the following way. We have N decision trees as part of a RF regressor. Each decision tree T_i (where $i = 1, 2, \ldots, N$) makes a prediction $\hat{y}_i$ for a given input vector x. The prediction error for each decision tree is calculated as follows, using the absolute error:

$$e_i = |\hat{y}_i - y| \quad (1)$$

where y is the true value.

The prediction of a RF regressor is the average of the predictions made by each individual decision tree in the forest:

$$\hat{y} = \frac{1}{N} \sum_{i=1}^{N} \hat{y}_i \quad (2)$$

We select the smallest of these errors:

$$e_{\min} = \min_{i=1,\ldots,N} (|\hat{y}_i - y|, |\hat{y} - y|) \quad (3)$$

If the smallest error exceeds a threshold value of 6, a signal is made for the irregular coordinate:

$$\text{Coordinate} = \begin{cases} \text{Irregular Coordinate} & \text{if } e_{min} > 6 \\ \text{Regular Coordinate} & \text{if } e_{min} \leq 6 \end{cases} \quad (4)$$

Each ship is then given a degree of irregularity (DOI), which is a percentage of coordinates signaled as irregular over all the coordinates in the ship's trajectory. This can be expressed as follows:

$$\text{DOI} = \left(\frac{N_{\text{irregular}}}{N_{\text{total}}}\right) \times 100 \quad (5)$$

where $N_{\text{irregular}}$ is the number of coordinates signaled as irregular, and N_{total} is the total number of coordinates in the ship's trajectory.

If the DOI value exceeds a threshold value of 1, a ship is clustered as irregular:

$$\text{Ship} = \begin{cases} \text{Irregular Ship} & \text{if } DOI > 1 \\ \text{Regular Ship} & \text{if } DOI \leq 1 \end{cases} \quad (6)$$

The DOI values are used in later stages to limit prediction only to ships, where prediction is meaningful. The distribution of the ships' DOI values between 0 and 20% is illustrated in Fig. 3 with 120-min forecasts. As one can see, most of the ships have a DOI value at most 1% and are clustered as regular.

2.4 Estimate of the Time Prediction Error

Time prediction was not included in the RF model, but its error for a hypothetical third RF model predicting it (two RF models are used for latitude and longitude prediction) can be approximated from the location error. Figure 4 shows the initial position as $p_0(t_0)$, leading to the observed point $p_1(t_1)$ and to a predicted position $b = \hat{p}(t_b)$. The triangle occurs as a projection to a plane, which osculates Earth sphere at the location p_0. Planar projection was deemed sufficient, since the largest distances involved with 2h prediction are approx. 80 km, leading to a rather insignificant relative horizontal geometric error 53×10^{-6}.

We use a point $a = \hat{p}(t_a)$ on the predicted path for the derivations. The point a is closest to the actual position $p_1(t_1)$.

The distances $l_a = \|a - p_0\|$, $l_b = \|b - p_0\|$ and $l_1 = \|p_1 - p_0\|$ are the lengths of the triangle shown in Fig. 4 The distance l_a has:

$$l_a = l_1 \cos \alpha = (p_1 - p_0) \cdot (b - p_0)/\|b - p_0\|. \quad (7)$$

We define the relative time prediction error e as:

$$e = \frac{|t_a - t_0 - (t_1 - t_0)|}{(t_1 - t_0)}. \quad (8)$$

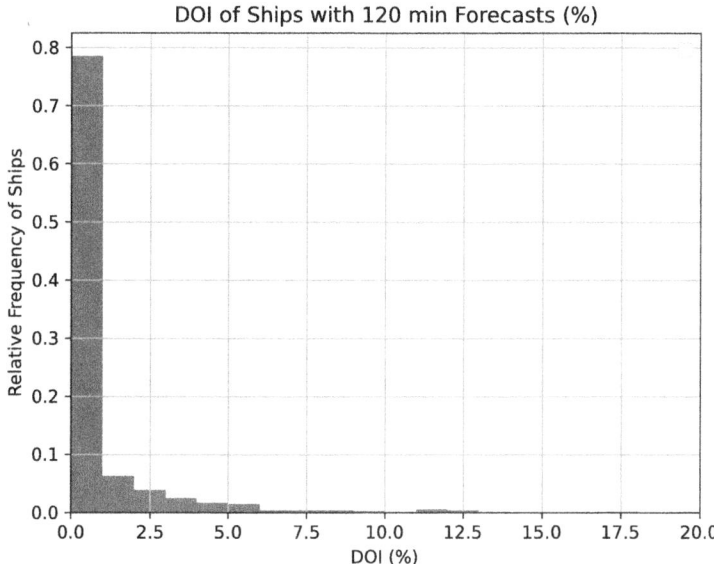

Fig. 3. The degree of irregularity (DOI) calculated for the ships with 120 min forecasts between 0 and 20%.

Since the velocity $v(t)$ is not directly predicted by the model, we consider an assumption that the predicted time t_b matches the actual time t_1, and the speed $\hat{v}(t)$ scales accordingly: $t_b = t_1$ and $\hat{v} = v_1 l_b / l_1$. Under this assumption, the predicted time t_a is: $t_a = (t_1 - t_0) l_a / l_b$ and the relative time error is:

$$e = \left|\frac{l_a}{l_b} - 1\right| \tag{9}$$

Naturally, this assumption does not hold, and it is not even an upper limit estimate of the time error, but it still approximates the capabilities of such systems, where time is a part of the learning and prediction process. Largest errors do occur over longer time intervals, drastic course errors, or abrupt velocity changes.

3 Results

The 5 min prediction was very accurate since no large errors can occur in such a small interval. The predicted (red) and realized (blue) trajectories are shown at 15 min and 120 min to future in Fig. 5. The detail (a) shows most of the predictions being accurate, and big errors occur at arrival to a port, anchoring or stopping, or drastic change of course nearby a destination port. The increase of red length seems to be approximately linear over time (30 min and 60 min images, which are not shown, were also used to derive this conclusion), which means a chance to deviate from prediction at any moment is almost constant.

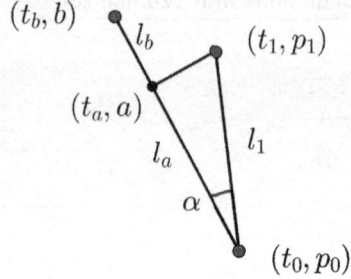

Fig. 4. The scheme of estimating the time error. The observed and predicted paths are represented as line segments p_0–p_1 and p_0–b, where $b = \hat{p}_1$ is the prediction result corresponding to the ground truth $p_1(t_1)$, and $p_a(t_a)$ is a point nearest to the ground truth $p_1(t_1)$.

The constant probability of deviation from predicted route is best explained by local geographic features like islands and shapes of the common ship lanes. Again, this effect can be countered by adding more teaching data until the geometric resolution is adequate.

Qualitatively, errors of 15 min prediction in Fig. 5 are of two kinds: the data loss between 55.5...57° lat, and ships stopping or anchoring to wait for the approach window to a port. The 120 min prediction brings two more sources for error: ships choosing an alternative port or making another adaptation, occurrences of which are not covered by the data, and natural variations in navigation, see e.g. surroundings of location 60° N, 25° E.

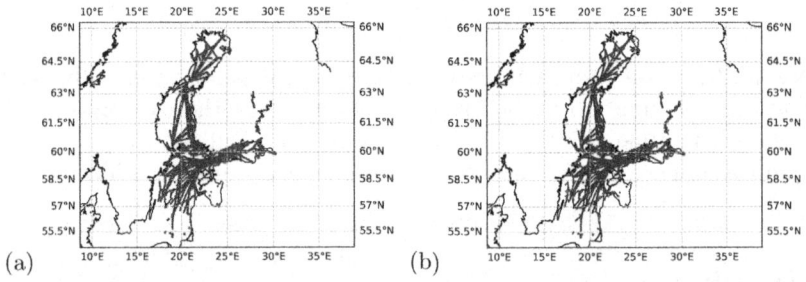

Fig. 5. The forecasted (red) and realized (blue) trajectories at (a) 15 min and (b) 120 min to future. (Color figure online)

The prediction accuracy in terms of MAE over different prediction times 5, 15, 30, 60 and 120 min is summarized in Fig. 6. Forecasting errors were averaged per ship and summarized as distribution curves. As seen, 5 min is a too small time interval for this kind of large area prediction. Also, the result is very close to the simple constant velocity vector assumption. The prediction error is not cumulative as the erroneous path lengths in Fig. 5. This means, that when taking the 5 min distribution in Fig. 6 and multiplying it by factor 3 along the abscissa, one gets larger errors than actualized by the 15 min prediction. This means the model has learned the behavior of ships beyond the shorter interval models. The same phenomenon can be seen till the 120 min prediction. This gives hopes that predictions over longer durations (e.g. 240 min) might be meaningful after there is more teaching data.

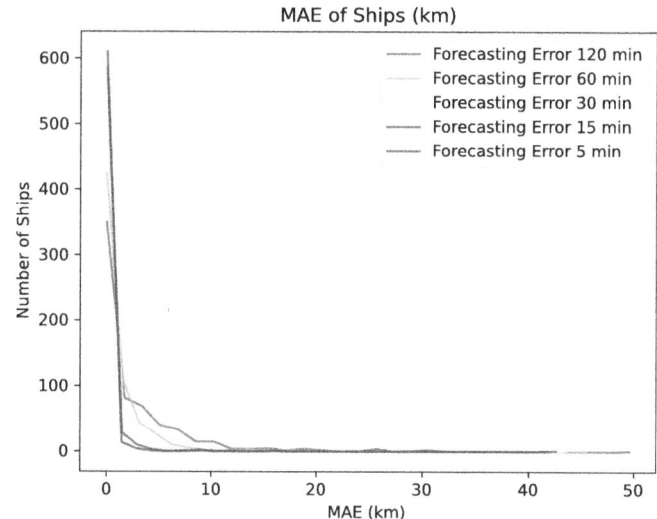

Fig. 6. MAE of the position prediction (km) over different time intervals: blue (5 min), green (15 min), yellow (30 min), orange (60 min), and red (120 min). (Color figure online)

The mean of each error distribution is listed in Table 5 for two categories, irregular and regular ships. One has to note, that the mean of MAE over all ships is a pessimistic measure since irregular ships like fishing ships do require specific methods and they should not in this comparison. The MAE of all ships is accompanied with the MAE of regular ships to make the distinction clear. A comparison study is VesNet [10], which is attentional multitask learning (MTL) model over both the Baltic Sea and Danish sea areas. Another baseline are the MAE values of deep learning methods (encoder-decoder) reported in [5]. The area of this result is surroundings of the Danish straits. The labeled results (in gray) have the ship destinations used in the prediction.

The ships are ranked by their DOI to two categories (irregular and regular ships). There is no definite cut-off MAE value between the two classes by MAE value, but the cut-off is set at $DOI = 1$. Probabilities of MAE values in 120 min forecasts are compared between regular and irregular ships in Fig. 7. The distribution of the regular and irregular ships' MAE values have been normalized

Table 5. Mean ship prediction error MAE for each prediction time interval, all the ships and only regular ships included. Latitude and longitude prediction are separated because the major part of traffic is longitudinal. Results are also compared to Ves-Net [10] and encoder-decoder model [5]. Labeled results of Encoder-decoder (in gray) have the ship destination known.

Pred. time (min)		5	15	30	60	120
RF MAE (km)	All	**0.40**	**0.58**	**0.92**	1.7	**3.2**
RF MAE (km)	Reg.	0.23	0.36	0.59	1.1	2.1
VesNet MAE (km)		0.50	0.73	1.0	1.6	3.3
Enc.-dec. MAE (km)	Unlabeled				1.4	3.6
Enc.-dec. MAE (km)	Labeled				1.0	2.0

separately (divided by the amount of regular and irregular ships respectively). The ships clustered as regular are clearly more likely to have smaller MAE values. The major part of the irregular ships have slow speeds, and 120 min is too small time period to demonstrate their erratic movement.

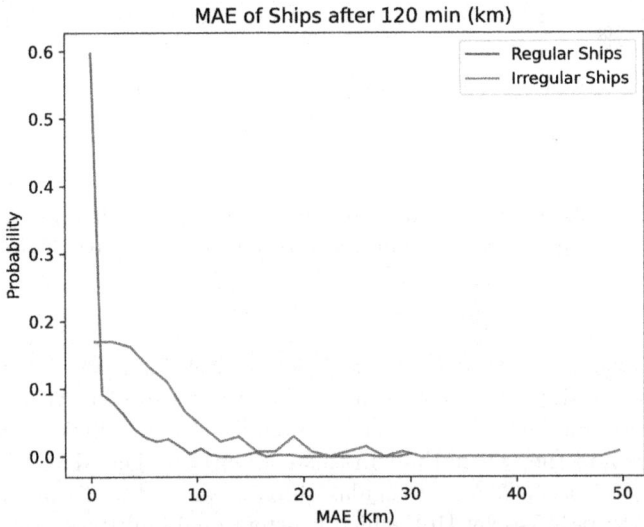

Fig. 7. MAE (km) of regular and irregular ships in 120 min predictions. The regular ships are shown in blue and irregular in red. (Color figure online)

The mean standard deviation (MSD) of the predicted values of all the decision trees is calculated for the ships to demonstrate the uncertainty of the forecasts. The dependency between MSD of the decision trees' predictions and MAE is depicted for each ship in Fig. 8. In many but not all of the cases the uncertainty of the forecasts increases when the MAE values increase. Four example ships shown in Fig. 9 are marked with a), b), c) and d) in Fig. 8. Example a) is about an anchored ship, ship b) is doing a tight turn while staying well predicted. The example c) has an intermittent visit to a port. The example d) represents one of rather uncertain predictions.

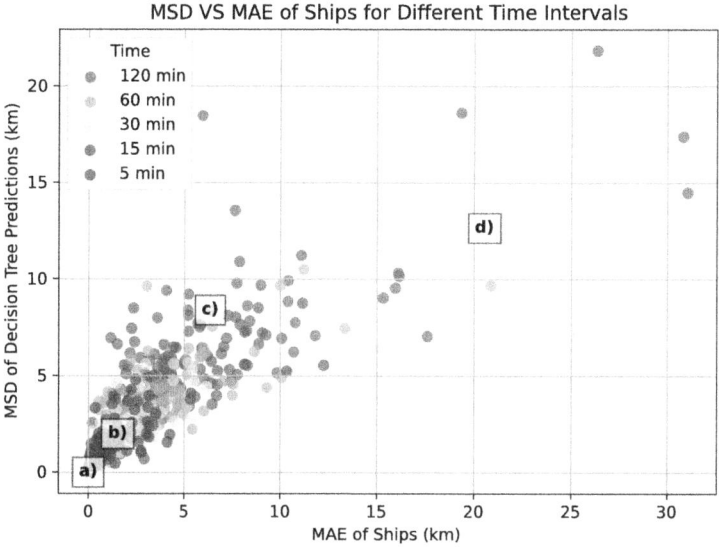

Fig. 8. The dependency between MSD of the decision trees' predictions and MAE depicted for each ship after 5 min (blue), 15 min (green), 30 min (yellow), 60 min (orange), and 120 min (red). (Color figure online)

The irregularity of the ships is illustrated in Fig. 9. Case a) is an anchored ship, its movement stays within a radius of 4 m. The actual movement is in blue, the prediction is red, and grey is the mass of decision trees. Case b) has a ship steering steadily towards the right, and the RF regressors track to movement rather well. The overall pattern would be similar even if the sampling moments were chosen differently. Case c) has a regular passenger ship with an established schedule. It actually performs two trajectories (Stockholm - Mariehamn and Mariehamn - Turku) but the data model considers it as one trajectory (Stockholm - Turku). Case d) has a ship shuttling between three ports, and without using the destination field information provided by AIS, the RF tree structure becomes indecisive. Yellow bars are observed parts of the trajectory, where no decision tree was able to match the previous movement and uncommon behaviour of the ship was signalled. The ship was also clustered as irregular.

Technically, an abnormality is an observation which is rarer than a predefined limit. One choice of rarity limit is that no decision tree fits both the observed data D and the observed trajectory T, see yellow sequences in detail d) of Fig. 9. This is similar to a probability of T to occur $p(T|D) \approx 0$, given a large and

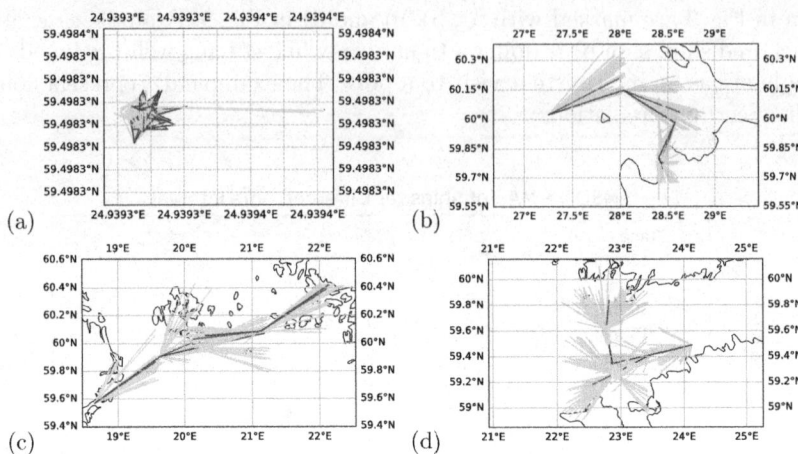

Fig. 9. Four 120 min prediction cases. Color codes blue, red, grey correspond to observed and predicted trajectories and to all the decision trees. Yellow is a trajectory not matching to any previous observation. (a) The ship is anchored. The predictions are very accurate and certain. The RF models and decision trees represent the route well. (b) The ship selects its route at two branching points of common shipping lanes. The branching structure at (lat, lon) points (27.5°, 60.1°) and (28.6°, 60.0°) are represented well. (c) The ship is on its regular route between 2–3 ports. (d) The ship stops, makes an uncommon turn and stops. Abnormal behavior condition is triggered several times and the ship is clustered as irregular. (Color figure online)

representative enough data D. Figure 10 has a color coding blue-red-yellow for predicted and realized ship movement, and exceptional movement not supported by any decision tree, respectively. The exceptional ship behavior occurs mostly nearby ports and archipelago islands. The waiting areas nearby ports are a common cause of yellow segments.

Feature importance according to the RF internal feature importance score (FIS) is depicted in Fig. 11. The visualization is split to two models, one predicting latitudes only, and one predicting longitudes. The recent ship trajectory (latitude and longitude at 0 min, −120 min, and −60 min), and heading and sog at 0 min are the most important features. One could assume that both latitude and longitude models would have equal order of importance of features, but the large variation between ranking of pairs like *lat_60* and *lon_60* indicate, that some other application area could have quite different ranking.

3.1 The Speed of Computing

To predict the trajectory at 5 future moments requires 0.11 s (0.05 s for longitude and 0.06 s for latitude values) per ship, which allows updating all the ships in the Baltic Sea at every 10 min (0.11 s × 1054 ships × 5 prediction intervals = 580 s). This number can be further reduced by utilizing a reduced set of features, using more efficient computing environment and improving the data representation.

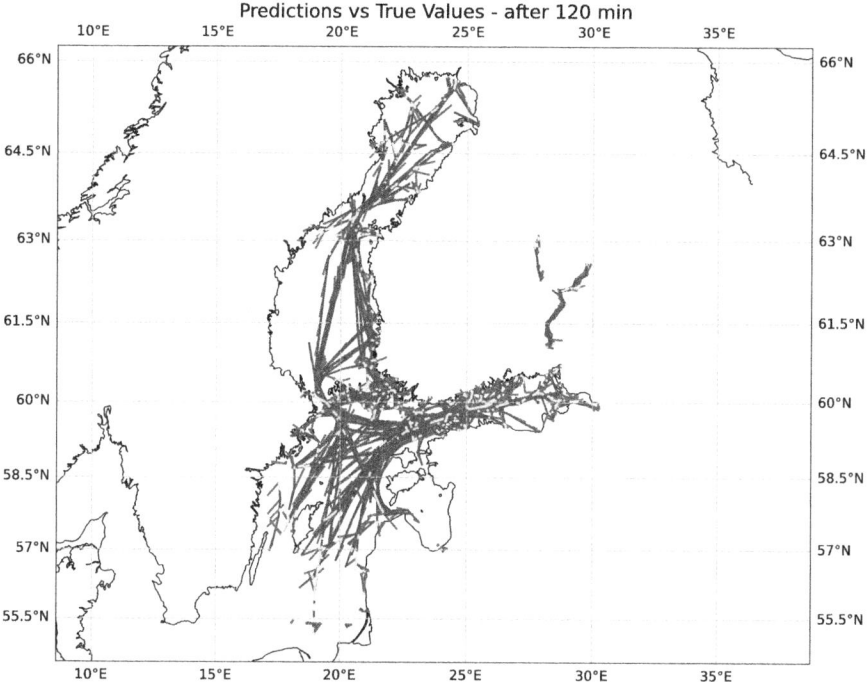

Fig. 10. Abnormal coordinates detected based on how the decision trees are able to predict the observed behavior. Blue: observed, red: predicted, yellow: not able to predict by any decision tree. (Color figure online)

4 Discussion

There are not many prediction attempts over as large area as the Baltic Sea. Danish Maritime Authority (DMA) data used in [5] covers a 1/3rd of the surface covered in our study, but the case is geographically nearby, and the traffic is from multiple origins to multiple destinations. Therefore is an excellent reference in this respect. The Danish case shows the importance and potential for including the AIS destination field to the prediction process.

Perhaps the most important goal in wide-area prediction is to direct the computing power and modelling attention to ships, which actually are predictable. Most ships maintain their status (regular, not regular) over time, and even there are fishing ship prediction models [34], they either require much longer data recording period, or they utilize models coupled to existence of fisheries [22]. Since RF can represent alternative paths, prediction result, and also perceived regularity of some ships it becomes better when the amount of data grows. The line between regular and irregular must be drawn according to the goals of the prediction: ship rendezvous and arrival time estimation concerns predictable

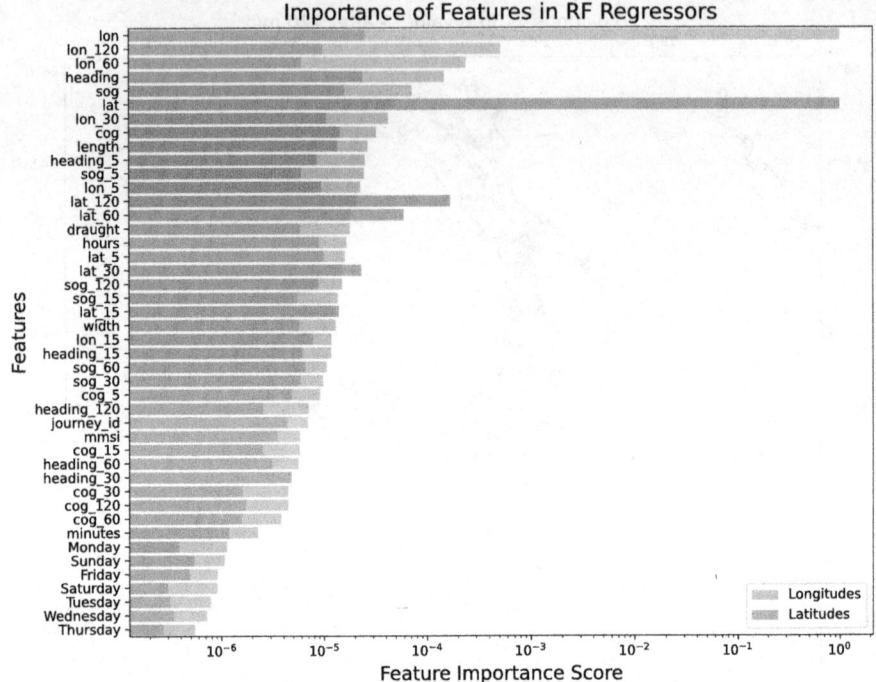

Fig. 11. Feature importance scores of the RF regressors predicting longitude and latitude coordinates.

ships, while large-scale short-term maritime awareness needs predictions of most of the ships.

Typical abnormalities in Fig. 10 are the following:

- Waiting queues and anchoring areas near large ports. Solving these requires some sort of port models [15] including the estimates of the internal load port is facing and of the outbound traffic.
- Zones with patchy AIS recording (southern part of Baltic Sea, for example). These zones require the fusion of other AIS sources, which was not attempted in this research.
- Archipelago with a network of route choices and avoidance of other traffic. The avoidance logic is outside of this research, but is indispensable at narrow ship lanes when attempting 60 min and 120 min predictions.
- Swapping a destination to a neighbouring port. This usually includes a modification of the AIS destination field, which could be added as a part of the prediction.

Examples of four of the above cases are depicted in Fig. 9. Red and blue lines are predicted and unfolded movements. Grey are relevant decision trees and yellow is a completely unique trajectory not matching with any earlier data. Detail (a) is an anchored ship with the radius of movement matching the effective range of a typical anchor chain. Detail (b) has a ship passing an island. That spot has a branching of ship lanes, and grey decision trees reveal this well. Detail (c) has a very regular ferry between Stockholm, Mariehamn and Turku. The decision trees cover all the traffic along that route. Detail (d) has ships making unusual stops while crossing a congested ship lane.

4.1 Computational Complexity

The computational complexity of usual RF models is explored in [2]. Having data $\mathcal{D} \subset \mathbb{R}^{n \times d}$ as n samples with d features[1], one builds a decision tree in time $\mathcal{O}(nd \log d)$, a forest of M trees in time $\mathcal{O}(Mnd \log d)$ and makes a prediction in time $\mathcal{O}(Mn \log n)$ for all n samples. Here, the prediction cost is the most important. It depends on the size n of data, and in this sense properly localized, seasonal data sets become important. The suggested steps in a maritime awareness project would be:

1. Select the prediction update interval 10 ... 30 min. This is a practical decision based on goals of the situational awareness process.
2. Use m AIS attributes suggested in this work.
3. Tune the number of trees M as a hyper-parameter of the model.
4. Select largest sample size n, which still allows the chosen prediction interval (see step 1). This includes practical decisions over seasonality and spatial scope of data.
5. Select the criterion for the category of irregular ships. This can be based on the number of available trees (this work).
6. Add an uncertainty measure provided to the end user. See the previous step.
7. Drop the support of longest duration predictions if the uncertainties grow too high.

RF regression models [4] are computationally more expensive models, and they cannot be used in a data stream situation. RF models could bring stability to prediction in cases where there is a clear deviation between the current ship position and its prediction. Regression models can better interpolate without exact fits in the data history and should be used as a backup of ordinary RF prediction.

4.2 Arrival and Randezvous Times

The current implementation has latitudes and longitudes predicted by separate RF models. Time could be covered by a similar model as in [25] (basically a 1D signal prediction over speed) requiring 3 RF models. An alternative is to use two independent models for the ship direction (cog) and speed.

Our RF model does not use the time nor velocity information. Figure 4 has the direction error α, which could serve in an integration of speed $\hat{v}(t)$ predicted using the AIS field *speed* and the orientation $\hat{\alpha}(t)$ used from the AIS field *cog*.

---
[1] $n = 1600$ trajectories with $d = 30$ features, see Table 1.

The location $\hat{p}(t)$ would be an indirect output by integrating velocity vector $\cos \hat{\alpha}(t), \sin \hat{\alpha}(t)$. An alternative is by coupling time and location by an error function, which expresses both errors either in time or distance domain. These alternatives remain to be experimented with.

It is of interest which kind of time errors one can assume, if we expanded the model to cover randezvous times. Table 6 summarizes the relative time error over 5 ... 120 min prediction intervals. One can see the exponential increase of the error, suggesting a certain practical limit to prediction time interval.

The basic assumption in producing the results is having time error not coupled to location error (see Section 2.3), which allows scaling the time error at the proximity of the ground truth location. This is not a true upper nor lower limit for error, since in reality location and timing will be coupled to some extent in models, which predict both at the same time.

Table 6. Estimation of the relative time error over 5 ... 120 min predictions. All ships have regular and irregular ships.

Pred. time (min)		5	15	30	60	120
Relative time error (1)	All	0.12	0.14	0.20	0.26	0.32
	Reg.	0.09	0.11	0.17	0.22	0.27

4.3 Stream Data RF Models

A stream data prediction model [21] has these properties: it inspects each sample only once, is ready to predict right from the beginning even data keeps evolving possibly over infinite time, while using only finite amount of processing time and memory.

There are problems in establishing the reliability of the predictions initially, since every new entry is basically an outlier or anomaly. Also, during the first 2 h of the new trajectory being observed, the prediction model (both RF and SRF) does not have full feature set available. This would require a short-term history model, which uses for example history of 5, 15 and 30 min.

The processing of a data entry would be this:

1. Preprocessing and possible rejection of the entry with a new ship mmsi.
2. Storing first 30 min of entries of the same mmsi. Only every 5 min are sampled totalling of 7 entries.
3. When 7 entries are in the queue (per ship), the first prediction of the short-term history model follows.
4. When 25 entries are in the queue, the first long-term history prediction occurs.
   The short-term prediction removes the mmsi from a list of active trajectories.
5. Both models are executed with the current data every 10 min.

A side effect of this arrangement is that ships, which are just leaving a port will be processed by a short-term model, which sees only the immediate coastal area. Of course, some ships are anchored and can change their destination field, igniting a new short-term prediction process. Popular anchoring areas will be covered by the short-term model over time.

5 Conclusions

A RF is used to predict ship location in the future with the mean prediction error (MAE) per ship being 1.7 ... 3.2 km for regular and irregular ships over the prediction time interval 60 ... 120 min. This corresponds well to [5] and [10], which span about the same geographic area and have multi-origin/multi-destination traffic, and We implemented also prediction time intervals 5 min, 15 min, and 30 min, which yielded errors 0.23–0.92 km.

The MAE of ships clustered as regular, 79 % of the ships, is approximately 66 % of the MAE of all the ships. The difference of the two ship categories is essential. Irregular ships can be divided to many categories (temporarily anchored, patrolling, fishing, loitering), which each would require a specialized predictor with additional inputs.

A set of dominant features was defined. For the relatively short data collection period (two weeks), the calendar features (weekend, office day etc.) were not relevant. The prediction time increases linearly over the data amount and decreases approximately inversely over the features used. Pruning approximately 40 % of the features while increasing the data recording window to 2 months seems to allow a large-area prediction every 10 min.

The proposed method adapts to large areas easily and requires no space partitioning. Also, separate seasonal models can be easily established. We managed to prove that this method suits to real-time online prediction.

Limitations and Further Studies: Limitations are related mainly to the research focus: 1) We omitted building models which utilize localized reliability (certain locations are prune to produce larger prediction errors than others), 2) We omitted the timing prediction, since we were mainly interested on spatial aspects having an application related to ice-breaker assistance on mind. 3) We did not utilize the AIS destination field and 4) we omitted the sensor fusion with meteorological data. Also 5) branching phenomena and 6) stream data models were omitted.

Further studies are needed to have 2) randezvous and arrival times predicted e.g. making changes described in the Sect. 4.2. AIS destination field [5] would require some extra textual pre-processing. A location and weather dependent model adaptation 1),4) is a possibility, but this would require much more balanced teaching data over different weather conditions. Work is needed in properly modeling 5) the branching situation, where two different routes are approximately as likely. A branching indication would be a useful property in a maritime situational decision-making.

We decided to omit 6) the SRF online models to make the research effort more manageable. This is an unfortunate limitation, which can be corrected by a further research.

There are possibilities to further improve the accuracy of the models with a larger training set, since it is sparse at some locations of the rather heterogeneous area. More suitable hyper-parameters, e.g., different timing for location sampling was not attempted. Categorisation of irregular ships should be controlled by a hyper-parameter, but we currently do not know how to formalize this. Although the existing literature suggests stream data SRF is as accurate as the batch version described in this work, the real-time load and arrangements needed for an online SRF service have to be tested separately, this especially during seasonal shifts.

Acknowledgments. This work is part of the AI-ARC project funded by the European Union's Horizon 2020 research and innovation programme under grant 96 agreement No. 101021271.

AIS data was gathered from https://www.digitraffic.fi/en/marine-traffic/.

Author contributions. CRediT Authorship Contribution Statement. Tanja Vähämäki: Data curation, Formal analysis, Investigation, Methodology, Software, Validation, Visualization. **Farshad Farahnakian**: Resources, Writing - Original Draft. **Paavo Nevalainen**: Conceptualization, Investigation, Writing - Original Draft, Writing - Review & Editing. **Jukka Heikkonen**: Conceptualization, Funding acquisition, Project administration, Supervision.

Availability of Data. Approx. 560 MB of the Baltic Sea AIS data from 26 July 2022 to 12 August 2022 as well as generated feature and label files are available at 10.5281/zenodo.12704422 in CSV format.

Disclosure of Interests. The authors declare that they have no known competing financial interests or personal relationships that could have appeared to influence the work reported in this paper.

References

1. Abdulsalam, H., Skillicorn, D.B., Martin, P.: Streaming random forests. In: 11th International Database Engineering and Applications Symposium (IDEAS 2007), pp. 225–232 (2007). https://doi.org/10.1109/IDEAS.2007.4318108
2. Biau, G.: Analysis of a random forests model. arXiv e-prints arXiv:1005.0208 (2010)
3. Breiman, L.: Bagging predictors. Mach. Learn. **24** (1996). https://doi.org/10.1007/BF00058655
4. Breiman, L.: Random forests. Mach. Learn. **45** (2001). https://doi.org/10.1023/A:1010933404324
5. Capobianco, S., Millefiori, L.M., Forti, N., Braca, P., Willett, P.: Deep learning methods for vessel trajectory prediction based on recurrent neural networks. IEEE Trans. Aerosp. Electron. Syst. **57**(6) (2021). https://doi.org/10.1109/TAES.2021.3096873

6. Duca, A.L., Bacciu, C., Marchetti, A.: A k-nearest neighbor classifier for ship route prediction. In: OCEANS 2017 - Aberdeen, pp. 1–6 (2017). https://doi.org/10.1109/OCEANSE.2017.8084635
7. Farahnakian, F., et al.: A comprehensive study of clustering-based techniques for detecting abnormal vessel behavior. Remote Sens. **15**(6) (2023). https://doi.org/10.3390/rs15061477
8. Gan, S., Liang, S., Li, K., Deng, J., Cheng, T.: Ship trajectory prediction for intelligent traffic management using clustering and ann. In: 2016 UKACC 11th International Conference on Control (CONTROL), pp. 1–6 (2016). https://doi.org/10.1109/CONTROL.2016.7737569
9. Jeong, J.S., Park, G.K., Kim, J.S.: Prediction table to improve maritime situation awareness by VTS operator. In: 2013 International Conference on Fuzzy Theory and Its Applications (iFUZZY), pp. 479–481 (2013). https://doi.org/10.1109/iFuzzy.2013.6825487
10. Jiang, F., Wang, H., Li, Y.: Vesnet: a vessel network for jointly learning route pattern and future trajectory. ACM Trans. Intell. Syst. Technol. **15**(2) (2024). https://doi.org/10.1145/3639370
11. Kang, H.S., et al.: Prediction of ship collision risk on Singapore strait using AIS data. J. Transp. Syst. Eng. (2021). https://api.semanticscholar.org/CorpusID:258840037
12. Karatas, G., Senkul, P., Ayran, O.: Trajectory prediction for maritime vessels using AIS data. In: Proceedings of the 12th International Conference on Management of Digital EcoSystems (2020). https://api.semanticscholar.org/CorpusID:227179742
13. Lakshminarayanan, B., Roy, D.M., Whye Teh, Y.: Mondrian forests: efficient online random forests. arXiv e-prints arXiv:1406.2673 (2014). https://doi.org/10.48550/arXiv.1406.2673
14. Li, H., Jiao, H., Yang, Z.: AIS data-driven ship trajectory prediction modelling and analysis based on machine learning and deep learning methods. Transp. Res. Part E: Logist. Transp. Rev. **175**, 103152 (2023). https://doi.org/10.1016/j.tre.2023.103152. https://www.sciencedirect.com/science/article/pii/S1366554523001400
15. Li, Y., et al.: Research on multi-port ship traffic prediction method based on spatiotemporal graph neural networks. J. Mar. Sci. Eng. **11**(7) (2023). https://doi.org/10.3390/jmse11071379
16. Liu, F.T., Ting, K.M., Zhou, Z.H.: Isolation forest. In: 2008 Eighth IEEE International Conference on Data Mining, pp. 413–422 (2008). https://doi.org/10.1109/ICDM.2008.17
17. Lundberg, S., Lee, S.I.: A unified approach to interpreting model predictions. arXiv e-prints arXiv:1705.07874 (2017)
18. Ma, Q., et al.: A hybrid deep learning method for the prediction of ship time headway using automatic identification system data. Eng. Appl. Artif. Intell. **133**, 108172 (2024). https://doi.org/10.1016/j.engappai.2024.108172. https://www.sciencedirect.com/science/article/pii/S0952197624003300
19. Manapragada, C., Webb, G.I., Salehi, M.: Extremely fast decision tree. In: Proceedings of the 24th ACM SIGKDD International Conference on Knowledge Discovery & Data Mining, KDD 2018, pp. 1953–1962. Association for Computing Machinery, New York (2018). https://doi.org/10.1145/3219819.3220005
20. Nishizaki, C., Terayama, M., Okazaki, T., Shoji, R.: Development of navigation support system to predict new course of ship. In: 2018 World Automation Congress (WAC), pp. 1–5 (2018). https://doi.org/10.23919/WAC.2018.8430436

21. Read, J., Bifet, A., Holmes, G., Pfahringer, B.: Scalable and efficient multi-label classification for evolving data streams. Mach. Learn. **88** (2012). https://doi.org/10.1007/s10994-012-5279-6
22. Tania, T.M., Smout, S., Photopoulou, T., Mark, J.: Identifying fishing grounds from vessel tracks: model-based inference for small scale fisheries. R. Soc. Open Sci. **6** (2019). https://doi.org/10.1098/rsos.191161
23. Uney, M., Millefiori, L.M., Braca, P.: Prediction of rendezvous in maritime situational awareness. In: 2018 21st International Conference on Information Fusion (FUSION), pp. 622–628 (2018). https://doi.org/10.23919/ICIF.2018.8455816
24. Virjonen, P., Nevalainen, P., Pahikkala, T., Heikkonen, J.: Ship movement prediction using k-NN method. In: 2018 Baltic Geodetic Congress (BGC Geomatics), pp. 304–309 (2018). https://api.semanticscholar.org/CorpusID:52161077
25. Wang, J., Guo, Y., Wang, Y.: A sequential random forest for short-term vessel speed prediction. Ocean Eng. **248**, 110691 (2022). https://doi.org/10.1016/j.oceaneng.2022.110691. https://www.sciencedirect.com/science/article/pii/S0029801822001470
26. Wang, Y., Wu, L.: A functional model of AIS data fusion. In: Shi, Z., Goertzel, B., Feng, J. (eds.) 2nd International Conference on Intelligence Science (ICIS). Intelligence Science I, vol. AICT-510, Shanghai, China, pp. 191–199. Springer, Cham (2017). https://doi.org/10.1007/978-3-319-68121-4_20. https://inria.hal.science/hal-01820899, Part 3: Big Data Analysis and Machine Learning
27. Weber, J.H., Häder, H., Nielsen, J.: Automatic identification system (AIS): the state of the art. J. Navig. **61**, 257–268 (2008). https://doi.org/10.1017/S0373463308004744
28. Xiao, Z., Zhang, L., Fu, X., Zhang, W., Zhou, J.T., Goh, R.S.M.: Concurrent processing cluster design to empower simultaneous prediction for hundreds of vessels' trajectories in near real-time. IEEE Trans. Syst. Man Cybern. Syst. **51**(3), 1830–1843 (2021). https://doi.org/10.1109/TSMC.2019.2906381
29. Yang, D., Wu, L., Wang, S.: Can we trust the AIS destination port information for bulk ships?-implications for shipping policy and practice. Transp. Res. Part E: Logist. Transp. Rev. **149**, 102308 (2021). https://doi.org/10.1016/j.tre.2021.102308. https://www.sciencedirect.com/science/article/pii/S136655452100082X
30. Yang, D., Wu, L., Wang, S., Jia, H., Li, K.X.: How big data enriches maritime research – a critical review of automatic identification system (AIS) data applications. Transp. Rev. **39**(6), 755–773 (2019). https://doi.org/10.1080/01441647.2019.1649315
31. Zhang, C., et al.: AIS data driven general vessel destination prediction: a random forest based approach. Transp. Res. Part C: Emerg. Technol. **118**, 102729 (2020). https://doi.org/10.1016/j.trc.2020.102729. https://www.sciencedirect.com/science/article/pii/S0968090X20306446
32. Zhang, D., Chu, X., Liu, C., He, Z., Zhang, P., Wu, W.: A review on motion prediction for intelligent ship navigation. J. Mar. Sci. Eng. **12**(1) (2024). https://doi.org/10.3390/jmse12010107
33. Zhang, M., Kujala, P., Musharraf, M., Zhang, J., Hirdaris, S.: A machine learning method for the prediction of ship motion trajectories in real operational conditions. Ocean Eng. **283**, 114905 (2023). https://doi.org/10.1016/j.oceaneng.2023.114905
34. Zhang, M., Kujala, P., Musharraf, M., Zhang, J., Hirdaris, S.: A machine learning method for the prediction of ship motion trajectories in real operational conditions. Ocean Eng. **283** (2023). https://doi.org/10.1016/j.oceaneng.2023.114905

35. Zhou, Y., Daamen, W., Vellinga, T., Hoogendoorn, S.: Review of maritime traffic models from vessel behavior modeling perspective. Transp. Res. Part C: Emerg. Technol. **105**, 323–345 (2019). https://doi.org/10.1016/j.trc.2019.06.004. https://www.sciencedirect.com/science/article/pii/S0968090X18318114
36. Zissis, D., Xidias, E.K., Lekkas, D.: Real-time vessel behavior prediction. Evol. Syst. **7** (2016). https://doi.org/10.1007/s12530-015-9133-5

Characteristics of Heterogeneous Traffic Flow Involving Different Intelligent Level Autonomous Vehicles

Xuan Wang, Junwei Zeng, Yongsheng Qian[✉], and Xu Wei

Lanzhou Jiaotong University, Lanzhou 730000, Gansu, China
qianyongsheng@mail.lzjtu.cn

Abstract. For a considerable period, the expressway will always be a mixture driving of conventional human-driven cars and autonomous cars due to the advancements in intelligent networking and autonomous driving technology. Based on the Gipps safety distance model, this study considers the time delay of L2 and L3 autonomous cars, and distinguishes between human-driven cars and autonomous cars. The STCA lane-changing model is enhanced and a hybrid traffic flow cellular automata model is developed based on their respective features. Then, further exploration and analysis are conducted to examine fundamental diagrams, space-time diagrams, congestion ratio (CR), and safety evaluation. The research indicates that self-driving vehicles can increase road capacity, alleviate congestion, and enhance road safety. Moreover, L3 autonomous cars have better effect than L2 autonomous cars on improving road traffic. This paper will provide a reference for studying the driving behavior of autonomous cars and the traffic flow characteristics associated with various intelligence levels of autonomous cars.

Keywords: Different levels of autonomous cars · Heterogeneous traffic flow · Cellular automata model · Expressway

1 Introduction

Advancements in artificial intelligence and autonomous driving tech have resulted in the growing prevalence of autonomous vehicles on our roads recently [1]. These vehicles are essentially machines that utilize various technologies, such as AI, visual computing, radar, monitoring equipment and GPS, working together to allow computers to drive cars automatically and safely, without needing active control from humans. The Society of Automotive Engineers (SAE) has established a framework to understand the different levels of driving automation, categorizing them into six distinct levels. These stages range from Level 0, which is manual driving where the human driver is fully in control, to Level 1, where the vehicle can assist the driver in some tasks. Level 2 represents partial automation, which can take over certain driving functions but still requires the driver to be engaged. Level 3 is conditional automation, which can manage the majority of driving tasks under specific conditions but may still need occasional human intervention.

Level 4 represents high automation, allowing the vehicle to function autonomously in certain conditions, while Level 5 indicates complete automation, meaning the vehicle can navigate itself in all circumstances without any drivers' involvement. According to the Intelligent Connected Vehicle Technology Roadmap 2.0, it is projected that by 2025, L2 autonomous vehicles and L3 autonomous vehicles will comprise about 50% of the market. This proportion is expected to grow to 70% by 2030. Additionally, it is anticipated that L4 autonomous vehicles will account for around 20% of the market by 2030. As we look further ahead to 2035, it is expected that L5 autonomous vehicles will also start to make their way into the market, marking a significant shift in how we think about transportation [2].

As technology advances and driver acceptance increases, mixed traffic flows are expected to be very common [3]. As a result, the capacity containing autonomous vehicles is expected to increase. In addition, L3 autonomous vehicles can also access the operating status of neighboring vehicles and have automatic lane-changing function to make more informed decisions when changing lanes. This might also improve road capacity. In addition to superior driving performance, autonomous vehicles have potential to have a dramatic impact on vehicle energy consumption, the safety of traffic system operation, and traffic efficiency [4]. Autonomous vehicles are gaining recognition and acceptance among the public, with a growing expectation that they will eventually supplant human-driven vehicles as the primary mode of transportation in the future [5].

The mixture of autonomous vehicles and conventional human-driven vehicles will complicate traffic flow features that would otherwise only involve autonomous vehicles. So far, there are lots of researches on autonomous cars and human-driven cars. Yang et al. proposed an on-ramp cellular automata (CA) model considering safety distance and interconnection, and simulated the traffic flow under different CAV permeability and CAV distribution [6]. Sala et al. used a macroscopic model to estimate the average length of CAV platoons based on specific traffic demand and the level of CAV adoption [7]. Perraki et al. introduced a model predictive control (MPC) strategy for managing highway traffic, integrating traditional control methods alongside operations by vehicles using Vehicle Automation and Communication Systems (VACS), and assessed it through micro traffic simulations [8]. However, existing studies of mixed traffic flows that include autonomous vehicles do not take different levels of automation of autonomous vehicles into account. Therefore, from a practical point of view, further research into mixed traffic flows, especially those containing different levels of autonomous vehicles, is becoming increasingly necessary.

Currently, there have been multiple studies investigating how traffic flow with a combination of different levels of automation perform. Guan et al. classified vehicles according to the SAE automation levels into human-driven vehicles, partially-automated vehicles, and fully-automated vehicles [9]. Miqdady et al. developed a hybrid simulation scenario (L0-L4) employing the Gipps model to assess how automation affects traffic safety, modifying parameters for each level of automation [10]. The findings indicate that higher levels of automation enhance traffic safety [11]. Overall, autonomous vehicles contribute to greater stability in mixed traffic flows, with homogeneous flows exhibiting better stability compared to mixed flows.

Over the past fifty years, multiple theories have been put forward concerning traffic flow involving human-driven vehicles, resulting in development of diverse traffic flow and prediction models. Among these, the CA traffic flow model has gained significant traction among researchers for its capacity to depict complex phenomena within traffic systems. The NaSch model, introduced by Nagel and Schreckenberg, was the first CA model applied to traffic simulation [12]. CA serves as a discrete spatial-temporal dynamic system governed by local rules. By utilizing CA, it is possible to simulate and model the complexity and emergence of traffic systems through the application of specific conditions and rules. Because of its effectiveness, simplicity and efficiency, numerous extended CA models have been created based on the NaSch model [13].

For the driving model involving autonomous vehicles, Jiang and Wu first proposed the CA model involving autonomous vehicles [14]. Motivated by Gipps safe distance rule, Qiu et al. proposed a CA model for autonomous vehicles from safe distance [15]. Qin et al. learned that as the proportion of autonomous vehicles increases, traditional traffic flows become more stable [16]. Numerous simulations indicate that autonomous vehicles integrated into traffic significantly enhances both traffic flow and free flow speed, with improvements becoming more pronounced as penetration rate of autonomous vehicles increases [17].

There have been numerous studies on the attributes of hybrid traffic flow that include autonomous vehicles. Zhong et al. quantified the potential impact of CAV on human-driven vehicles (HVS) by examining high-resolution vehicle trajectory data obtained from microsimulations [18]. Raphael et al. experimentally demonstrated that intelligent control of self-driving cars can suppress time-travel stop waves that can occur even in the absence of geometric or lane-change triggers [19]. Ghiasi et al. put forward a model that can manage the smooth deceleration wave of CAVs based on the idea of CAV trajectory smoothing, and the findings of their research indicated that it could dampen traffic oscillations while also decreasing fuel consumption and emission [20]. Drawing on the idea of smoothing CAVs' trajectories, Ghiasi et al. developed a model aimed at controlling the backward deceleration waves of these vehicles and found that it effectively reduces traffic fluctuations, leading to lower fuel consumption and fewer emissions of pollutants [20].

There is currently a range of studies about how different levels of autonomous vehicles mixing on the road. Kavas-Torris et al. utilized SUMO software to create autonomous vehicles by adjusting current car-following model and lane-changing model. Besides, they also investigated the impact of both autonomous and non-autonomous vehicles on traffic flow through real-world studies [21]. Focused on Level 0, Level 2 and Level 4 autonomous vehicles, Zhang et al. adjusted following rules based on drivers' personalities and a variable time headway strategy, introducing parameters to assess lane-changing abilities, and evaluated traffic flow using fundamental graphical models and average travel time [22]. Ma et al. initiated their study by examining the distribution of vehicles across various automation levels and assessed the impact of these vehicles on traffic flow using microscopic simulations [23].

Compared with the existing studies, this paper also considers the time delay and takeover mode of autonomous vehicles to establish traffic flow cellular automaton model, aiming to explore the characteristics of hybrid traffic flow between conventional human-driven vehicles and different levels of autonomous vehicles, including capacity, space-time diagrams, congestion ratio(CR), safety evaluation and lane-changing times. Since L4 autonomous vehicles have not yet entered the market, only L2 and L3 autonomous vehicles are considered in this study. In terms of the selection of simulation tools, since this study is to discuss the influence on various types of vehicles on the overall traffic flow, and CA model has the ability to simulate complex nonlinear traffic flow mechanics, CA model is chosen as the simulation model.

2 Traffic Flow Model

2.1 Functions of Autonomous Cars

According to the SAE International document on driving automation systems [2], L2 autonomous cars have a system that controls the vehicle's operation. However, the driver still needs to monitor the surroundings and take control if necessary. The vehicle helps with lane keeping, steering, braking, and acceleration. L3 autonomous cars, in addition to these features, can also change lanes automatically.

2.2 Following Model

2.2.1 Safe Distance

In this paper, considering the intelligence level of the autonomous vehicle, the Gipps safe distance rule is introduced into CA model to enhance the accuracy, and the established model is used for deep analysis of safe driving behavior in traffic flow. The safe distance [24] is meant to make sure when the front vehicle suddenly brakes, the vehicle behind can react in time to avoid a collision. In this paper, we will define the safe distance for human-driven vehicles as follows:

$$d_{safe,n}(t) = x_{n+1}(t) - x_n(t) - l_{n+1} = v_n(t)\tau_n + \frac{v_n(t)^2}{2b_n} - \frac{v_{n+1}(t)^2}{2b_n} \quad (1)$$

where, $d_{safe,n}$ refers to safe distance for the vehicle n; $x_n(t)$ refers to position of the vehicle n at t moment; $x_{n+1}(t)$ refers to that front vehicle of the vehicle n; l_{n+1} refers to body length of vehicle ahead; $v_n(t)$ refers to velocity of the vehicle n at t moment; $v_{n+1}(t)$ refers to that of the front vehicle of the vehicle n; τ_n refers to reaction time of the vehicle n; b_n refers to deceleration of vehicle.

Since the safety assistance systems in autonomous vehicles are not as advanced as those in fully autonomous cars and there is a time delay [11], we can express the safe distance for autonomous vehicles as follows:

$$d_{safe,n}(t) = x_{n+1}(t) - x_n(t) - l_{n+1} = v_n(t)(\tau_n + t_d) + \frac{v_n(t)^2}{2b_n} - \frac{v_{n+1}(t)^2}{2b_n} \quad (2)$$

2.2.2 Safe Driving Speed

The safe driving speed $v_{safe,n}$ for conventional human-driven cars based on the previously mentioned safe distance is:

$$v_{safe,n}(t) = -b_n \tau_n + \sqrt{b_n^2 \tau_n^2 + b_n[2d_{safe,n} - v_n(t)\tau_n + (\frac{v_{n+1}^2(t)}{b_n})]} \qquad (3)$$

Since autonomous car can keep its lanes and have features for automatic acceleration and braking, the reaction time of it is the system's reaction time. However, there is a time delay which differs from that of driver's. Therefore, we can define the safe driving speed for autonomous vehicles as follows:

$$v_{safe,n}(t) = -b_n(\tau_n + t_d) + \sqrt{b_n^2(\tau_n + t_d)^2 + b_n[2d_{safe,n} - v_n(t)(\tau_n + t_d) + (\frac{v_{n+1}^2(t)}{b_n})]} \qquad (4)$$

When analyzing the characteristics of vehicle driving behavior, the operational characteristics on different types of vehicles are fully considered to determine the difference between autonomous cars and conventional human-driven cars in terms of safety distance, reaction time, and random slow down probability, and to establish a hybrid traffic flow CA model. Our research develops a CA model for hybrid traffic flow involving human-driven cars and autonomous cars based on the literature [25], the following rules for human-driven cars are as follows:

Step 1. Acceleration (if $d_n > d_{safe,n}$)

$$v_n(t+1) = min(v_n(t) + 1, v_{max}) \qquad (5)$$

Step 2. Deterministic deceleration (if $d_n < d_{safe,n}$)

$$v_n(t+1) = min(v_n(t) + a_n, v_{max}, v_{safe,n}(t), d_n) \qquad (6)$$

Step 3. Uniform velocity (if $d_n = d_{safe,n}$)

$$v_n(t+1) = min(v_n(t), d_n) \qquad (7)$$

Step 4. Probabilistic random deceleration

$$v_n(t+1) = max(v_n(t) - b_n, 0) \qquad (8)$$

Step 5. Position updates

$$x_n(t+1) = x_n(t) + v_n(t+1) \qquad (9)$$

where, a refers to acceleration of vehicle; v_{max} refers to maximum velocity of vehicle.

2.2.3 Following Rules for Autonomous Cars

According to the current literature on switching modes [26], the autonomous car does not consider the probability of random deceleration, the second step is deleted. The Step 2 of following rules for autonomous cars is as follows:

Step 2. Deterministic deceleration

$$v_n(t+1) = min(v_n(t) + a_n, v_{max}, v_{safe,n}(t), d_{safe,n}) \qquad (10)$$

2.3 Lane-Changing Model

For both autonomous and non-autonomous cars, lane changes should be performed in a manner that minimizes the need for adjacent vehicles in the targeted lane to significantly reduce their speed. With reference to existing literature [27], this paper establishes lane change rules for autonomous vehicles. L3 autonomous vehicle automatically changes lanes when the conditions for lane change are met.

(1) Polite Lane Change (PLC) rule

The basic assumption regarding lane-changing behavior PLC is when human drivers try to change lanes, their actions should not disrupt the movement of adjacent vehicles in the target lane. The rule supports altruistic behavior and does not promote self-centered behavior. The lane-changing conditions are as follows:

$$\begin{cases} d_n(t) < min(v_n + 1, v_{max}) \\ d_{n,other}(t) > d_n(t) \\ d_{n,back}(t) > V_{max}(t) \\ V_{max}(t) = d_{safe}(t) \end{cases} \quad (11)$$

where,

$d_n(t)$ – the distance between the vehicle n and the front vehicle in current lane at t moment;

$d_{n,other}(t)$ – the distance between the vehicle and its front vehicle in target lane at the same time;

$d_{n,back}(t)$ – the distance between the vehicle and that behind it in target lane at the same moment.

(2) Aggressive Lane Change (ALC) rule

ALC is closer to reality than PLC for human-driven cars. Existing research [28, 29] on the lane-changing behavior of human-driven cars has shown that, in many cases, the slower front vehicle induces the driver behind to consider overtaking.

$$\begin{cases} d_{n,other,back}(t) \geq d_0(t) \\ v_n(t) \geq d_{n,other,back}(t) \end{cases} \quad (12)$$

(3) PLC or ALC for an AV

Autonomous vehicles will choose PLC if it meets the PLC conditions, whereas ALC will be chosen with a probability P_c if it meets the ALC conditions, indicating an impact on the speeds of nearby vehicles. If the lane-changing condition is not feasible, the vehicle will stay in its current lane.

2.4 L3 Autonomous Vehicles Takeover Model

According to the Automatic Lane Keeping System Regulations [30] (ALKS) proposed and adopted by the United Nations Economic Commission for Europe (UNECE) in the

World Forum on the Harmonization of Automobile Regulations, the special attributes of L3 autonomous vehicles are as below.

The system asks driver to take over the vehicle under certain conditions. According to the provisions in ALKS, the following conditions are set for takeover request: (1) If driver fails to take over in time, the minimum risk control shall begin 10s after the takeover request begins at the earliest; (2) During minimum risk maneuver (MRM), the vehicle decelerates in the lane at a velocity of no more than 4 m/s until vehicle stops or driver takes over vehicle, and the vehicle minimum risk control begins. When using MATLAB for simulation analysis, the takeover model is designed as follows.

$$v_n(t+1) = \max(v_n(t) - 4, 0) \tag{13}$$

$$d_{safe,n}(t) = x_{n+1}(t) - x_n(t) - l_{n+1} = v_n(t)(\tau_n + 10) + \frac{v_n(t)^2}{2b_n} - \frac{v_{n+1}(t)^2}{2b_n} \tag{14}$$

Formula (13) represents the velocity evolution rule after the minimum risk control starts, and formula (14) represents the minimum distance from the obstacle ahead before the takeover request starts.

3 Simulation Analysis

3.1 Simulation Setting

The simulation is carried out using MATLAB 2019b. The Simulation model parameters and values are as Table 1 shown. At the beginning, there is a specific quantity of vehicles N, which their maximum speed is 33 m/s. The traffic density is $\rho = N/2L$, average speed is $\bar{v} = \frac{1}{N} \sum_{n-1}^{n} v_n$, and traffic flow is $Q = \rho \bar{v}$. Initially, conventional human-driven cars, L2 autonomous cars, and L3 autonomous cars are spread along the road based on their respective proportions. We provide the simulation model parameters and corresponding values in Table 1, which aligns with previous research [31, 32].

Table 1. Simulation model parameters and values [31, 32]

Variable	Parameter value
Unit cell length	1 m
Time step	1 s
Number of lanes	2
Length of each lane	2 km
HDV length	5 cells

(*continued*)

Table 1. (*continued*)

Variable	Parameter value
AV length	5 cells
Maximum speed (HDV/AV)	33/22 cell/s
Maximum acceleration (HDV/AV)	1.5/2 cell/s2
Maximum deceleration (HDV/AV)	2/2.5 cell/s2

3.2 Influence of AV Mixing Ratio on Capacity

To more clearly illustrate the influence on different levels of autonomous vehicles, Fig. 1 shows the corresponding flow changes at different levels of autonomous vehicles under different penetration rate of autonomous cars, where the legend represents different proportions of autonomous cars. From the illustrations, it is evident that with the same density, the traffic capacity grows as the percentage of autonomous vehicles rises. When the ratio of autonomous vehicles remains constant, the traffic capacity initially rises and then declines as the density increases. This is because at low and medium density, an increase in density means an increase in vehicles on the road, and therefore an increase in traffic. However, due to road capacity constraints, traffic will not continue to increase even if vehicle density continues to increase. At this time, individual vehicles may change lanes in pursuit of higher speeds, which will cause certain interference with traffic flow. This statement aligns with our overall comprehension of the capabilities of autonomous cars. However, there is a difference in traffic capacity (maximum traffic volume) between L2 and L3 autonomous cars. Under the same proportion of autonomous vehicles, the traffic capacity corresponding to L2 and L3 autonomous cars is different, and the traffic capacity corresponding to L3 autonomous cars is greater than that of L2 autonomous cars. This is because the delay time of L2 autonomous cars is greater than that of L3 autonomous cars, so the required safety distance of L2 autonomous cars is greater than that of L3 autonomous cars, and the quantity of L2 autonomous cars accommodated by the same lane length is greater than that of L3 autonomous cars. Obviously, the addition of L3 autonomous cars has greatly increased road capacity, as shown in Fig. 1.

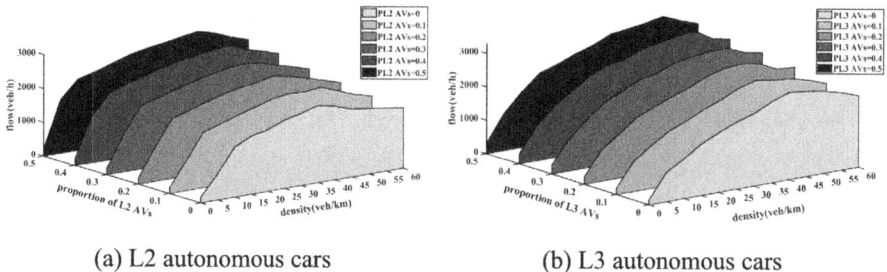

(a) L2 autonomous cars   (b) L3 autonomous cars

Fig. 1. Capacity at different densities with different proportions of autonomous vehicles

3.3 Time-Space Diagrams

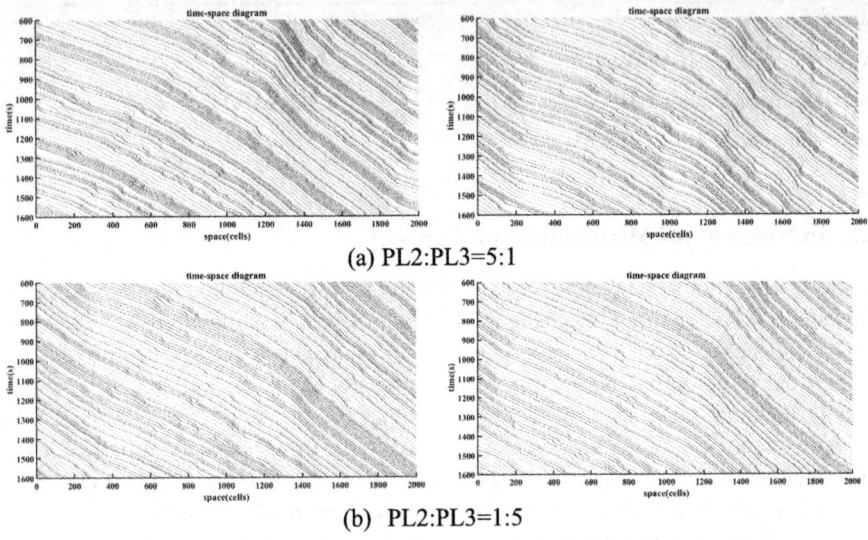

Fig. 2. Space-time trajectory diagrams (density = 40 veh/km)

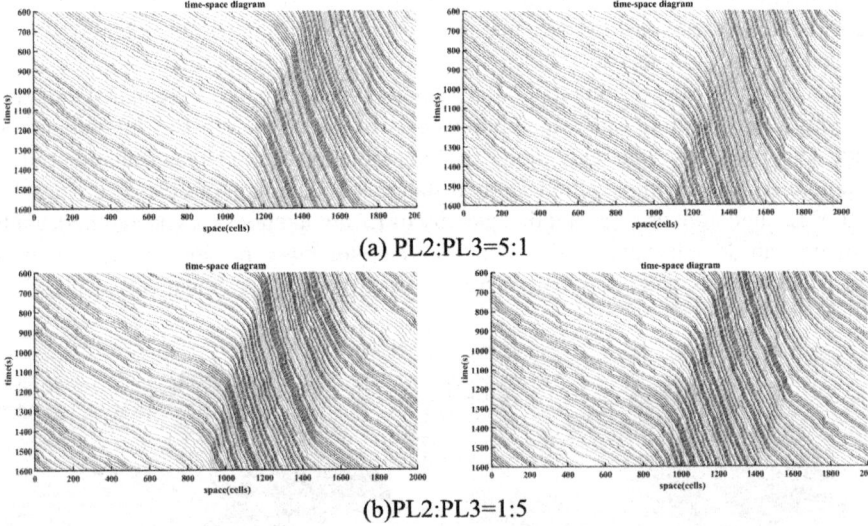

Fig. 3. Space-time trajectory diagrams (density = 60 veh/km)

To further investigate the impact of autonomous cars on traffic flow characteristics, we then studied space-time trajectory diagrams. Figure 2 and Fig. 3 show the space-time trajectory diagrams under different proportions of autonomous cars at different densities, in which the red line denotes trajectory of L2 autonomous cars, the purple one represents that of L3 autonomous cars and the green one denotes that of conventional human-driven cars. The slope of vehicles' trajectory in the space-time diagrams rep indicates corresponding velocity, and the bending of vehicle trajectory reflects the change of velocity. Compared with the space-time trajectory diagrams under high density (Fig. 3), that under medium density (Fig. 2) changed is not obvious, indicating that vehicles on the road at this time are not affected by density. This is because there is enough free space at low and medium density, the flow is under free state, vehicles basically run on the current lane, vehicles almost do not change lanes, and basically maintain the same velocity. With the rise in density, the number of vehicles changing lanes also increases at high density. However, in this scenario, the increase in vehicle density prevents them from achieving higher speed through lane-changing. It is evident that L2 autonomous cars and L3 autonomous cars experience distinct changes in velocity over time by comparing Fig. 3(a) and Fig. 3(b), despite having the same percentage representation.

3.4 Influence of AVs on Congestion Ratio

To understand how the proportion of autonomous cars affects traffic congestion, we introduce the congestion ratio (CR) [33]. This helps us analyze how traffic congestion changes as the number of autonomous cars varies. The congestion ratio is defined the vehicle with a velocity of less than 10 km/h at a certain time as the vehicle in congestion, and the congestion ratio can be expressed by the ratio of vehicles experiencing congestion, its formula is as follows:

$$CR = \frac{n}{\Delta TN} \quad (15)$$

where, n denotes number of vehicles under congestion; ΔT is simulation time; N refers to total number of vehicles during simulation period;

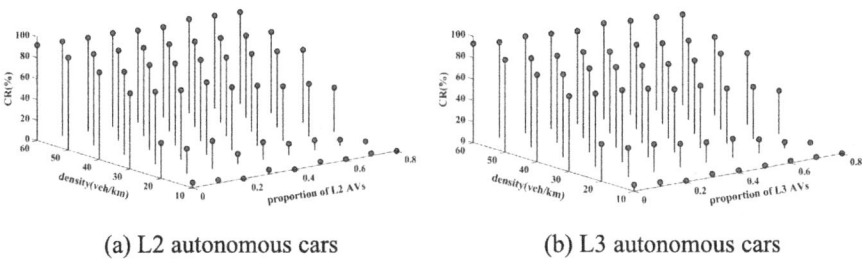

(a) L2 autonomous cars   (b) L3 autonomous cars

Fig. 4. Congestion ratio under different proportions of autonomous vehicles at different densities

Figure 4 shows that whether they are L2 or L3 autonomous cars, the CR decreases with higher density as the proportion of autonomous vehicles increases under the same density. Under the same proportion of autonomous cars, CR increases with density. When the proportion of autonomous cars is the same, the CR of L3 autonomous cars decreases more with the increase of density than that of L2 autonomous cars. Studies have shown that L2 and L3 autonomous cars can reduce road congestion, but the relief is less obvious at high densities. This is because the safety assistance function equipped with autonomous cars can always monitor the surrounding driving environment information, which includes tracking the speed of nearby vehicles, the distance between front vehicle and rear one, and so on. As density increases, autonomous vehicles exhibit improved acceleration and deceleration performance. This enables them to influence the movement of other vehicles, allowing the overall traffic flow to operate at a faster speed and delaying appearance of congestion, so the congestion is not high. However, when density reaches 40 veh/km, the reduction of CR becomes smaller and smaller.

3.5 Safety Evaluation

This paper intends to use TET (Time Exposed TTC) and TTC (Time to Collison) as safety evaluation indicators to explore the safety change rules of autonomous vehicles [34]. TTC represents the remaining time before collision. If the speed discrepancy between the front vehicle and rear one is positive and remains the same, we can express the TTC for the vehicle as follows

$$TTC_n(t) = \frac{x_{n+1}(t) - x_n(t) - l}{v_n(t) - v_{n+1}(t)}, v_n(t) > v_{n+1}(t) \tag{16}$$

The smaller the TTC value, the lower the risk of vehicle collision at that moment. Based on this formula (16), the total time spent in critical traffic conditions TET is calculated as follows:

$$TET(t) = \sum_{n=1}^{N} \sum_{r=1}^{T} \delta \times \Delta t \tag{17}$$

$$\delta_t = \begin{cases} 1, & 0 < TTC_n(t) < TTC^* \\ 0, & other \end{cases} \tag{18}$$

where, Δt refers to simulation time.

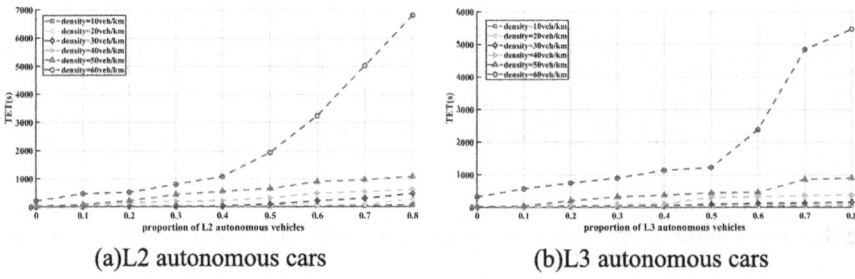

(a) L2 autonomous cars   (b) L3 autonomous cars

Fig. 5. TET values under different proportions of autonomous vehicles at different densities

Vehicles safety performance with different proportions of L2 and L3 autonomous cars under different densities are shown in Fig. 5, which demonstrates the relationship between TET values and the proportions of autonomous cars and density. From the figures, it can be seen that TET value rises as density of traffic flow increases. Meanwhile, with proportion of autonomous cars increases, TET value fluctuates within a certain range, but the overall trend is consistent and decreases with the increase of proportion, which indicates that the traffic safety risk is gradually reduced, and it has a mitigating effect on the congestion, which greatly improves road safety.

4 Conclusions

This paper considers both the time delay and takeover model of autonomous vehicles, and establishes a CA model including different levels of autonomous vehicles, and then studies their influence on the characteristics of hybrid traffic flow. Utilizing an enhanced hybrid traffic flow CA model, simulations are conducted to study the traffic flow behaviors of varying levels of autonomous vehicles under different proportions. The following findings are derived through numerical simulation analysis: (1) Autonomous vehicles can improve capacity, reduce congestion, and elevate road safety; L3 autonomous vehicles can enhance capacity, alleviate congestion more effectively, and exhibit greater ease in lane-changing behavior compared to L2 autonomous vehicles. The higher the intelligence level of autonomous cars, the more conducive to stability of traffic flow. The autonomous cars on the expressway in the future will also have L4 autonomous cars and L5 autonomous cars, and their functions are different from L2 and L3 autonomous cars, so the next step needs further research. In addition, the dynamic acceleration of the vehicle can also be considered, and the acceleration process of the vehicle can be regarded as non-constant acceleration. We can also use other models to explore and analyze its traffic flow characteristics in the future.

Acknowledgements. This work is supported by the National Natural Science Foundation of China (Grant No. 72361017, Grant No. 52362047, Grant No. 71861024), the Major Research Plan of Gansu Province (Grant No. 21YF5GA052), the 2021 Gansu Higher Education Industry Support Plan (Grant No. 2021CYZC-60), the Natural Science Foundation of Gansu Province (Grant No. 18JR3RA119), the Excellent Doctoral Program of Gansu Province (Grant No. 23JRRA906), and the Double—First Class Major Research Programs, Educational Department of Gansu Province (Grant No. GSSYLXM—04).

References

1. Shang, M., Wang, S., Stern, R.E.: Extending ramp metering control to mixed autonomy traffic flow with varying degrees of automation. Transp. Res. Part C Emerg. Technol. **151**, 104119 (2023)
2. J3016_202104: Taxonomy and Definitions for Terms Related to Driving Automation Systems for On-Road Motor Vehicles - SAE International (n.d.). https://www.sae.org/standards/content/j3016_202104/. Accessed 12 July 2023
3. Van Arem, B., Van Driel, C.J., Visser, R.: The impact of cooperative adaptive cruise control on traffic-flow characteristics. IEEE Trans. Intell. Transp. Syst. **7**(4), 429–436 (2006)
4. Talebpour, A., Mahmassani, H.S.: Influence of connected and autonomous vehicles on traffic flow stability and throughput. Transp. Res. Part C: Emerg. Technol. **71**, 143–163 (2016)

5. Elliott, D., Keen, W., Miao, L.: Recent advances in connected and automated vehicles. J. Traffic Transp. Eng. (Engl. Ed.) **6**(2), 109–131 (2019)
6. Yang, S., Du, M., Chen, Q.: Impact of connected and autonomous vehicles on traffic efficiency and safety of an on-ramp. Simul. Model. Pract. Theory **113**, 102374 (2021)
7. Sala, M., Soriguera, F.: Capacity of a freeway lane with platoons of autonomous vehicles mixed with regular traffic. Transp. Res. Part B: Methodol. **147**, 116–131 (2021)
8. Perraki, G., Roncoli, C., Papamichail, I., Papageorgiou, M.: Evaluation of a model predictive control framework for motorway traffic involving conventional and automated vehicles. Transp. Res. Part C: Emerg. Technol. **92**, 456–471 (2018)
9. S.O.-R.A.V.S. Committee, et al.: Taxonomy and definitions for terms related to on-road motor vehicle automated driving systems. SAE Stand. J 3016 (2014)
10. Guan, H., Wang, H., Meng, Q., Mak, C.L.: Markov chain-based traffic analysis on platooning effect among mixed semi-and fully-autonomous vehicles in a freeway lane. Transp. Res. Part B: Methodol. **173**, 176–202 (2023)
11. Miqdady, T., de Oña, R., Casas, J., de Oña, J.: Studying traffic safety during the transition period between manual driving and autonomous driving: a simulation-based approach. IEEE Trans. Intell. Transp. Syst. **24**(6), 6690–6710 (2023)
12. Barlovic, R., Santen, L., Schadschneider, A., Schreckenberg, M.: Metastable states in cellular automata for traffic flow. Eur. Phys. J. B-Condensed Matter Complex Syst. **5**, 793–800 (1998)
13. Maerivoet, S., De Moor, B.: Cellular automata models of road traffic. Phys. Rep. **419**(1), 1–64 (2005)
14. Jiang, R., Wu, Q.S.: The adaptive cruise control vehicles in the cellular automata model. Phys. Lett. A **359**(2), 99–102 (2006)
15. Qiu, X.P., Ma, L.N., Zhou, X.X., Yang, D.: The mixed traffic flow of manual-automated driving based on safety distance. J. Transp. Syst. Eng. Inf. Technol. **16**(4), 101–108 (2016)
16. Yan-Yan, Q., Hao, W., Wei, W., Qian, W.: Stability analysis and fundamental diagram of heterogeneous traffic flow mixed with cooperative adaptive cruise control vehicles. Acta Physica Sinica **66**(9) (2017)
17. Gong, S., Du, L.: Cooperative platoon control for a mixed traffic flow including human drive vehicles and connected and autonomous vehicles. Transp. Res. Part B: Methodol. **116**, 25–61 (2018)
18. Zhong, Z., Lee, E.E., Nejad, M., Lee, J.: Influence of CAV clustering strategies on mixed traffic flow characteristics: an analysis of vehicle trajectory data. Transp. Res. Part C: Emerg. Technol. **115**, 102611 (2020)
19. Stern, R.E., et al.: Dissipation of stop-and-go waves via control of autonomous vehicles: Field experiments. Transp. Res. Part C: Emerg. Technol. **89**, 205–221 (2018)
20. Ghiasi, A., Li, X., Ma, J.: A mixed traffic speed harmonization model with connected autonomous vehicles. Transp. Res. Part C: Emerg. Technol. **104**, 210–233 (2019)
21. Kavas-Torris, O., Lackey, N., Guvenc, L.: Simulating the effect of autonomous vehicles on roadway mobility in a microscopic traffic simulator. Int. J. Automot. Technol. **22**(3), 713–733 (2021)
22. Zhang, L.Y., Duan, X.K., Ma, J., Zhang, M., Wen, Y., Wang, Y.: Mechanism of road capacity under different penetration scenarios of autonomous vehicles. Int. J. Simul. Model. **21**(1), 172–183 (2022)
23. Ma, X., Hu, X., Weber, T., Schramm, D.: Effects of automated vehicles on traffic flow with different levels of automation. IEEE Access **9**, 3630–3637 (2020)
24. Gipps, P.G.: A behavioural car-following model for computer simulation. Transp. Res. Part B: Methodol. **15**(2), 105–111 (1981)
25. Hu, X., Huang, M., Guo, J.: Feature analysis on mixed traffic flow of manually driven and autonomous vehicles based on cellular automata. Math. Probl. Eng. **2020**(1), 7210547 (2020)

26. Hu, L., Cai, H., Huang, J., Cao, D., Zhang, X.: The challenges of driving mode switching in automated vehicles: a review. IEEE Trans. Veh. Technol. (2023)
27. Wang, X., Zeng, J., Qian, Y., Wei, X., Zhang, F.: Heterogeneous traffic flow of expressway with Level 2 autonomous vehicles considering moving bottlenecks. Physica A: Stat. Mech. Appl. 129991 (2024)
28. Lee, S.E., Olsen, E.C., Wierwille, W.W.: A comprehensive examination of naturalistic lane-changes (No. FHWA-JPO-04-092). United States. Department of Transportation. National Highway Traffic Safety Administration (2004)
29. George Oketch, T.: New modeling approach for mixed-traffic streams with nonmotorized vehicles. Transp. Res. Rec. **1705**(1), 61–69 (2000)
30. UNECE. UN regulation on Automated Lane Keeping Systems (ALKS) extended to trucks, buses and coaches | UNECE, 26 November 2021. https://unece.org/media/Transport/press/362551
31. Jiang, Y., Wang, S., Gao, K., Liu, M., Yao, Z.: Cellular automata model of mixed traffic flow composed of intelligent connected vehicles' platoon. J. Syst. Simul. **34**(5), 1025–1032 (2022)
32. Guo, Y., Xiang, Q., Li, S., Zhang, T., Yao, R.: Impacts of large vehicles on traffic safety in freeway interchange merging areas and improvement measures. In: MATEC Web of Conferences, vol. 124, p. 04004. EDP Sciences (2017)
33. Jiang, Y., Wang, S., Yao, Z., Zhao, B., Wang, Y.: A cellular automata model for mixed traffic flow considering the driving behavior of connected automated vehicle platoons. Physica A **582**, 126262 (2021)
34. Minderhoud, M.M., Bovy, P.H.: Extended time-to-collision measures for road traffic safety assessment. Accid. Anal. Prev.. Anal. Prev. **33**(1), 89–97 (2001)

Exploring the Potential Application of Ramp Metering Systems to Improve the Performances of Roundabout Corridors

Lorenzo Brocchini[1(✉)] and Antonio Pratelli[2]

[1] University of Florence, 50121 Firenze, Italy
lorenzo.brocchini@unifi.it
[2] University of Pisa, 56126 Pisa, Italy

Abstract. This work is part of broader research concerning some innovative topics such as the study of new dynamic simulation-based approaches to improve traffic efficiency and road safety in Roundabout Corridors. In particular, the present paper examines Ramp Metering systems focusing on their application to roundabouts, to improve their efficiency and safety. Ramp Metering, a traffic management strategy traditionally used on freeway on-ramps to control the rate at which vehicles enter the mainline, has been thorough in various contexts but is relatively novel in its application to roundabouts. The first part of the paper aims to analyse how Ramp Metering can mitigate congestion and enhance overall traffic efficiency and safety in roundabouts. Instead in the second part, a new potential application for Roundabout Corridors is introduced, where coordinated Ramp Metering could optimize traffic flow across multiple roundabouts in succession, leading to improvements throughout the artery. The case study analyzed is the congested Roundabout Corridor located in Pisa, Italy. The corridor was analyzed through the Aimsun simulation software, comparing the current state and a possible future project state with the addition of Ramp Metering systems. Finally, this paper also suggests directions for future research, including the development of further Intelligent Transportation Systems (ITS) such as smart cameras that could further improve the effectiveness of Ramp Metering on Roundabout Corridors. The article is divided as follows: Introduction, Ramp Metering systems operations, Roundabout Corridors characteristics, Application of Ramp Metering in Roundabouts, New Implementation for Roundabout Corridors, Conclusions and Future Research works.

Keywords: Ramp Metering · Roundabout Corridors · Intelligent Transportation Systems · Dynamic Simulations · Traffic Congestion

1 Introduction

The present paper examines Ramp Metering systems with a focus on their application to roundabouts, emphasizing their potential to improve traffic flow and safety in Roundabout Corridors.

The first part of the paper synthesizes existing studies, highlighting the effectiveness of integrating Ramp Metering with roundabout intersections, analysing how this integration can mitigate congestion and enhance overall traffic efficiency and safety. Instead in the second part, a new potential application for Roundabout Corridors is introduced, where coordinated Ramp Metering could optimise traffic flow across multiple roundabouts in succession, leading to improvements throughout the artery. The case study analysed is the congested Roundabout Corridor located on the Via Aurelia Nord, Pisa, Italy. The corridor was analysed through the Aimsun dynamic simulation software, making a comparison between the current state and a possible future project state with the addition of Ramp Metering systems. The key contributions and innovations of this work emphasize the effectiveness of Ramp Metering Systems as a type of Intelligent Transport System that is particularly well-suited for both urban and suburban areas. Due to their low cost and ease of installation, these systems offer a practical solution for enhancing critical infrastructure, such as Roundabout Corridors, with minimal effort and significant benefits. Finally, in the conclusion, this paper also suggests directions for future research, including the development of further Intelligent Transportation Systems (ITS) that could further improve the effectiveness of Ramp Metering on Roundabout Corridors. The paragraphs will be as follows: Ramp Metering systems operations, Roundabout Corridors characteristics, Application of Ramp Metering in Roundabouts, New Implementation for Roundabout Corridors (Italian Case Study), Conclusions and Future Research works.

2 Ramp Metering System Operation

Ramp Metering systems, initially developed as a traffic management strategy, aim to control the flow of vehicles entering freeways and highways to improve traffic flow, reduce congestion, and enhance safety [1]. By regulating vehicle entry speeds, these systems help alleviate the impact on-road performance. This chapter covers the main concepts of Ramp Metering systems, including their main components, their typologies, their implementation and operation, their benefits and finally some of their possible future developments [2]. Starting with the key components, are: Ramp Metering Signals (traffic signals to control vehicle entry), Detection Systems (loop detectors, cameras, etc.), Control Algorithms (software that determines the timing of the signals based on real-time traffic data), Communication Systems (infrastructure for data exchange between other components), User Interface and Monitoring Tools (systems for traffic operators to monitor and adjust them). There are various types of Ramp Metering, among which are: Fixed-Time Metering (operates on pre-determined schedules with fixed cycle times); Traffic-Responsive Metering (adjusts metering rates in real-time based on current traffic data from detection systems); Adaptive Metering (uses advanced algorithms to predict traffic conditions and adjust metering rates proactively). Regardless of the type chosen, the implementation and operation of Ramp Metering systems are, in order: the Site Selection (identifying ramps where metering will be most effective involves analysing traffic patterns, congestion levels, and safety concerns); the System Design (designing the ramp metering system, including the placement of signals, detection systems, and communication infrastructure); the Testing and Calibration (before full deployment, the system undergoes testing and calibration to ensure it operates correctly under various traffic conditions); the Operation and Maintenance (continuous monitoring and maintenance are crucial for effective operation; traffic operators must adjust settings and respond to issues as they arise) [3–5]. One of the main concepts that pushed the authors to delve deeper into this research topic is the benefits linked to Ramp Metering systems. Among the advantages they are certainly worth mentioning: are improved traffic flow, reduced congestion, enhanced safety, and environmental advantages. Starting from these benefits, it is possible to affirm that the future of Ramp Metering lies in its integration with smart city initiatives, leveraging advanced data analytics and connected vehicle technologies. Artificial intelligence and machine learning algorithms will enhance the predictive capabilities of ramp metering systems, making them more responsive and efficient. Additionally, vehicle-to-infrastructure (V2I) communication will enable future systems to optimize metering based on direct feedback from vehicles. Despite challenges, advancements in technology and strategic planning will continue to enhance the effectiveness and integration of Ramp Metering systems in the future [6, 7]. The present research started from these last concepts.

3 Roundabout Corridors Characteristics

Roundabout Corridors represent a new research topic in traffic engineering and could represent one of the new technologies for future transportation. Roundabout corridors offer a robust solution for managing traffic flow, improving safety and road aesthetic and environmental quality. By understanding the key characteristics, design considerations, benefits and challenges associated with roundabout corridors, designers and engineers can create effective and sustainable transportation solutions that meet the needs of modern urban and suburban environments. A Roundabout Corridor is defined as: "an infrastructure that includes a series of three or more roundabouts that function independently along an artery" [8]. The essential characteristics to define a road corridor as a Roundabout Corridor are: the corridor must have from 3 to 6 roundabouts; branches must have 2 or 4 lanes, typically in suburban areas; roundabouts must have 1 or 2 lanes on the ring; the speed limit must be between 25 mph (~40 km/h) and 50 mph (~80 km/h); total length must be between 0.5 miles (~800 m) and 4.5 miles (~7200 m); the distance between two consecutive roundabouts must be between 650 feet (~200 m) and 6465 feet (~1970 m); the characteristics of the lateral arrangements may vary (presence or absence of sidewalks, pedestrian crossings, cycle paths, rest areas, etc..) [9–11]. In this research work, the Roundabout Corridor will be analyzed: "SS1 (State Steet 1) - Via Aurelia Nord" in Pisa (Tuscany, Italy), which presents the characteristics indicated above and therefore can be considered as such.

4 Application of Ramp Metering in Roundabouts

In this last theoretical chapter, we will explore the application of Ramp Metering systems to roundabouts. Existing research already shows improvements in isolated roundabouts using Ramp Metering. The objective of this research aims to summarize these studies, highlighting the effectiveness of integrating Ramp Metering with roundabouts, before discussing its potential application to Roundabout Corridors [12]. As already said, Ramp

Metering is a traffic management technique that regulates the flow of vehicles entering a roadway to reduce congestion and improve safety. When applied to roundabouts, they aim to manage the influx of vehicles at critical entry points, enhancing the roundabout's efficiency and reducing the potential for traffic conflicts. This chapter explores their application, benefits, considerations and resulting challenges that need to be overcome. Regarding the applications, Ramp Metering in roundabouts involves using traffic signals at roundabout entry points to control the rate at which vehicles enter the circulatory roadway. The goal is to balance traffic flow, ensuring the roundabout operates smoothly without excessive queuing or congestion at any entry point. The traffic signals are installed at the approaches to the roundabout and as said, they control vehicle entry, allowing only a specified number of vehicles to enter the roundabout at a time. They work in combination with detection systems, i.e., sensors that detect the presence and flow of vehicles both at the entry points and within the roundabout. These sensors provide real-time data to adjust the metering rate. This is all based on control algorithms that determine the optimal metering rate based on current traffic conditions, balancing the entry flow to prevent overloading the circulatory roadway. Moving on to the benefits of Ramp Metering applied to Roundabouts, taking up some of the concepts already exposed in previous chapters [13, 14]. Their application surely improves traffic flow: by regulating entry, ramp metering ensures a more balanced distribution of vehicles within the roundabout, reducing the risk of congestion. This brings several benefits: enhanced safety (controlled entry reduces conflict points and potential collisions, especially during peak traffic periods); reduced delays (preventing excessive queuing at roundabout approaches helps minimize delays and ensures smoother overall traffic movement); optimized roundabout capacity (effective metering allows roundabouts to handle higher volumes of traffic without compromising performance). With this in mind, we can move on to some important operational considerations. Ramp metering devices are systems of dynamic adjustment, capable of adapting in real-time to changes in traffic volume and flow patterns to maintain optimal roundabout performance. Additionally, they can function in coordination with local traffic management systems, ensuring synchronized control and information sharing, especially in urban areas with multiple roundabouts. Finally, they uphold the principles of public awareness and compliance: educating drivers about the purpose and operation of ramp metering at roundabouts is crucial for ensuring compliance and effectiveness [15, 16]. The topics just explored will be taken up again in the next chapter of this research work, concerning a new possible and innovative application of Ramp Metering.

5 New Possible Implementation for Roundabout Corridors

In this paragraph, we will examine the Roundabout Corridor of Pisa. Specifically, in Pisa (Tuscany, Italy), the "SS1 (State Street 1) - Via Aurelia Nord" was once a road characterized by a series of traditional intersections controlled by traffic lights. Today, however, it has been transformed into a Roundabout Corridor with five roundabouts. The "SS1 (State Street 1) - Via Aurelia" is a crucial Italian state road linking Rome to France. This route follows the scenic coastline of the Tyrrhenian and Ligurian Seas, passing through nine provincial capitals and numerous significant tourist destinations.

The infrastructure under consideration is situated in the western part of Pisa, Tuscany, Italy. It spans approximately 2.5 kms and features a single thoroughfare interrupted by five roundabout intersections as can be seen in Fig. 1.

Fig. 1. Roundabout Corridor in Pisa (source: Google Earth Pro).

As previously said, today, each intersection is represented by a roundabout, making this road corridor fully compliant with the characteristics of a Roundabout Corridor. Previously, these intersections were traditional intersections controlled by traffic lights. Over the past 15 years, they have all been redesigned into roundabouts. The transformation of the infrastructure from a series of traffic-lighted intersections to a Roundabout Corridor led to increased traffic despite significant investment, creating a paradox. This has been demonstrated starting from another fundamental concept, namely that today road intersections have no longer be studied as isolated, but as part of a system. The mistake that was made in this case was in fact not to consider the five intersections as a single infrastructure (a road corridor), but only as separate intersections. In fact, by carrying out timely analyses of individual intersections, the transition from traffic lights (previous state) to roundabouts (current state) certainly led to improvements in terms of efficiency and safety. However, it is also true that experimental evidence has demonstrated the fact that there has been an increase in costs in terms of delay in the main direction, leading to overall driver discontent. Two possible reasons led to this and can be summarized as follows. First of all, the fact that roundabouts in themselves are more "attractive" for road users than traffic lights and therefore traffic on the infrastructure has had an unexpected increase due to these transformations. Furthermore, even more importantly, roundabouts, unlike traffic lights, destroy any road hierarchy; better said, thanks to traffic lights it is possible to hierarchize the flow on the main road by giving it longer green times compared to secondary roads, while with roundabouts this is not possible. Starting from this last consideration, the authors then thought about how to solve the Roundabout Corridor problem without returning to the previous configuration. The possible answer was to try applying Ramp Metering systems to the Roundabout Corridor. The simulation results obtained using Aimsun software are presented for both the "standard conditions" and the "Ramp Metering application". It is important to note that standard values for key simulation parameters were used in both cases to maintain the integrity of the comparison. Altering these parameters could have compromised the results. The parameters used were: Simulation Step $= 0.80$ s; Reaction Time at Stop $= 1.20$ s; and Reaction Time at Traffic Light $= 1.60$ s. Furthermore, as regard the "Ramp Metering application", various tests were conducted applying them to the different roundabouts,, but only the best scenario is presented. In this scenario, Ramp Metering was applied to roundabouts 3 and 4 using a fixed cycle time, as shown in the Fig. 2.

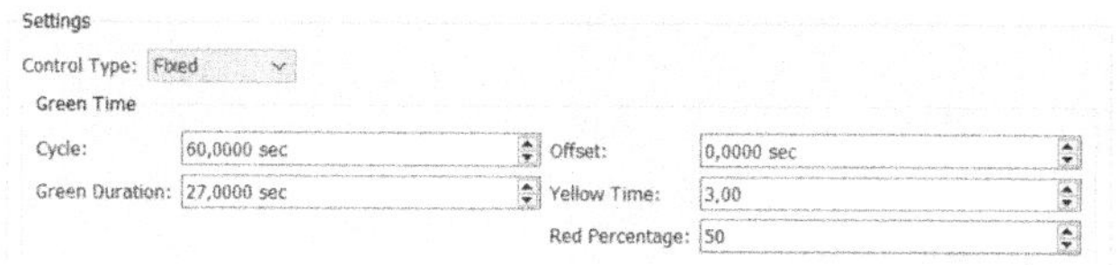

Fig. 2. Parameters of Ramp Metering application utilizing in Aimsun software.

In addition to the fundamental outputs of Aimsun (specifically, the time series data highlighting delays, queue lengths, and speeds), the number of conflicts was also calculated using the SSAM application [17, 18]. Over the past decade, the Federal Highway Administration (FHWA) has developed and released the Surrogate Safety Assessment Model (SSAM) program. This tool allows designers, researchers, and road design companies to assess intersection safety by estimating the frequency of conflicts using the.trj trajectory file, an output of dynamic simulation software. It is also noteworthy that roundabout 4 is an innovative, unconventional roundabout, which makes exploring innovative solutions like Ramp Metering even more relevant. As shown in Fig. 3, roundabout 4 is a Two-Geometry Roundabout, characterized by an elliptical shape with a circular central island [19, 20]. The dynamic simulations consider the unique geometry of such intersections, even if they fall outside standards.

Fig. 3. Intersection 4 – Unconventional Two-Geometry Roundabout (source: Google Earth Pro).

Below various images of roundabout 3 are provided. The first image (Fig. 4a), extracted from Google Earth, shows Via Aurelia Nord (the main road of the corridor) and Via Fosso Ducaria (the secondary road). The other images depict the roundabout reconstructed in Aimsun under both "standard conditions" (Fig. 4b) and with "Ramp Metering application", both in the situation of the green lantern being lit (Fig. 5a) and the red lantern being lit (Fig. 5b).

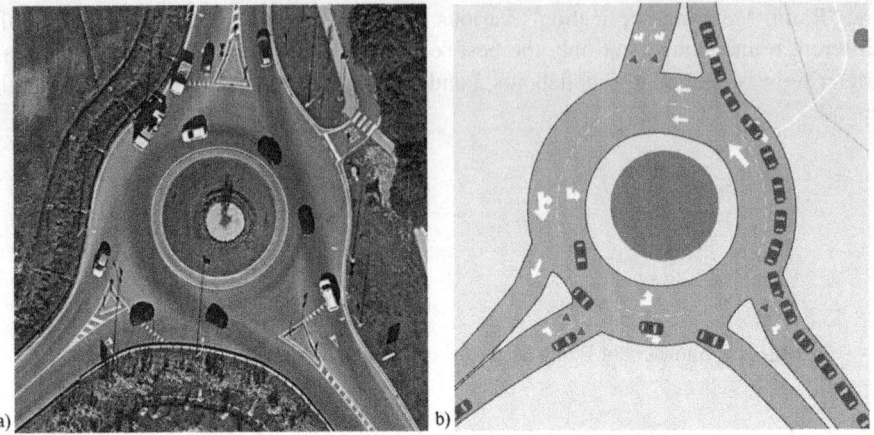

Fig. 4. a) Roundabout 3 (source: Google Earth Pro); b) Roundabout 3 in Aimsun software ("standard conditions").

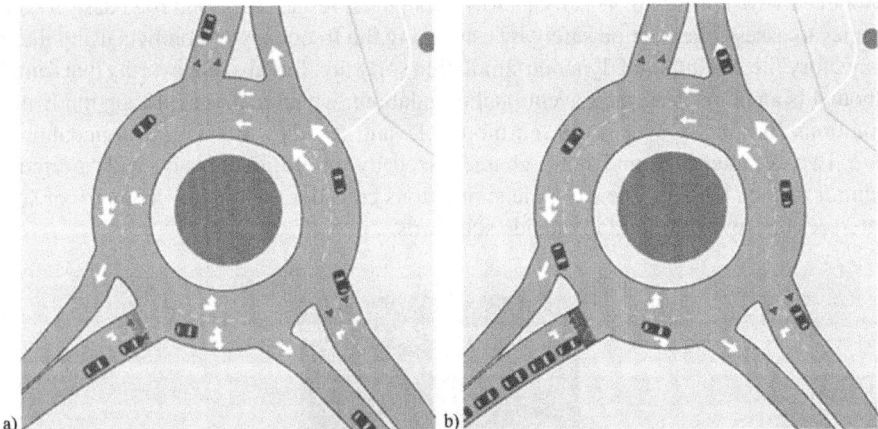

Fig. 5. a) Roundabout 3 in Aimsun software ("Ramp Metering application", green lantern); b) Roundabout 3 in Aimsun software ("Ramp Metering application", red lantern). (Colour figure online)

Even a quick glance at these images reveals that ramp metering allows vehicles on the secondary road to be stopped, prioritizing the main road and resulting in shorter queues.

However, the tables with the results of the time series and SSAM analyses, as previously mentioned, are reported. In detail, three tables are reported: 1) representing the results of the "standard conditions" (Fig. 6); 2) representing the results of the "Ramp Metering applications" (Fig. 7); 3) the percentage difference of the most important parameters (Fig. 8).

"STANDARD CONDITIONS"			
Time Series	Value	Standard Deviation	Units
Delay Time - All	72	22	sec/km
Density - All	24	5	veh/km
Flow - All	4617	223	veh/h
Harmonic Speed - All	26	4	km/h
Input Count - All	4735	172	veh
Input Flow - All	4735	172	veh/h
Max. Virtual Queue - All	474	202	veh
Mean Queue - All	120	54	veh
Mean Virtual Queue - All	212	74,01	veh
Missed Turns - All	7,42	3,00	
Number of Lane Changes - All	1,90E+02	8,91E+00	#/km
Number of Stops - All	0,16	0,02	#/veh/km
Speed - All	36	2	km/h
Stop Time - All	57	22	sec/km
Total Distance Traveled - All	5.969	349	km
Total Distance Traveled (Vehicles Inside) - All	260,03	31,09	km
Total Number of Lane Changes - All	2339,75	109,56	
Total Number of Stops - All	9179,5	809,07	
Total Travel Time - All	256,1	39,03	h
Total Travel Time (Vehicles Inside) - All	67,66	41,63	h
Total Travel Time (Waiting Out) - All	180,58	57,44	h
Travel Time - All	139,04	22,5	sec/km
Vehicles Inside - All	338,42	78,64	veh
Vehicles Lost Inside - All	0	0	veh
Vehicles Lost Outside - All	0	0	veh
Vehicles Outside - All	4617,17	223,33	veh
Vehicles Waiting to Enter - All	473,08	202,37	veh
Waiting Time in Virtual Queue - All	116,5	38,83	sec
SSAM: Total Number of Conflicts	Rear End	Lane Change	Crossing
2495	2049	344	102

Fig. 6. Time Series and SSAM output of "standard conditions".

6 Conclusions and Future Research Works

This article is divided into two main parts. The first theoretical part examines Ramp Metering systems, focusing on their application to roundabouts, reconstructing the state of the art. The second part, more substantial and innovative, proposes their new potential application to Roundabout Corridors, where Ramp Metering systems could optimise traffic flow, leading to improvements throughout the artery. The case study analyzed is the congested Roundabout Corridor located on the Via Aurelia Nord, Pisa, Italy. The corridor was analyzed through the Aimsun dynamic simulation software, comparing the current state (named "standard conditions") and a possible future state of the project (named "Ramp Metering applications") with, obviously, the addition of the Ramp Metering systems. The conclusions that can be drawn from the study carried out are almost

"RAMP METERING APPLICATION"			
Time Series	Value	Standard Deviation	Units
Delay Time - All	56	16	sec/km
Density - All	21	4	veh/km
Flow - All	4682	199	veh/h
Harmonic Speed - All	29	3	km/h
Input Count - All	4849	126	veh
Input Flow - All	4849	126	veh/h
Max. Virtual Queue - All	339	113	veh
Mean Queue - All	91	41	veh
Mean Virtual Queue - All	145	36,67	veh
Missed Turns - All	8,00	2,80	
Number of Lane Changes - All	1,87E+02	1,09E+01	#/km
Number of Stops - All	0,13	0,01	#/veh/km
Speed - All	37	1	km/h
Stop Time - All	43	16	sec/km
Total Distance Traveled - All	6.041	340	km
Total Distance Traveled (Vehicles Inside) - All	262,02	18,94	km
Total Number of Lane Changes - All	2300,25	134,54	
Total Number of Stops - All	7642	563,31	
Total Travel Time - All	218,77	25,37	h
Total Travel Time (Vehicles Inside) - All	54,94	33,85	h
Total Travel Time (Waiting Out) - All	135,88	28,86	h
Travel Time - All	123,06	15,54	sec/km
Vehicles Inside - All	346,58	95,66	veh
Vehicles Lost Inside - All	0	0	veh
Vehicles Lost Outside - All	0	0	veh
Vehicles Outside - All	4681,67	199,36	veh
Vehicles Waiting to Enter - All	337,5	113,41	veh
Waiting Time in Virtual Queue - All	85,94	19,95	sec
SSAM: Total Number of Conflicts	Rear End	Lane Change	Crossing
2095	1641	354	100

Fig. 7. Time Series and SSAM output of "Ramp Metering application".

Time Series	Value Difference %		
Delay Time - All	-22,19%		
Mean Queue - All	-24,22%		
Speed - All	4,43%		
SSAM: Total Number of Conflicts	Rear End	Lane Change	Crossing
-16,03%	-19,91%	2,91%	-1,96%

Fig. 8. Percentage difference of the most important parameters.

explained in the Excel tables relating to the "Time Series" and the "SSAM" applications. In fact, in Fig. 8 you can see the percentage difference between the current situation and the project situation with the addition of the Ramp Metering systems. Both in terms of Delay Time, Average Queue, Speed and Total Number of Conflicts, improvements of approximately 20% are recorded. This data is very interesting because it demonstrates that applying Ramp Metering systems to Roundabout Corridors is an excellent solution for improving traffic, especially on the main route where the greatest number of vehicles is concentrated and the highest degree of dissatisfaction on the part of the drivers. Furthermore, it should not be overlooked that the installation of Ramp Metering systems in this context would have very low costs and quick setup times. Therefore, such an intervention could be taken into consideration by the bodies that own the road to improve the efficiency of the infrastructure. Before concluding, the authors also suggest directions for future research. Currently, the study uses a fixed cycle time for distributing red and green signals, referred to as "Fixed-Time Metering" (as illustrated in Fig. 2). However, given the variability of real traffic conditions, a crucial point for future research could be exploring how to adapt these systems to unpredictable scenarios involving diverse traffic dynamics or infrastructure layouts. About that, there are plans to develop advanced Intelligent Transportation Systems (ITS), such as smart cameras, to enhance the effectiveness of Ramp Metering on Roundabout Corridors, through the use of the now widespread artificial intelligence. This advancement would evolve the system from "Fixed-Time Metering" to "Traffic-Responsive Metering" or "Adaptive Metering," or even to a hybrid model that dynamically switches between the two based on time of day and corresponding traffic conditions, such as the differences between rush hour and off-peak periods.. Specifically, installing intelligent cameras near the roundabout branches could monitor vehicle queues and send real-time data to traffic lights at secondary branch entrances. This would allow the traffic lights to adjust red and green times dynamically based on current traffic conditions [21]. In any case, these aspects will be further explored in the continuation of the research.

References

1. Division of traffic operations: ramp metering design manual. Division of Traffic Operations (2022)
2. Johnson, F., Bajenov, M.: SCATS ramp metering – from North American origins to autonomous vehicle readiness. In: 25th ITS World Congress, Copenhagen, Denmark, 17–21 September 2018
3. Trubia, S., Curto, S., Barberi, S., Severino, A., Arena, F., Pau, G.: Analysis and evaluation of ramp metering: from historical evolution to the application of new algorithms and engineering principles. Sustainability **13**(850) (2021)
4. Shaaban, K., Khan, M.A., Hamila, R.: Literature review of advancements in adaptive ramp metering. Procedia Comput. Sci. **83**, 203–211 (2016)
5. Join, C., Abouaïssa, H., Fliess, M.: Ramp metering: modeling, simulations and control issues. In: Advances in Distributed Parameter Systems, pp. 227–242 (2022)
6. Shang, M., Wang, S., Stern, R.E.: Extending ramp metering control to mixed autonomy traffic flow with varying degrees of automation. Transp. Res. Part C **151**, 104119 (2023)
7. Cheng, Y., Chang, G.-L.: Arterial-friendly local ramp metering control strategy. Transp. Res. Rec. **2675**(7), 67–80 (2021)
8. Bugg, Z., Schroeder, B.J., Jenior, P., Brewer, M., Rodegerdts, L.: Methodology to compute travel time of a roundabout corridor. Transp. Res. Record: J. Transp. Res. Board **2483**, 20–29 (2015)

9. Fernandes, P., Fontes, T., Pereira, S.R., Rouphail, N.M., Coelho, M.C.: Multicriteria assessment of crosswalk location in urban roundabout corridors. Transp. Res. Record: J. Transp. Res. Board **2517**, 37–47 (2015)
10. Fernandes, P., Guarnaccia, C., Teixeira, J., Sousa, A., Coelho, M.C.: Multi-criteria assessment of crosswalk location on a corridor with roundabouts: incorporating a noise related criterion. Transp. Res. Procedia **27**, 460–467 (2017)
11. Fernandes, P., Salamati, K.B., Rouphail, N.M., Coelho, M.C.: The effect of a roundabout corridor's design on selecting the optimal crosswalk location: a multi-objective impact analysis. Int. J. Sustain. Transp. **11**(3), 206–220 (2016)
12. Salem, H.H., Papageorgiou, M.: Ramp metering impact on urban corridor traffic: field results. Transp. Res. Part A: Policy Pract. **29**(4), 303–319 (1995)
13. An, H.K., Yue, W.L., Stazic, B.: Estimation of vehicle queuing lengths at metering roundabouts. J. Traffic Transp. Eng. **4**(6), 545–554 (2017)
14. Martin, M., García, A., Moreno, A.T., Llorca, C.: Capacity and operational improvements of metering roundabouts in Spain. Transp. Res. Procedia **15**, 295–307 (2016)
15. Mosslemi, M.: Using metering signals at roundabouts with unbalanced flows to improve the traffic condition: The Case Study of Kannik Area in Stavanger (2008)
16. Hummer, J.E., Milazzo, J.S., Schroeder, B., Salamati, K.: Potential for metering to help roundabouts manage peak period demands in the united states. Transp. Res. Record: J. Transp. Res. Board **2402**, 56–66 (2014)
17. Vasconcelos, L., Neto, L., Seco, A.M., Silva, A.B.: Validation of the surrogate safety assessment model for assessment of intersection safety. Transp. Res. Board **2432**, 1–9 (2014)
18. Pratelli, A., Brocchini, L., Leandri, P., Aiello, R.: Comparison between existing accident models and surrogate safety assessment models (SSAM) on unconventional roundabouts, with focused applications of the latter to some real study cases. Int. J. Adv. Intell. Syst. **17**(1&2), 61–72 (2024)
19. Pratelli, A., Souleyrette, R.R., Brocchini, L.: Two-geometry roundabouts: design principles. Transp. Res. Procedia **64**, 299–307 (2022)
20. Pratelli, A., Brocchini, L.: Two-geometry roundabouts: estimation of capacity. Transp. Res. Procedia **64**, 232–239 (2022)
21. Taale, H., Hoogendoorn, S., Legius, P.: Metering with traffic signal control - development and evaluation of an algorithm. Trans. Res. Procedia **8**, 204–214 (2015)

The Research on Customer Demand of Asia-Europe Liner Shipping Companies Based on Kano Model

Yiyang Chen[✉]

Shanghai Maritime University, Shanghai 201306, China
202230610043@stu.shmtu.edu.cn

Abstract. Against the backdrop of the sudden Red Sea crisis, the Asia-Europe container shipping route was forced to detour to the Cape of Good Hope route, which brought about problems such as supply chain disruption, tightening of the supply side, and adjustment of ship schedules. So, this paper does research on the customer demands of Asia-Europe container liner companies. This paper firstly establishes a reasonable customer demand indicators system, and then builds the Kano model, and applies the entropy weight method and TOPSIS method to obtain the key factors to meet customer demands. The findings reveal that: Emergency handling capability, schedule reliability and schedule flexibility have the greatest impact on improving customer satisfaction. Their comprehensive importance has been greatly improved and they belong to expected demands according to the quantitative Kano model. Therefore, these factors are the primary direction for liner companies to improve service quality in the background of the Red Sea crisis.

Keywords: Kano model · Customer demand · liner shipping company

1 Introduction

We know that container shipping is an important part of shipping trade, and the Asia-Europe route is one of the world's major ocean routes [1], and its seaborne trade volume accounts for around 10% of the container trade. However, the Red Sea crisis that broke out at the end of 2023 forced all container vessels on the Asia-Europe route to sail around the Cape of Good Hope, greatly increasing the route distance. This paper is of great significance to the customer demand research of liner shipping companies on the Asia-Europe line. The Kano model divides customer needs into five categories based on the relationship between different customer needs and customer satisfaction. Among them, A represents attractive demand, M represents essential demand, O represents expected demand, and I represent nondifference demand. Based on the customer demands of container liner companies, this paper uses the Kano model to investigate customer satisfaction and then uses the TOPSIS method to make decisions on the model results, and then uses the entropy weight method to calculate the importance of customer needs, and finally obtains the comprehensive importance function of user demands.

© The Author(s), under exclusive license to Springer Nature Switzerland AG 2025
A. Razminia and D. H. Nguyen (Eds.): ITFT 2024, CCIS 2378, pp. 271–281, 2025.
https://doi.org/10.1007/978-3-031-84148-4_21

2 Indicators System for Customer Demand of Asia-Europe Liner Shipping Companies

2.1 Analysis of Customer Value of Asia-Europe Liner Shipping Companies

According to the customer value, this paper analyzes the customer demands of liner shipping companies. Liner shipping company services generally refer to container shipping logistics [2, 3]. Based on the market environment and shipping logistics service process, the customer expectations of shipping logistics services are followed: Service informatization. Since the Asia-Europe line across multiple oceans and has a long voyage, customers can obtain real-time dynamics of their goods. In addition, logistics inquiries, order tracking, information feedback and other demands during the service period can be achieved through the Internet. Service customization. With the upgrading of the manufacturing industry and the steady development of high-tech industries, the value of single-container goods on the Asia-Europe line has increased significantly. Customers hope to use customized services for the valuable goods to ensure safety. Service stabilization. With the continuous occurrence of global geopolitical factors and supply chain disruptions [4], the service stability expected by customers is particularly important.

2.2 Determination of Customer Demands of Asia-Europe Liner Shipping Companies

This paper draws on relevant literature to explore the characteristics of liner shipping company customer demands [5, 6]. Based on the SEVRQUAL and the LSQ model, this paper builds evaluation indicators system taking into account five factors: tangibility, reliability, responsiveness, assurance, and empathy, as shown in Table 1.

Table 1. Indicators system for customer demand of Asia-Europe liner shipping companies

Customer demands	Notes
Schedule reliability	V1
Schedule flexibility	V2
Schedule resumption ability	V3
Freight rates	V4
Shipping guarantee	V5
Route coverage	V6
Digital services	V7
Document services	V8
Emergency handling capability	V9

3 The Model

3.1 Establishment of Fuzzy Kano Model

Combining the advantages of fuzzy theory in dealing with uncertain information, this paper establishes a relatively fuzzy Kano model for quantitative analysis and demand classification. This model can use fuzzy sets to reflect the complexity and uncertainty of customer demands, and at the same time use the advantages of the Kano model to classify demand factors [7].

Design of Fuzzy Kano Questionnaire. There is no essential difference between the fuzzy Kano questionnaire and the traditional Kano questionnaire in terms of form. The difference between the two questionnaires lies in the design of the options. When answering each question in the traditional Kano questionnaire, customers are only allowed to fill in a unique option, while the fuzzy Kano questionnaire allows customers to fill in multiple answers to a question fuzzily based on the membership assignment, and assign values in [0,1] to the options according to their own situation, ensuring that the sum of the values filled in each row is 1. The traditional Kano questionnaire and the fuzzy Kano questionnaire are shown in Tables 2 and 3.

Table 2. Traditional Kano questionnaire example

Demands	Like	Naturally	Never mind	Passable	Dislike
Provides this demand	√				
This demand is not provided					√

Table 3. Fuzzy Kano questionnaire example

Demands	Like	Naturally	Never mind	Passable	Dislike
Provides this demand	0.6	0.4	0	0	0
This demand is not provided	0	0	0.2	0.5	0.3

Constructing Fuzzy Evaluation Matrix. The fuzzy matrix is constructed using the obtained customer demand factor information. Taking Table 3 as an example, it is assumed that the matrix with insufficient service demand factors is D, and the matrix with sufficient service demand factors is F. So: F = [0.6 0.4 0 0 0], D = [0 0 0.2 0.5 0.3]. Based on the fuzzy relationship, the sufficient and insufficient matrices are fuzzy transformed. Through orthogonal transformation, the two matrices can be calculated to obtain the evaluation matrix S. As shown in formula (1).

$$S = F^T D = \begin{bmatrix} 0 & 0 & 0.12 & 0.3 & 0.18 \\ 0 & 0 & 0.08 & 0.2 & 0.12 \\ 0 & 0 & 0 & 0 & 0 \\ 0 & 0 & 0 & 0 & 0 \\ 0 & 0 & 0 & 0 & 0 \end{bmatrix} \quad (1)$$

Through the Kano classification table, the elements in the matrix S are matched with the categories in the table, and the preliminary results of the classification of the service demand element can be obtained. Then their (A, O, M, Q, R, I) membership vectors are: $t_A = s_{12}+s_{13} + s_{14} = 0.42$, $t_I = 0.28$, $t_O = 0.18$, $t_M = 0.12$, $t_Q = t_R = 0$. The membership vector of the service demand is T and the calculation formula of T is formula (2).

$$T = (\frac{M}{0.12}, \frac{O}{0.18}, \frac{I}{0.28}, \frac{A}{0.42}, \frac{Q}{0}, \frac{R}{0}) \quad (2)$$

Determine the Threshold. To make the obtained data more reliable and credible, the threshold α is introduced, and the threshold is used to filter the data in the membership vector [8]. This paper by reading literature to determine the value of $\alpha = 0.4$.

3.2 Calculation of Customer Satisfaction Based on TOPSIS Method

In calculating the demand classification based on the fuzzy Kano model, this section uses the TOPSIS method to solve the satisfaction and dissatisfaction of the demand [9, 10], and calculates the relative closeness of each solution to the ideal solution by constructing the positive and negative ideal solutions of the evaluation problem. This paper regards the customer demand in the Kano model as the evaluation solution, and SI and DSI are regarded as evaluation indicators. The weights of the two indicators are equally important. Among them, SI belongs to the benefit attribute, and DSI belongs to the cost attribute. The calculation formula is formula (3) and (4):

$$SI = \frac{F_A + F_O}{F_A + F_O + F_M + F_I} \quad (3)$$

$$DSI = \frac{F_O + F_M}{F_A + F_O + F_M + F_I} \quad (4)$$

Then, the relative closeness value of each customer demand satisfaction is calculated using the TOPSIS formula principle. The steps of customer demand satisfaction decision based on TOPSIS are:

1. Construct a customer demand satisfaction evaluation matrix. Assuming that there are n options and m decision attributes in the multi-attribute decision problem, the evaluation decision matrix formula is:

$$X_{ij} = \begin{bmatrix} X_{11} & \cdots & X_{1m} \\ \vdots & \ddots & \vdots \\ X_{n1} & \cdots & X_{nm} \end{bmatrix}_{n \times m} \tag{5}$$

2. Calculate the standardized matrix and weighted normalized decision matrix after weighted normalization, denoted as Z_{ij}, As shown in formula (6):

$$Z_{ij} = \frac{X_{ij}}{\sqrt{\sum_{i=1}^{n} X_{ij}^2}} \tag{6}$$

3. Calculate the positive and negative ideal solutions of the weighted normalized evaluation matrix, where "J" is the benefit attribute and "j" is the cost attribute.

$$Z_j^+ = \{\max_{1 \leq i \leq m}(\{Z_{ij}\}_{i=1}^m) | j \in J, \min_{1 \leq i \leq m}(\{Z_{ij}\}_{i=1}^m | j \in J)\} \tag{7}$$

$$Z_j^- = \{\min_{1 \leq i \leq m}(\{Z_{ij}\}_{i=1}^m) | j \in J, \max_{1 \leq i \leq m}(\{Z_{ij}\}_{i=1}^m | j \in J)\}. \tag{8}$$

4. Calculate the Euclid distance between the positive and negative ideal solutions.

$$D_i^+ = \sqrt{\sum_{j=1}^{n}(Z_{ij} - Z_j^+)^2}, i = 1,2,3\cdots,m \tag{9}$$

$$D_i^- = \sqrt{\sum_{j=1}^{n}(Z_{ij} - Z_j^-)^2}, i = 1,2,3\cdots,m \tag{10}$$

5. Calculate the relative closeness value of each target. As shown in formula (11).

$$V_i = \frac{D_i^-}{D_i^- + D_i^+}, 0 \leq V_i \leq 1, i = 1, 2, 3, \cdots, m \tag{11}$$

3.3 Determination of the Comprehensive Importance of Customer Demand

The calculation of the importance of customer demand usually relies on the subjective experience of experts. This paper uses the entropy weight method. The entropy weight method relies less on subjectivity and can make an evaluation value close to the truth and the calculation steps are:

1. Construct a customer demand primary importance decision matrix. Based on the 5-point Likert scale. The larger the value, the higher the importance of the demand. If n respondents evaluate the importance of m demand factors, the formula is:

$$X_{ij} = \begin{bmatrix} X_{11} & \cdots & X_{1m} \\ \vdots & \ddots & \vdots \\ X_{n1} & \cdots & X_{nm} \end{bmatrix}_{n \times m} \tag{12}$$

2. Normalize the first step evaluation matrix and record it as matrix P.

$$P_{ij} = \frac{X_{ij}}{\sum_{i=1}^{n} X_{ij}} \tag{13}$$

3. Calculate the entropy weight of the demand index. As shown in formula (14).

$$H_j = -k \sum_{i=1}^{n} P_{ij} \ln P_{ij}, j = 1, 2, 3 \cdots m, k = 1/\ln n \tag{14}$$

4. Calculate the degree of consistency and the weight value. As shown below:

$$W_j = \frac{d_j}{\sum_{j=1}^{m} d_j}, j = 1, 2, 3 \cdots m, d_j = 1 - H_j \tag{15}$$

5. Combine the primary importance value of customer demand with the satisfaction value of customer demand to obtain the comprehensive importance function of customer demand. The calculation formula is formula (16).

$$f_j = v_i \times w_j \bigg/ \sum_{j=1}^{m} v_i \times w_j \tag{16}$$

Finally, the comprehensive importance vector of customer demands $f_j = (f_1, f_2, \cdots f_m)$ is obtained, and each customer demand is ranked according to the standardized comprehensive importance to determine the company's reasonable service direction.

4 Empirical Study

This paper selects C liner shipping company on the Asia-Europe route as the object of empirical research, conducts in-depth research on the demands of C company's customers, and improves C company's decision-making ability based on importance analysis. Since the port calls at Shanghai on the Asia-Europe route is relatively high, and Rotterdam port is a representative hub port in Europe, this paper selects C liner shipping company's Shanghai-Rotterdam route as the empirical study object.

4.1 Questionnaire Design and Collection

In order to enhance the reliability of the data obtained from the questionnaire survey, this paper conducted a Kano and importance survey on the same group of customers. Therefore, this study designed a comprehensive questionnaire, which combines the Kano part and the importance part, including three parts: (1) Basic information of customers. The basic information part includes the types exported goods and the quantity of goods. (2) Kano questionnaire part. According to the Kano theory, two positive and negative questions are designed for each of the nine needs. Taking "Schedule reliability" as an example, the positive question is designed as "If the "Schedule reliability is very high, how do you think?" The reverse question is designed as "If the "Schedule reliability" is not high, how do you think?" (3) Importance questionnaire part. According to the 5-point

Likert scale. The scores reflected in the questionnaire responses were: "very important" - 5 points, "important" - 4 points, "average" - 3 points, "unimportant" - 2 points, and "very unimportant" - 1 point.

The questionnaire officially started on 3rd March 2024, and the questionnaires were collected on 5th April 2024. The questionnaire was mainly conducted through a combination of offline survey questionnaires and online platforms. A total of 132 paper questionnaires were distributed, 101 were collected, and 96 were valid.

4.2 Demand Factor Classification

Taking V1 "Schedule reliability" as an example, we found through the fuzzy Kano model questionnaire survey that the first customer's answers to the positive question are shown in Table 4. Therefore, the matrix composed of the positive and reverse questions can be obtained. The matrix composed of the positive question and the matrix composed of the reverse question are $X = [0.9\ 0.1\ 0\ 0\ 0]$, $Y = [0\ 0\ 0\ 0.5\ 0.5]$. Then the fuzzy interaction matrix is shown below:

$$X^T Y = \begin{bmatrix} 0 & 0 & 0 & 0.45 & 0.45 \\ 0 & 0 & 0 & 0.05 & 0.05 \\ 0 & 0 & 0 & 0 & 0 \\ 0 & 0 & 0 & 0 & 0 \\ 0 & 0 & 0 & 0 & 0 \end{bmatrix}$$

Table 4. First Customer's response to schedule reliability

Demands	Like	Naturally	Never mind	Passable	Dislike
If the "Schedule reliability is very high, how do you think?	0.9	0.1	0	0	0
If the "Schedule reliability" is not high, how do you think?	0	0	0	0.5	0.5

According to the comparison between the above fuzzy matrix and the Kano classification table, the membership vector T=$(\frac{M}{0.05}, \frac{O}{0.45}, \frac{I}{0.05}, \frac{A}{0.45}, \frac{R}{0}, \frac{Q}{0})$ of V1 schedule reliability can be obtained, the threshold α= 0.4 is used to filter the data in the membership vector T, V1 membership vector is T=$(\frac{M}{0}, \frac{O}{1}, \frac{I}{0}, \frac{A}{1}, \frac{R}{0}, \frac{Q}{0})$. According to the priority of Kano attribute categories, the first customer believes that the demand for V1 is expected demand O. After counting all customers' demand classifications for V1, 60 people think it is an expected demand, 24 people think it is an attractive demand, and 12 people think it is an essential demand. Therefore, it can be obtained that customers think that V1 demand is an expected demand O. Then the demand classification of all indicators is counted, and the kano classification of 9 demands can be obtained. The classification results are shown in Table 5.

Table 5. Kano classification of liner shipping company customer demand indicators

Customer demands	Kano classification						Result
	A	O	M	I	R	Q	
V1	24	60	12				O
V2	24	60	12				O
V3	12	36	48				M
V4	12	48	24	12			O
V5		60	24	12			O
V6	36	24	12	24			A
V7	60	12		24			A
V8	12	24	48	12			M
V9	20	64	12				O

4.3 Satisfaction Calculations

According to formula (3) and (4), customer satisfaction and customer dissatisfaction can be obtained. The results are shown in Table 6.

Table 6. Liner shipping company customer demand satisfaction calculation

Customer demands	SI	DSI
V1	0.875	0.75
V2	0.875	0.75
V3	0.5	0.875
V4	0.625	0.75
V5	0.625	0.875
V6	0.625	0.375
V7	0.75	0.125
V8	0.375	0.75
V9	0.875	0.792

According to formulas (7–8), with the weighted normalized matrix Z as the target, the positive and negative ideal solutions can be obtained: Z^+=(0.252591, 0.266464), Z^-=(0.108253, 0.038066). According to formulas (9), (10) and 11, D +, D-, and v are calculated. The results are shown in Table 7.

Table 7. Relative closeness to customer demands and Euclid distance

Customer demands	D+	D-	v
V1	0.038066	0.238871	0.862545
V2	0.038066	0.238871	0.862545
V3	0.108253	0.231231	0.681124
V4	0.081593	0.203554	0.713857
V5	0.072169	0.239528	0.768465
V6	0.168502	0.104902	0.383689
V7	0.231231	0.108253	0.318876
V8	0.149273	0.190331	0.560451
V9	0.025378	0.249099	0.907542

Table 8. Entropy weight value and comprehensive importance of customer demand indicators

Customer demands	Hj	fj	Ranking
V1	1.757844	0.143046	2
V2	1.754917	0.142493	3
V3	1.758546	0.113063	6
V4	1.753981	0.117783	5
V5	1.755007	0.126966	4
V6	1.752247	0.063161	8
V7	1.750009	0.052336	9
V8	1.751996	0.092229	7
V9	1.749863	0.148923	1

4.4 Comprehensive Importance Calculations

This paper uses the entropy weight method to analyze all the collected questionnaire results and calculates the comprehensive importance of the 9 demands. The decision matrix X as shown below:

$$X = \begin{bmatrix} 5 & 4 & \cdots & 4 & 5 \\ 5 & 4 & \cdots & 3 & 5 \\ \vdots & \vdots & \ddots & \vdots & \vdots \\ 3 & 5 & \cdots & 4 & 3 \\ 5 & 3 & \cdots & 4 & 3 \end{bmatrix}_{96 \times 9}$$

According to formula (14), the entropy weights of the 9 demands can be obtained, and according to formula (16), the comprehensive importance of the customer can be obtained, and the results are shown in the following table.

As can be seen, the comprehensive priority considering the comprehensive importance of customer satisfaction is: V9 > V1 > V2 > V5 > V4 > V3 > V8 > V6 > V7. Among them, V9, V1 and V2 have the greatest impact on improving demand satisfaction, their comprehensive importance has been greatly improved and they belong to the expected demand. Therefore, emergency handling capability, schedule reliability and schedule flexibility are the primary service improvement directions of liner shipping companies. Although route coverage and digital services are attractive demands, their importance is relatively low. In the background of the Red Sea crisis, liner shipping companies can only meet basic demands.

4.5 Implications for Liner Shipping Companies

Firstly, it is necessary to strengthen cooperation with port enterprises to further improve the schedule reliability, the liner shipping companies can sign MOU to link the schedule reliability with the freight volume, and also strengthen the monitoring and prediction of specific terminals. Secondly, establish accountability system in order to improve emergency handling capability, it is recommended to arrange one person to connect with the customer throughout the process, this solution avoid unclear responsibilities and multi-headed external problems caused by complex processes. Finally, develop a more flexible freight rate mechanism, suggest that liner shipping companies appropriately relax the price concession range that can be given, develop a more flexible approval process, and also improve the employee's responsiveness.

5 Conclusion

Determining the comprehensive importance of customer demands is an important way for liner shipping companies to improve service quality under the special background, and it is also a key step to increase the target market share. This paper constructs an integrated model based on Kano, TOPSIS and entropy weight method, and establishes a comprehensive evaluation function for the basic importance of customer demands. This method uses TOPSIS to quantify the satisfaction questionnaire in the Kano model, which effectively ensures the accuracy of quantification. Secondly, the entropy weight method is used to determine the importance of demands, which can effectively weaken the customer's subjective judgment on the importance of demands, so that the ranking is more reasonable. Through examples, it is proved that the method in this paper effectively combines customer satisfaction and importance of demands and increases the efficiency and guarantee of liner services. However, the determination of the comprehensive importance of liner shipping company customer demands changes with the situation. Therefore, the focus of the next stage of research is to conduct a comprehensive dynamic analysis of the importance of customer demands.

References

1. Liu, Q., Yang, Y.: Liner shipping network vulnerability to component disruptions: a China-Europe container flow analysis. Transp. Res. Part D **131**, 104232 (2024)
2. Sun, Q., Li, W., Meng, Q.: Single-leg shipping revenue management for expedited services with ambiguous elasticity in transit-time-sensitive demand. Transp. Res. Part B **180**, 102886 (2024)
3. Anderl, C., Caporale, G.M.: Shipping cost uncertainty, endogenous regime switching and the global drivers of inflation. Int. Econ. **178**, 100500 (2024)
4. Zhou, Y., Yuen, K.F.: A Bayesian network model for container shipping companies' organisational sustainability risk management. Transp. Res. Part D **126**, 103999 (2024)
5. Guo, S., Jing, L., Qin, Y.: Analysis of the coupled spatial and temporal development characteristics of global liner shipping connectivity driven by trade. Ocean Coast. Manag. **251**, 107071 (2024)
6. Chao, S.-L., Ming-Miin, Y., Wei, S.-Y.: Ascertaining the impact of e-service quality on e-loyalty for the e-commerce platform of liner shipping companies. Transp. Res. Part E **184**, 103491 (2024)
7. Yang, Y., Li, Q., Li, C., Qin, Q.: User requirements analysis of new energy vehicles based on improved Kano model. Energy 133134 (2024)
8. Becker, W., Saisana, M., Paruolo, P., Vandecasteele, I.: Weights and importance in composite indicators: closing the gap. Ecol. Indicat. 12–22 (2017)
9. Wang, Y., Ye, G.-T., Ng, A.K.Y.: Choosing optimal bunkering ports for liner shipping companies: a hybrid fuzzy-Delphi–TOPSIS approach. Transp. Policy **35**, 358–365 (2024)
10. Shan, D.: Hybrid Kano-DEMATEL-TOPSIS model based benefit distribution of multiple logistics service providers considering consumer service evaluation of segmented task. Expert Syst. Appl. Part C **213**, 119292 (2023)

The Death and Life of Free-Floating Car Sharing in China: Case Study of Chengdu and Changchun

Hongjie Wang[1,2,3], Xia Luo[1,2,3](✉), Qiming Su[1,2,3], and Hongqing Bao[1,2,3]

[1] School of Transportation and Logistics, Southwest Jiaotong University, Chengdu, People's Republic of China
xia.luo@263.net
[2] National Engineering Laboratory of Integrated Transportation Big Data Application Technology, Southwest Jiaotong University, Chengdu, People's Republic of China
[3] National United Engineering Laboratory of Integrated and Intelligent Transportation, Southwest Jiaotong University, Chengdu, People's Republic of China

Abstract. In the last decades, free-floating car sharing (FFCS) has become a fast-growing mode of mobility service and promoted in many cities across the world. While lots of researchers have paid increasing attention to FFCS in recent years, little is known regarding the characterization and evolution of FFCS in Chinese cities, where car sharing services are gaining more and more popularity. In this work, we focus on two Chinese cities: Chengdu and Changchun, both of which have witnessed the shutdown of FFCS service and its replacement with station-based car sharing (SBCS). We rebuild 743,796 trips using public data from the smartphone application of a car sharing service, then explore the characteristics and spatiotemporal distribution of FFCS usage, and identify the differences compared to those in Europe and North America. Furthermore, we conduct a comparative analysis of the service coverage, demand spatiotemporal distribution and usage patterns of car sharing services before and after the shutdown of FFCS. Results show that, the shutdown of FFCS did not cause a severe disruption to the car sharing service. Instead, the introduction of SBCS effectively inherited and accommodated the travel demand. This work underscores the importance of considering the unique characteristics of China's FFCS services when develop operation strategies and demonstrates that a well-designed SBCS could emerge as a viable alternative.

Keywords: Shared Mobility · Big Data · Operation Scheme Transition

1 Introduction

As an innovative urban mobility service, car sharing has sparked interest in the academic community in recent years. The operating scheme of car sharing can be categorized into FFCS (Free-floating Car Sharing) and SBCS (Station-based Car Sharing) [1]. SBCS requires users drop off the car at one of the stations defined by the service provider, while

© The Author(s), under exclusive license to Springer Nature Switzerland AG 2025
A. Razminia and D. H. Nguyen (Eds.): ITFT 2024, CCIS 2378, pp. 282–292, 2025.
https://doi.org/10.1007/978-3-031-84148-4_22

FFCS allow them to drop off the car at any legal parking space within the designated area. FFCS has gained popularity among travelers for its flexibility and been promoted globally, becoming an essential part of urban transportation system.

Car sharing services have experienced rapid development in the Chinese market since their introduction in 2010, where SBCS is still the norm [2]. According to GoFun, one of the main SBCS operators in China, its car sharing services have launched in over 260 Chinese cities, with a fleet of over 10,000 cars and more than 20 million users [3]. However, the development of FFCS in China is not smooth, and this popular operating scheme, prevalent in European and American cities, has only appeared in a few Chinese cities. For instance, Car2go, one of the world's largest FFCS operators, launched China's first FFCS service in 2016. It provided services in 7 Chinese cities at its peak, but finally exited the Chinese market in 2019. The significant variation in market performance reflects the unique nature of car sharing demand in China, which further underscores the importance of conducting empirical research on it.

Understanding how the FFCS services are used is crucial for improving the system performance [4]. Several works have been done to capture the characterization of FFCS usage. Giordano et al. conducted a comprehensive analysis in 23 cities in Europe and North America, investigated the characteristics of FFCS services [5]. María et al. focused on the city of Madrid, obtaining FFCS usage patterns by analyzing the temporal and spatial distribution of FFCS flow [6]. Based on the characterization of FFCS service in 4 European cities, Cocca et al. studied the impact of different operation strategies to support the design of FFCS system [7]. Until recently, knowledge about car sharing usage in Chinese cities has been mostly acquired through SBCS services. Feng et al. calculated the number of transactions to represent the carsharing demand in Beijing, and explored the factors associated with one-way and round-trip FFCS usage [8]. Huo et al. analyzed the historical order data of 30 stations in Beijing to explore the characterization of SBCS trips [9]. Jin et al. combined a nested logit mode with trip survey data to estimate the demand of SBCS in Beijing. Though FFCS services had their origins in SBCS, results of the literature on SBCS cannot be directly transfer to the FFCS [6].

Regarding the comparative analysis of the two operation schemes, existing knowledge has been mostly acquired through surveys. Becker et al. used survey data to compare FFCS and SBCS, results indicated that there are indeed differences in user groups and usage patterns between these two schemes [10]. Heilig et al. integrated SBCS and FFCS in a multiple-day travel demand model basing on the data from a household travel survey, and analyzed the characterization of customers, trip purposes, trip lengths, number of trip in the two operation schemes [11].

To the best of our knowledge, few studies have analyzed the characteristics of FFCS in China with actual usage data, hindering the development of targeted strategies for its sustainable growth. In this work, we track the car sharing usage data in two Chinese cities: Chengdu and Changchun, both of which have previously provided large-scale FFCS services and witnessed a transition from FFCS to SBCS. With the help of this long-time span and high-precision dataset, we perform a comparative empirical analysis of the service coverage, demand spatiotemporal distribution and usage patterns of FFCS/SBCS services. The contributions of this work are summarized into: (1) rebuild and analysis the trip information basing on public data, explore the characteristics of

FFCS usage and spatiotemporal distribution in Chinese cities; (2) employ an event-based social experiment (the shutdown of FFCS and the transition to SBCS) to examine the impact of the transition; (3) provide data-driven insights for car sharing operators to develop the appropriate operation strategies.

2 Data

The inclusion of ICT (Information and Communication Technology) enables car sharing operators to provide convenient service and generate large amounts of data that can help researchers to capture the characterization of car sharing usage. Recently, several works have been devoted to take advantage of the data from shared mobility operator's websites for characterization analysis [5, 6, 12]. Inspired by these works, we designed an app-based platform tailored to the characteristics of Chinese car sharing services to track the status of shared vehicles.

Car sharing operators present information through smartphone apps, including details on available FFCS vehicles (vehicle ID, energy level, and location), stations (station ID, vehicles' information within it, and location), and FFCS service zones boundary. When a user reserves a particular vehicle, it disappears from the page and becomes unavailable for other users. Once returned, it reappears and becomes available for users to choose. Thus, we can rebuild the historic trip information through the vehicle's availability status. We collect data every 60s using the available public API from a major car sharing operator in Chengdu and Changchun during September 2022 to June 2023 (264 days). According to the actual using process of car sharing services, we infer the reservation periods through the unavailability of a vehicle and identify whether the vehicle has moved by comparing the pick-up and drop-off locations. After cleaning up the erroneous data caused by technical problems such as network interruption, platform server maintenance, and the trips for charging or refueling the vehicles (which are basically done by the operators), we obtained 743,796 valid trips for analysis. It is worth mentioning that the dataset does not contain any user's personal information, in this paper there is no risk to violate the users' privacy.

3 Analysis and Results

We scrutinize the shutdown of FFCS services through a series of analyses, treating it as a social experiment. To keep the length of the paper in a reasonable extent, a selection of representative analysis results is presented in this section. These analyses aim to provide a more comprehensive view of FFCS services in Chinese cities, and to figure out the differences between them with the SBCS launched in the same cities.

3.1 Study Area and Supply-Side Changes

In Chengdu and Changchun, the scheme shift occurred on 2023-04-01 and 2022-12-20, respectively. Prior to this, both cities provided FFCS services, users could drop off the car at any legal parking space within the FFCS area, while they were also encouraged to drop off the car to specified stations. After the FFCS services shutdown, the operators deployed a considerable number of additional stations that were densely scattered throughout the original FFCS zones, as depicted in Fig. 1.

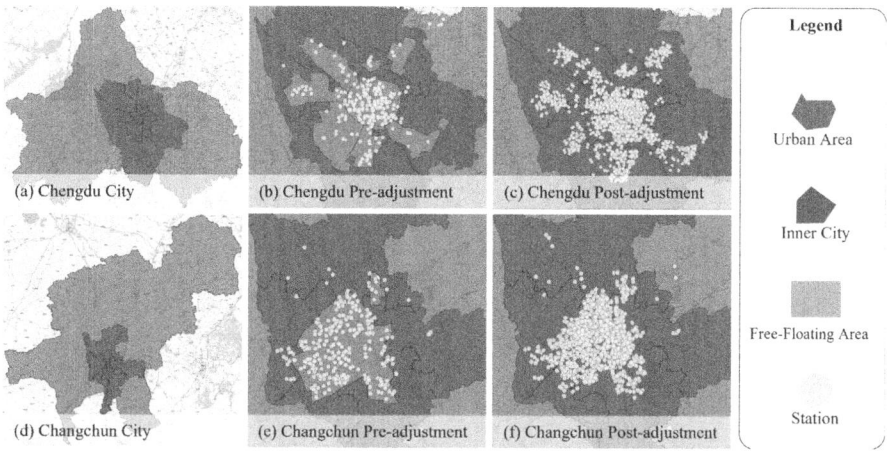

Fig. 1. Distribution of FFCS zones/SBCS stations before/after the operation scheme transition.

Figure 1 (a) and (d) represent the schematic maps of Chengdu and Changchun, respectively. We specifically emphasize the inner city as the primary research area of this work, where concentrates the population, socio-economic activities, as well as the carsharing services. Figure 1 (b) and (e) show the car sharing services supply before the scheme shift. Figure 1 (c) and (f) show the site distribution after the scheme shift. It is worth noting that during the selected period, the fleet size and usage fees for car sharing services in both cities have not significantly changed, thus, it can be inferred that the changes in service supply primarily stem from the shift of operating scheme (See Table 1 "Service Supply Information" section for detailed statistics).

3.2 Study Period and Usage Changes

Usage during the study period is shown in Fig. 2, in which trips are classified into four types: FFCS trips(FF to FF), SBCS trips(SB to SB), mixed trips(SB to FF, and FF to SB). For instance, a "SB to FF" trip indicates the user pick up a shared car at a station and drop off it at the free-floating area. Two special periods were included in the study period: the Spring Festival (when car sharing service suspended in Chengdu) and the impact of the pandemic (days when there were lockdowns or outbreak alerts are culled in both cities). The analysis of these two special periods is beyond the scope of this study.

Combined with the detailed statistics in Table 1 "Trip Information" section, we can see that: (1) Before the scheme shift, not all trips were "strictly FFCS" trips. In Chengdu (Changchun), the percentage of FFCS trips, SBCS trips and mixed trips are 44.6%(60.5%), 19.7%(7.1%) and 35.7%(32.4%), respectively. (2) The scheme shift led to an approximately 14% decline in of the total daily number of trips in both cities. (3) The daily number of trips basically hold a general fluctuation pattern with a weekly cycle, with weekends showing significantly higher trip volumes compared to weekdays. A slightly increased of the ratio of weekend to weekday trip volumes can be observed in both cities.

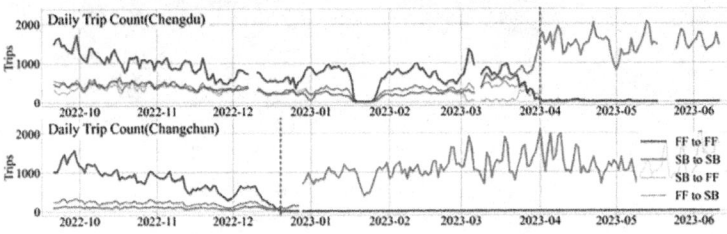

Fig. 2. Break-down of trips per day into type of starting and ending locations.

3.3 Usage Characterization

Relying solely on the number of trips is insufficient to accurately assess the service level of car sharing systems [13]. We conducted in-depth analyses of car sharing usage data from multiple perspectives. For ease of conducting comparative analysis, we present a summary of various statistical results in Table 1.

In the "Service Supply Information" section, we present the changes in car sharing service supply before and after the scheme shift, some of which have already been discussed earlier (see Fig. 1). Here we introduce an indicator: service coverage level, which measures the extent to which the new SBCS service covers the original FFCS service. To unify the calculation units, we create a buffer of 500 m around each station (the tolerance access distance of car sharing users [5, 12, 14]) to form a virtual operational area and perform spatial operations using QGIS. The service coverage level (area) is then determined by the proportion of overlap between the virtual area and the FFCS area. And the service coverage level (trip) is represented by the proportion of FFCS trips that fall within the virtual operational areas. Results show that, Chengdu, with just 55% area coverage, covers more than 90% of the original trips. While Changchun covers more than 99% trips with the area coverage of 84%.

In the "Trip Information" section, we conduct a tally of the total count for each type of trip, which have been discussed earlier (see Fig. 2). Here we introduce the spatial dispersion index to measure the extent of spatial dispersion in trip distributions, defined as the ratio of the variance to the mean of the distances between all pairs of trip pick-up points. Intuitively, in a SBCS system, the pick-up and drop-off points of trips tend to concentrate within the stations, thus maintaining a relatively low level of spatial dispersion. However, results show that after the shift from FFCS to SBCS, the spatial dispersion of car sharing trips in Chengdu remains relatively unchanged, while a slight increase of which in Changchun mainly because the relatively large number of stations. Based on the previous analysis, we hypothesize that, before the shift to SBCS, a considerable number of trips were already concentrated in certain hotspots, leading to the spontaneous aggregation of FFCS trips.

Table 1. Statistics summary of the dataset.

City		Chengdu		Changchun	
Service Supply Information					
Operating Scheme		FFCS	SBCS	FFCS	SBCS
Operating Area of FF (km^2)		904	0	382	0
Number of Stations		250	1464(+486%)	283	1195(+322%)
Average Fleet Size		1236	1133(−8%)	1324	1317(−1%)
Service coverage level (area)		55.24%		84.27%	
Service coverage level (trip)		91.24%		99.05%	
Trip Information					
Trips	FF to FF	147,159	–	69,708	–
	SB to SB	64,919	–	8,155	–
	FF to SB	64,846	–	18,330	–
	SB to FF	52,669	–	18,990	–
	In total	329,593	106,403	115,183	192,617
	In total(Daily)	1717	1478(-13.9%)	1280	1107(−13.5%)
Spatial Dispersion Index		0.525	0.523(−0.4%)	0.548	0.571(+4.2%)
Performance Index					
Daily Trips per Car		1.41	1.34(−5%)	1.03	0.88(−15%)
Median Trip Duration (min)	FF to FF	46.42	–	33.86	–
	SB to SB	45.67	–	36.75	–
	FF to SB	48.33	–	39.83	–
	SB to FF	47.72	–	37.67	–
	In total	47.27	60.22(+27%)	34.87	41.07(+18%)
Median Trip Distance (km)	FF to FF	9.40	–	4.33	–
	SB to SB	5.93	–	4.86	–
	FF to SB	12.76	–	5.88	–
	SB to FF	11.96	–	5.76	–
	In total	10.03	14.33(+43%)	4.64	5.73(+23%)

In the "Performance Index" section, we use daily trips per car to reflect the intensity of car sharing usage, which is calculated by the average number of times each car is used per day, often being used to measure the success of car sharing [12]. Besides, median trip duration and median trip distance can help us to characterize the intensity of individual trip, providing a direct reflection of the overall profit of the car sharing trips. Results show that following the scheme shift, both cities witness a decline in daily trips per car, with Chengdu experiencing a 5% decrease and Changchun facing a 15% decrease. Conversely, the median trip duration and median trip distance are both higher than all categories of trips before the transition (except for the median trip distance in Changchun, which is slightly lower than the mixed trips). The findings reveal that, although the scheme shift led to an overall decrease in system usage intensity, both trip duration and distance significantly increased, resulting in a notable rise in the profitability of individual trips.

3.4 Spatiotemporal Analysis of Demand

To unify the units of analysis for both FFCS and SBCS services, the study area was divided into square zones measuring 500 m × 500 m, for each zone, the total number of trips before and after the scheme shift is loaded, and then normalized based on the ratio of its trip volume to the highest one.

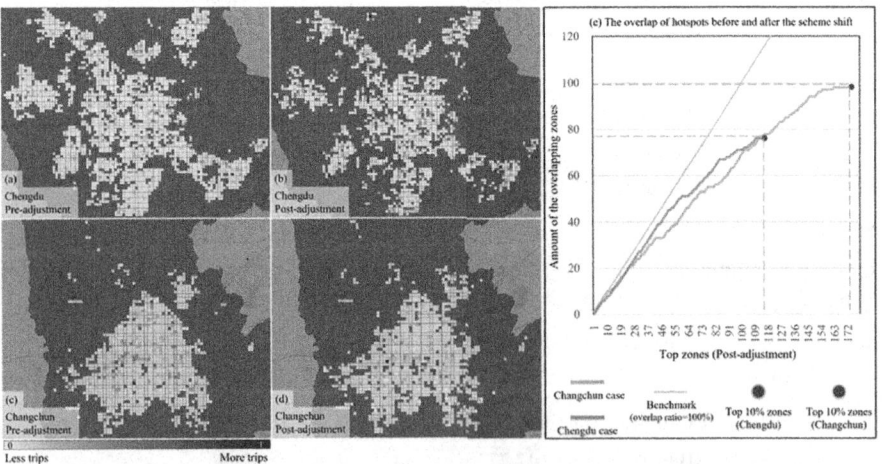

Fig. 3. Distribution of car sharing trips before and after the scheme shift.

Figure 3 (a-d) illustrate the distribution of car sharing trips in Chengdu and Changchun, in which the zones without trips have been excluded. Results show that: (1) the hotspots are dispersed across the operating areas, indicating significant spatial heterogeneity in the demand for car sharing service; (2) the shutdown of FFCS service has eliminated trips in some zones, but the original car sharing demand may be redirected to nearby stations; (3) combined with the previous analysis of service coverage, the original demand for FFCS is likely to have been absorbed by the new SBCS service.

We further analyze whether the hotspots have changed due to the scheme shift and to what extent. As shown in Fig. 3(e), we select the top 10% of zones based on trip counts before the scheme shift and figure up how many of them remain in the top 10% after the shift. The x-axis represents the top zones ranked by the number of trips, starting with the most popular zone on the left and ending with the top 10% of hotspots on the right. Zones that were also in the top 10% for trip count before the scheme shift are labeled as "overlapping zones", with their cumulative values displayed on the y-axis. The findings indicate that after the scheme shift in Chengdu, 56% of the top 10% hotspots (covering 39% of all trips) were previously among the top 10% hotspots. When considering the top 5% hotspots (covering 26% of all trips), this proportion rises to 67%. Furthermore, for the top 1% hotspots (comprising 8% of all trips), it increases to 81%. Similar patterns are observed in Changchun, with even higher proportions of the overlapping zones.

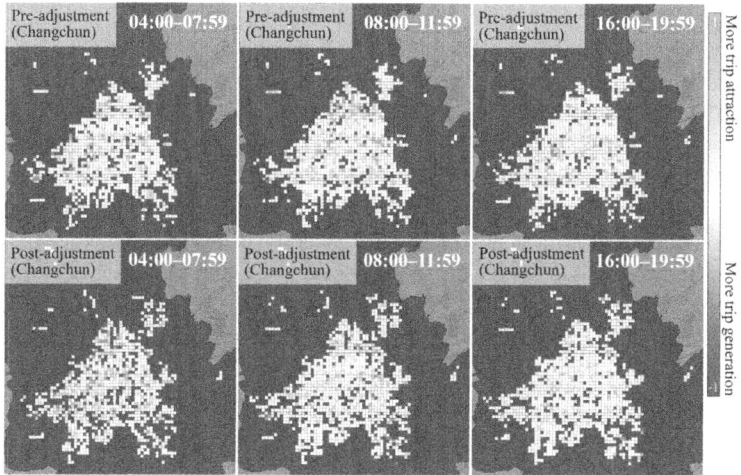

Fig. 4. Spatiotemporal distribution of net flow (selection of representative results).

To capture more detailed insights into the car sharing demand, we divide the day into six segments for analysis: 00:00–03:59, 04:00–07:59, 08:00–11:59, 12:00–15:59, 16:00–19:59, 20:00–23:59. Then, we assign the net flow to each grid for each segment, net flow is calculated as the number of arrival trips minus the number of departure trips, depicting the attractiveness or generativeness of car sharing trips.

Figure 4 depicts a selection of representative results, in which the heatmaps show the generative zones (with more departure trips) in red while attractive zones (with more arrival trips) in green. Results reveal that: (1) The overall distribution patterns of all time periods remain relatively consistent before and after the scheme shift, with the differences mainly found in localized regions due to changes in operating areas; (2) After the scheme shift, the aggregation of car sharing demand becomes more prominent across all time periods. Regions that originally had higher attractiveness/generativeness now concentrate in smaller areas, forming more distinct hotspots; (3) Many zones exhibiting generativeness in the morning shift to attractiveness zones in the afternoon and evening, which aligns with the spatiotemporal characteristics of commuting patterns. After the mode transition, this characteristic is preserved and even becomes more pronounced, suggesting that the higher certainty provided by SBCS (e.g., stable parking spots) is more attractive to commuters.

4 Discussion and Conclusion

In this work, we have collected the latest data of car sharing services in Chengdu and Changchun, China, both of which experienced a shift in the operation scheme of car sharing service. The nature of our dataset provides us a valuable opportunity to investigate some less explored yet meaningful academic issues. With a series of comparative analyses, usage characterization, and spatiotemporal analysis, we gradually outlined the characterization and evolution of FFCS services in these cities, providing detailed insights into this emerging mode of transportation.

In terms of characterization of FFCS in China. We gathered information on FFCS usage from existing literature for a total of 31 cities for comparison [5, 6, 12, 15]. In terms of supply, the largest supply scale among the compared cases is found in Berlin, with a FFCS area of 125 km^2 and 1009 vehicles. As a comparison, the FFCS area in Chengdu and Changchun is 904 km^2 and 382 km^2 respectively, while the fleet sizes are 1236 and 1324, respectively. In terms of usage characterization, median trip duration of Chengdu and Changchun is 46.42 min and 33.86 min respectively, while the median trip distance is 9.40 km and 4.33 km, respectively. As for these two indicators, their maximum values in the comparison sample are: 33.56 min in New York City, and 8.4km in Munich. The daily trips per car of Chengdu and Changchun are 1.41 and 1.03 respectively, which are relatively low. The minimum value of daily trips per car in the comparison sample is 1.6 in Stockholm. In summary, compared to European and American cities, Chinese cities' FFCS services have a larger supply scale (especially in FFCS areas), relatively longer trip durations and distances, as well as the lower daily trips per car, which may be attributed to low vehicle density. Such differences may be due to the travel behavior and trip purposes of FFCS users in different cities, as well as factors such as parking policies, public transport supply and vehicle types in different cities, which will be discussed in our future research.

In terms of the evolution of FFCS in China. We reveal that the shutdown of FFCS services did not cause a severe disruption to car sharing services. Instead, the introduction of SBCS effectively inherited and accommodated the travel demand. After the scheme shift, both cities witness a decrease in the intensity of car sharing usage, but meanwhile a significant increase in trip duration and distance. An interesting phenomenon is that the spatiotemporal distribution of car sharing trips did not show significant change. We speculate that the success of the new SBCS service in retaining car sharing demand is due to its high coverage of the previously served FFCS areas, particularly with respect to historical trip start and end points. In addition, a large proportion of FFCS users seem to be willing to tolerate the fixed pick-up and drop-off in station, even if it means longer trip durations, distances, and service fee. Given the heightened operational challenges (e.g., relocation and maintenance) and policy constraints (e.g., parking restrictions) faced by FFCS, a well-designed SBCS could emerge as a viable alternative.

Acknowledgments. This work was supported by the Science & Technology Project of Sichuan Province under Grant [No. 24NSFSC0356] and Program of China Scholarship Council [No. 202307000110].

Disclosure of Interests. The authors have no competing interests to declare that are relevant to the content of this article.

References

1. Nansubuga, B., Kowalkowski, C.: Carsharing: a systematic literature review and research agenda. J. Serv. Manag. **32**, 55–91 (2021). https://doi.org/10.1108/JOSM-10-2020-0344
2. Li, L., Zhang, Y.: An extended theory of planned behavior to explain the intention to use carsharing: a multi-group analysis of different sociodemographic characteristics. Transp. (Amst) **50**, 143–181 (2023). https://doi.org/10.1007/s11116-021-10240-1
3. BeiqiEV GoFun. https://www.shouqiev.com/. Accessed 29 Jul 2023
4. Ciociola, A., Giordano, D., Vassio, L., Mellia, M.: Data driven scalability and profitability analysis in free floating electric car sharing systems. Inf. Sci. (Ny) **621**, 545–561 (2023). https://doi.org/10.1016/J.INS.2022.11.116
5. Giordano, D., Vassio, L., Cagliero, L.: A multi-faceted characterization of free-floating car sharing service usage. Transp. Res. Part C. Emerg. Technol. **125**, 102966 (2021). https://doi.org/10.1016/J.TRC.2021.102966
6. Ampudia-Renuncio, M., Guirao, B., Molina-Sánchez, R., Engel de Álvarez, C.: Understanding the spatial distribution of free-floating carsharing in cities: analysis of the new Madrid experience through a web-based platform. Cities **98**, 102593 (2020). https://doi.org/10.1016/J.CITIES.2019.102593
7. Cocca, M., Giordano, D., Mellia, M., Vassio, L.: Free floating electric car sharing: a data driven approach for system design. IEEE Trans. Intell. Transp. Syst. **20**, 4691–4703 (2019). https://doi.org/10.1109/TITS.2019.2932809
8. Feng, X., Sun, H., Wu, J., Lv, Y.: Understanding the factors associated with one-way and round-trip carsharing usage based on a hybrid operation carsharing system: a case study in Beijing. Travel Behav. Soc. **30**, 74–91 (2023). https://doi.org/10.1016/j.tbs.2022.08.007
9. Huo, X., Wu, X., Li, M., et al.: The allocation problem of electric car-sharing system: a data-driven approach. Transp. Res. Part D Transp. Environ. **78**, 102192 (2020). https://doi.org/10.1016/J.TRD.2019.11.021
10. Becker, H., Ciari, F., Axhausen, K.W.: Comparing car-sharing schemes in Switzerland: user groups and usage patterns. Transp. Res. Part A Policy Pract. **97**, 17–29 (2017). https://doi.org/10.1016/j.tra.2017.01.004
11. Heilig, M., Mallig, N., Schröder, O., et al.: Implementation of free-floating and station-based carsharing in an agent-based travel demand model. Travel Behav. Soc. **12**, 151–158 (2018). https://doi.org/10.1016/j.tbs.2017.02.002
12. Boldrini, C., Bruno, R., Laarabi, M.H.: Weak signals in the mobility landscape: car sharing in ten European cities. EPJ. Data Sci. **8**, 100032 (2019). https://doi.org/10.1140/epjds/s13688-019-0186-8
13. González, A.B.R., Wilby, M.R., Díaz, J.J.V., et al.: Utilization rate of the fleet: a novel performance metric for a novel shared mobility. Transp. (Amst) **50**, 285–301 (2023). https://doi.org/10.1007/s11116-021-10244-x

14. Jin, F., An, K., Yao, E.: Mode choice analysis in urban transport with shared battery electric vehicles: a stated-preference case study in Beijing, China. Transp. Res. Part A Policy Pract. **133**, 95–108 (2020). https://doi.org/10.1016/j.tra.2020.01.009
15. Sprei, F., Habibi, S., Englund, C., et al.: Free-floating car-sharing electrification and mode displacement: travel time and usage patterns from 12 cities in Europe and the United States. Transp. Res. Part D Transp. Environ. **71**, 127–140 (2019). https://doi.org/10.1016/J.TRD.2018.12.018

Intelligent Transportation Infrastructure and Sustainable Transportation

Research on Coordinated Control of Multiple Energy Storage Systems Considering No-Load Voltage Differences

Shi Xiao[1], Jingwen Zheng[1], Luqing Jiang[2], Yajie Zhao[2], Bin Li[2(✉)], Zhihong Zhong[2], and Hu Sun[2]

[1] CRRC Qingdao Sifang Co., Ltd., QingDao, China
[2] Beijing Jiaotong University, Beijing, China
binxlee@163.com

Abstract. As the scale of urban railway transit is continuously enlarging, the issue of energy consumption has grown increasingly conspicuous. Installing hybrid energy storage systems in traction substations gives the energy recovery device the ability to possess both high power density and high energy density. Because the no-load voltage of the substation in the urban railway distributed power supply system is different and often fluctuates. If the energy management strategy of the energy storage system is improper, and the unfavorable situation of energy circulation between the substation and the energy storage and between the multiple energy storages will occur. In this paper, each part of the traction power supply system with stationary hybrid energy storage system is modeled first, and then three operating conditions based on different no-load voltages are analyzed. By increasing the charging threshold, some operating conditions can be avoided, but at the same time, the dynamic resistance is easier to start, so a control strategy needs to be proposed to solve this problem in the future.

Keywords: Urban railway system · Hybrid energy storage · no-load voltage · Energy circulation

1 Introduction

The total length of our country's urban railway transit operating lines reaches 9,206.8 km in 2021, of which the mileage of subway operating lines is 7,209.7 km, and the mileage of new operating lines is 1,237.1 km this year. Energy-saving technology is the key support for achieving the objective of carbon peaking and carbon neutrality. Therefore, while fulfilling the strategic development goals, urban railway transit should also fulfill the important task of carbon emission reduction. Its energy consumption and energy saving and emission reduction measures have also become a social issue main focus.

Trains are the "big consumers" of electricity in urban railway transit systems. The consumption of traction energy by trains constitutes about 50% of the total energy consumption of urban railway transit. Therefore, reducing traction energy consumption is of great significance to promoting the sustainable development of the urban railway transit industry and achieving the "dual carbon" goal. Urban railway transit has the characteristics of high train density and frequent start and stop, so the regenerative braking energy generated by trains is considerable [1].

At present, the recycling and application of regenerative braking energy mainly includes three ways: train operation map optimization [2], energy feedback [3] and energy storage. Compared with the onboard energy storage system, the stationary energy storage system does not increase the weight of the train, does not occupy the vehicle installation space, and has the benefits of flexible installation and low development cost [4]. With the rapid development of energy storage technology and the sharp drop in price, stationary energy storage systems have become one of the mainstream ways to solve the problem of renewable energy utilization [5]. Choosing to combine energy-oriented energy storage and power-oriented energy storage to form a hybrid energy storage system can fully exert the advantages of both, making it both high-power and high-energy, thereby improving its performance of recovering regenerative braking energy, peak clipping, voltage regulation and emergency traction.

Since it is not suitable to add main substations in densely populated areas in urban areas, the urban railway transit power supply system mostly adopts distributed power supply, and multiple power sources are directly introduced from the urban medium-voltage grid along the line to power each traction or step-down substation [6, 7]. Power supply shall ensure that each traction substation and step-down substation can obtain double-circuit power supply. In the decentralized power supply mode, since the transmission line of the traction substation connected to multi-channel power supply is longer than that of the substation that is not directly connected, the line voltage drop is smaller. The substation has a high no-load voltage value and a larger output power [8]. (Fig. 1).

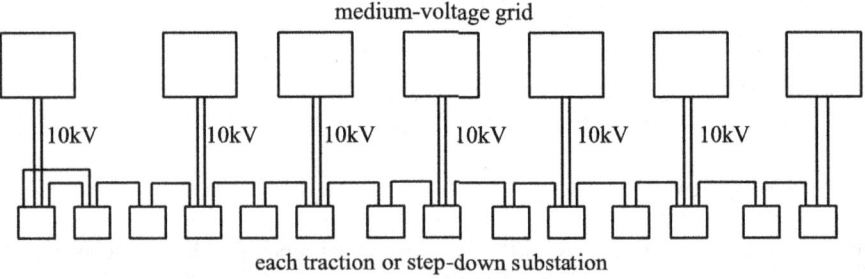

Fig. 1. Distributed power supply systems

Most of the energy management strategies proposed in the references only take a set of supercapacitor energy storage systems as the research object, and it is not necessary to take into account the impact of the no-load voltage difference resulting from the aforementioned decentralized power supply [9, 10]. However, the entire railway transit line is usually equipped with multiple sets of energy storage systems. At this time, the control parameters of the multi-energy storage system need to consider the comprehensive influence of the rectifier units of each station.

2 Traction Power Supply System with Hybrid Energy Storage System

As shown in Fig. 2, the structure of the traction power supply system including the stationary hybrid energy storage system. The hybrid energy storage system is installed in the substation. In this paper, a hybrid energy storage system made up of batteries and supercapacitors.

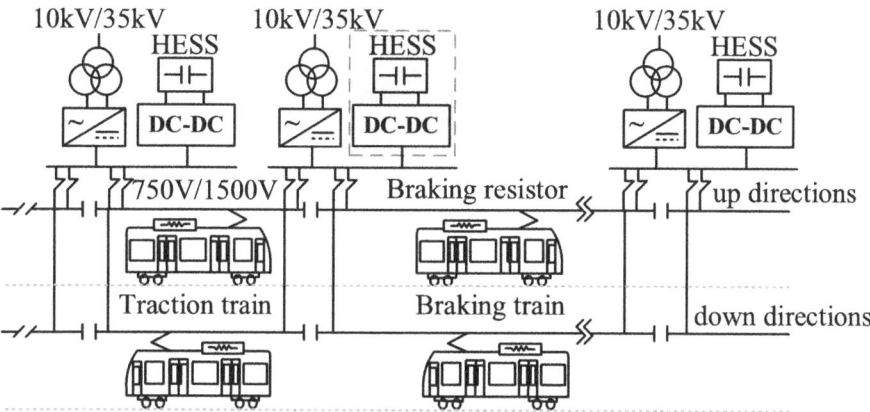

Fig. 2. Structure diagram of traction power supply system with stationary hybrid energy storage system

2.1 Substation

The equivalent model of the substation can be substituted by an ideal voltage source connected in series with an equivalent resistance [11], and its equivalent circuit diagram and voltage and current output characteristics are shown in Fig. 3. The output characteristics of the substation can be represented by formula (1):

$$u_{sub} = \begin{cases} u_{s0} - i_{sub} \cdot r_{s0}, & 0 < i_{sub} < i_g \\ u_{s1} - i_{sub} \cdot r_{s1}, & i_{sub} > i_g \end{cases} \qquad (1)$$

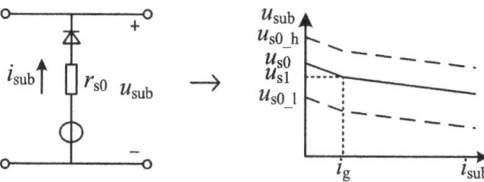

Fig. 3. Substation equivalent model and output characteristics

In the formula, u_{s0} is the no-load voltage of the substation, and u_{s1} is the equivalent voltage of the characteristic curve of the substation when $i_{sub} > i_g$; since the no-load voltage will fluctuate within the range of u_{s0_l} and u_{s0_h}, the characteristic curve of the substation will also follow the no-load voltage and translates up and down. r_{s0} and r_{s1} are the substation's equivalent resistance, i_g is the critical current value, u_{sub} and i_{sub} are the output voltage and the output current of the substation.

2.2 Train

When the train is braking, some of the regenerative braking energy produced by the train powers the on-board auxiliary system, while the remaining regenerative braking energy is returned to the traction network to be absorbed by adjacent traction trains and the energy storage system. At this time, the traction network voltage rises. Given the current-limiting feature of train regenerative braking, when the grid voltage goes beyond the allowable range, the on-board braking resistor begins to restrain the rise of traction grid voltage. The train model is illustrated in Fig. 4.

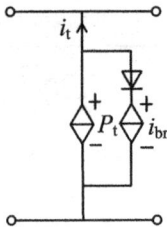

Fig. 4. Train equivalent model

2.3 Hybrid Energy Storage System

In this paper, an active topology in which the battery and the supercapacitor are independently controlled, that is, the battery and the supercapacitor are connected in parallel to the DC bus of the substation through bidirectional DC/DC converters, respectively. The battery module and the supercapacitor module respectively carry out the bidirectional flow of energy through the bidirectional DC/DC converter and the traction network to realize the recovery and utilization of the regenerative braking energy (Fig. 5).

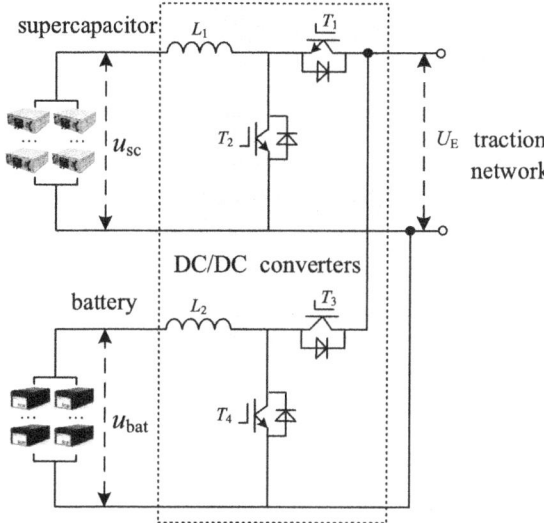

Fig. 5. Active topology of hybrid energy storage system

The control framework of the stationary hybrid energy storage system can be refined into four parts: the voltage threshold adjustment strategy, the voltage outer loop controller, the power distribution strategy and the current inner loop controller. The structure diagram is shown in Fig. 6. The voltage threshold adjustment strategy, the voltage outer loop controller and the power distribution strategy belong to the upper energy management system, and the current inner loop controller belongs to the converter control system.

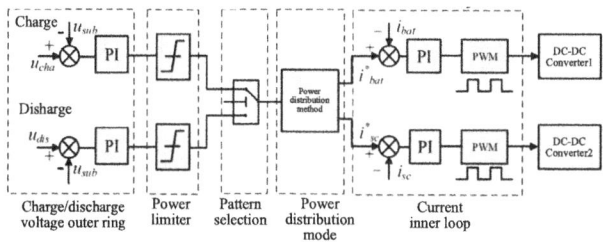

Fig. 6. Control structure diagram of hybrid energy storage system

According to the above-mentioned control structure of the hybrid energy storage system, the charge and discharge current of the battery and the current of the supercapacitor can be regulated. The hybrid energy storage system can be regarded as a controlled current source in the topology of the traction power supply system. Based on the energy conservation principle, the equivalent current source current of the traction grid-side energy storage system is

$$i_{hess} = \frac{u_b(t)i_b(t) + u_{sc}(t)i_{sc}(t)}{u_{sub}(t)} \quad (2)$$

As shown in Fig. 7, the equivalent circuit models of the low-voltage side and the traction grid side of the hybrid energy storage system.

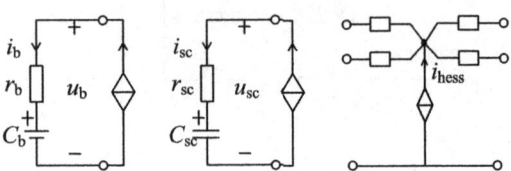

Fig. 7. Equivalent circuit model of energy storage system

The equivalent circuit topology of the traction power supply system is shown in Fig. 9. In the figure, the DC traction power supply system is a complex node network topology. Since the train runs in real time on the uplink and downlink lines, the topology has time-varying characteristics and parallel multi-conductors. Characteristic. According to the circuit topology in Fig. 8, the DC power flow of the traction power supply system can be solved.

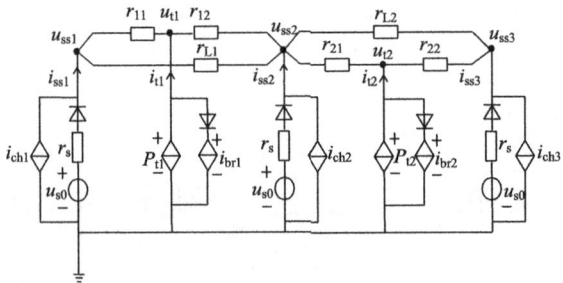

Fig. 8. Equivalent circuit topology of traction power supply system

3 The Effect of Different No-Load Voltage on Configuration

This paper studies the line conditions of the certain line of Beijing Subway, which has a total of 13 stations. Among them, there are nine traction substations.

Input the train power curve between each station into the power flow analysis model of the traction power supply system of this line, and the input train curve is shown in the following Fig. 9.

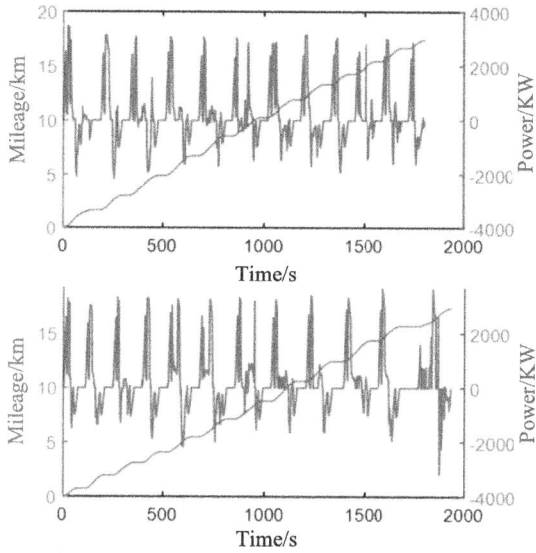

Fig. 9. Up and down train input curve

From Sihuidong to Tuqiao Station, energy storage devices are installed at each station. The hybrid energy storage configuration of each station is as shown in Table 1 below.

Table 1. Hybrid energy storage configuration

Substation	Supercapacitor (MW)	Battery (MW)
Sihui	0	0
Sihuidong	1.5	0
Gaobeidian	0	0
Communication University of China	1.5	0
Shuangqiao	0.8	0.2
Guanzhuang	1.5	0
Baliqiao	0.8	0.2
Tongzhoubeiyuan	1.5	0
Guoyuan	1.5	0
Liyuan	0.8	0.2
Tuqiao	1.5	0

According to the current configuration of the subway line, analyze the four groups of simulations without energy storage, with energy storage, the same no-load voltage, and different no-load voltages at the 300s departure interval, calculate the change of the energy-saving rate, and analyze the fluctuation of the no-load voltage due to this configuration.

When the no-load voltage settings of the substations on the whole line are identical, the output energy of the substations from Sihui Station to Baliqiao Station is relatively average, while the output energy of substations from Guoyuan Station to Tuqiao Station is higher. There is an ordinary substation between them, and the train is equivalent to starting and stopping twice, so the output energy of these substations is high.

After the energy storage is installed, when there are different no-load voltage settings across the line, stations with higher no-load voltage, such as Gaobeidian Station and Guoyuan Station, whether the no-load voltage settings on the entire line are identical or not, the output energy of substations will differ greatly. Among them, when the no-load voltages of the whole line are different, the Gaobeidian substation outputs 45.55 kWh more electricity compared with when the no-load voltages are the same (Fig. 10).

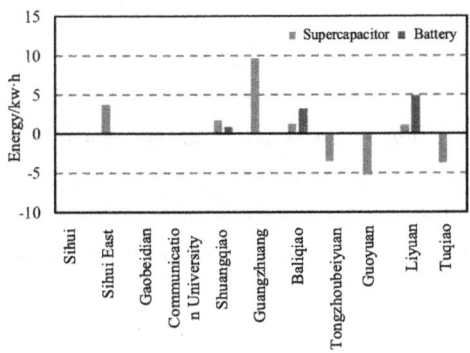

Fig. 10. Difference of energy storage charging energy

After considering the variation in no-load voltage, some substations whose no-load voltages are lower than those of adjacent stations, such as Sihui East Station, Guanzhuang Station, and Liyuan Station, have more charging energy than when the no-load voltage is the same; at the same time, the no-load voltage higher than adjacent stations, such as Tongzhou Beiyuan, Guoyuan, and Tuqiao, have less charging energy than when the no-load voltage is the same. Since the control strategy adopted by the energy storage system is a fixed threshold control approach based on no-load voltage, it is greatly affected by the no-load voltage of adjacent substations.

The Table 2 makes a comparison of the energy consumption of each part of the system and the change of the energy saving rate when the no-load voltage is the same or different. Compared with the same no-load voltage setting, the system's energy consumption and the loss of the braking resistor have increased, and the line loss after installing the energy storage has also increased. The energy saving rate dropped from 16.22% to 15.11%. For the same configuration scheme, different no-load voltage settings will also have different evaluations of the configuration results. Therefore, the actual no-load voltage difference should be considered during the configuration process.

Table 2. Energy consumption of each part of the system

Energy/kWh	Without ESS		With ESS	
	same no-load voltage	different no-load voltage	same no-load voltage	different no-load voltage
System energy consumption	317.46	320.78	265.95	272.30
Braking resistance loss	50.24	53.84	3.61	8.69
Line loss	28.62	28.12	24.01	25.10
Energy saving rate	–	–	16.22%	15.11%

4 The Effect of Different No-Load Voltage on the Control Strategy

Three working conditions were analyzed: the substation charging the accumulators, the accumulators charging and discharging each other, and the no-load voltage of the substation approaching the starting voltage of the braking resistor. On the basis of the first two working conditions, the method of increasing the charging threshold is used to avoid this working condition as much as possible occurs, but at the same time, it will also make the braking resistor easier to start. In the following, a control strategy is proposed to solve the above problems.

4.1 Case1: Substation Charging Energy Storage

In case 1, three stations of Sihui, Sihuidong and Gaobeidian are selected. This energy storage system is installed in Sihuidong, and there is a braking train in the interval (Fig. 11).

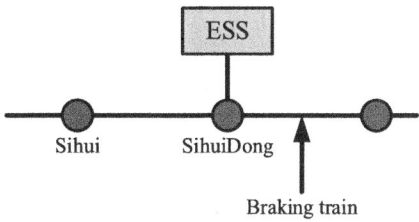

Fig. 11. Case 1

The parameter settings are as follows (Table 3 and Fig. 12).

The Table 4 summarizes the energy of each part of the system under this working condition. Since the no-load voltage of Sihui and Sihui East Stations is relatively low, the two substations hardly output energy. Ideally, when there is only train braking in the line, the substation does not output energy, moreover, the regenerative braking energy produced by the train is absorbed by the energy storage to the greatest extent possible. In the actual simulation, the Gaobeidian station outputs nearly 1 kWh of energy. Among the

Table 3. The parameter settings of case 1

	no-load voltage/V	Charge threshold/V	Disharge threshold/V
Sihui	828.81	–	–
Sihuidong	825.45	830	820
Gaobeidian	863.5	–	–

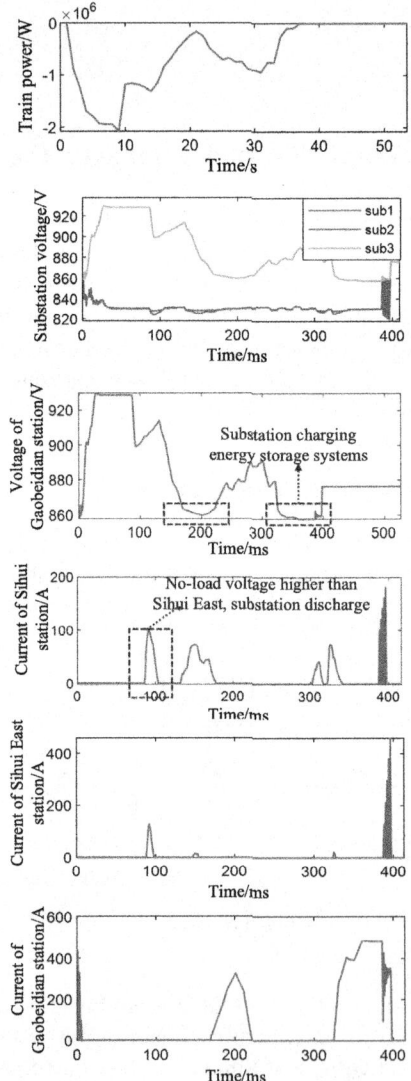

Fig. 12. Input waveform and simulation results

8.988 kWh of the energy produced by the train's braking, 0.473 kWh of energy is dissipated in the braking resistor, and the remaining energy is charged into the energy storage; and part of the energy in the energy storage comes from the Gaobeidian substation. This is because the no-load voltage of Gaobeidian is higher than the charge threshold related to the Sihuidong energy storage system. When the grid voltage is between the two, the substation charges the energy storage. Therefore, when the no-load voltage of the substation is very different, the improper selection of the threshold value of the energy storage system will cause the substation to charge the energy storage.

Table 4. Energy consumption of each part of the system

Energy	Value/V
Sihui output energy	0.1066
Sihuidong output energy	0.0518
Gaobeidian output energy	0.9316
Energy storage charge	8.905
Energy storage discharge	0
Train input braking energy	8.988
Braking resistor loss	0.473
Train regenerative braking energy recovered from energy storage	8.515
Energy from substation charging to storage	0.39

In view of the above situation, the threshold for charging of the Sihuidong energy storage system is boosted to be above the no-load voltage of Gaobeidian, and 868.5 V is selected as the charging threshold. At this time, the energy statistics of each part are as follows.

Compared with the charging threshold before the charging threshold is not increased, the charging threshold is increased until it is exceeds the no-load voltage of Gaobeidian, and the substation almost does not charge the energy storage. However, increasing the charging threshold makes some regenerative braking energy not absorbed in time. As the braking power of the train increases again, tthere is a high possibility that the grid voltage will be higher than the starting voltage of the braking resistor, leading to the loss of part of the regenerative braking energy on the braking resistor. Therefore, although increasing the charging threshold can avoid the unfavorable situation of charge the substation's energy storage to a certain extent, the excessively high voltage threshold also prevents the regenerative braking energy from being absorbed in time and is consumed on the braking resistor (Table 5).

Table 5. Energy statistics after raising the threshold

Energy	Value/V
Substation output total energy	0.011
Energy storage charge	6.986
Energy storage discharge	0
Train input braking energy	8.988
Braking resistor loss	1.563
Train regenerative braking energy recovered from energy storage	6.975

4.2 Case2: Mutual Charging Between Energy Storage

In case 2, three stations of Sihui, Sihuidong and Gaobeidian are selected. The energy storage system are installed in Sihuidong and Gaobeidian, and there is a traction train in the interval (Fig. 13).

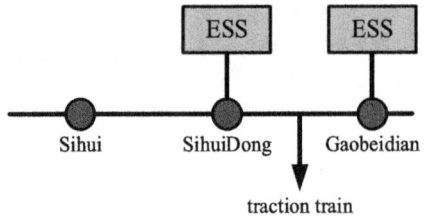

Fig. 13. Case 2

The parameter settings are as follows (Table 6 and Fig. 14).

Table 6. The parameter settings of case 2

	no-load voltage/V	Charge threshold/V	Disharge threshold/V
Sihui	828.81	–	–
Sihuidong	825.45	830	820
Gaobeidian	863.5	868.5	858.5

The Table 7 summarizes the energy of each part of the system under this working condition. Since the no-load voltage of Sihui and Sihui East Stations is relatively low, the two substations hardly output energy. Ideally, when there is only train traction in the line, the energy storage system or substation supplies power to the traction train at the same time, and the energy storage will not be charged. In the actual simulation, Sihuidong energy storage is charged with 4.625 kWh of energy. The actual energy flow path is

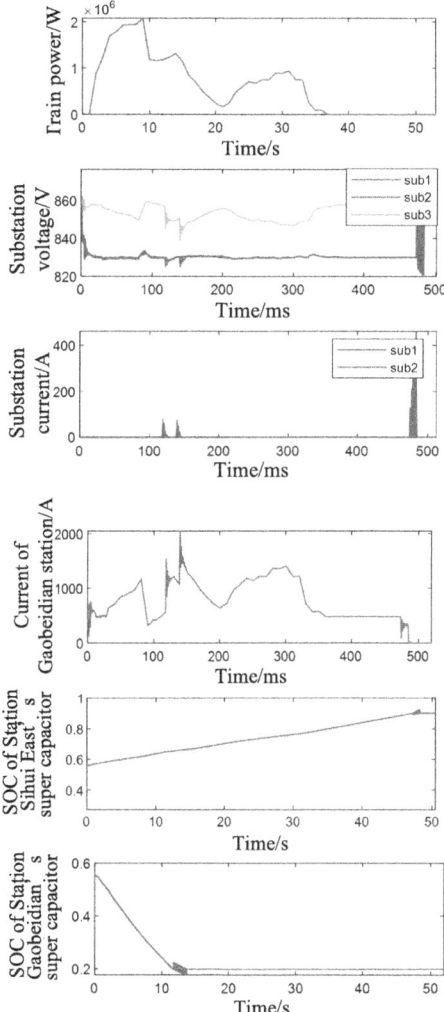

Fig. 14. Input waveform and simulation results

the Gaobeidian substation and energy storage to the traction train and the Sihuidong energy storage system, which increases the line loss. The energy flow path that is most conducive to energy saving is the flow of the Sihuidong substation and energy storage system to the traction train.

Threshold value for initiating charge of the Sihuidong energy storage system is increased to be higher than the no-load voltage of Gaobeidian, and 868.5 V is selected as the charging threshold. The energy of 8.988 kWh required for train traction is provided by the Gaobeidian substation, and there is no relationship between the energy storage and the substation circulation. Therefore, in the second work condition, by increasing

Table 7. Energy consumption of each part of the system

Energy	Value/V
Sihui output energy	0.0241
Sihuidong output energy	0.0355
Gaobeidian output energy	9.1984
Sihuidong energy storage charging energy	4.625
Gaobeidian energy storage and discharge energy	4.485
Train input traction energy	8.988
Braking resistor loss	0

the charging threshold of the energy storage system, the disadvantage of mutual charging among energy storage systems can be considerably avoided.

4.3 Case3: The No-Load Voltage of the Substation is Close to the Starting Voltage of the Braking Resistor

In case 3, three stations of Sihui, Sihuidong and Gaobeidian are selected. The energy storage system are installed in Sihuidong and Gaobeidian, and there is a braking train in the interval (Fig. 15).

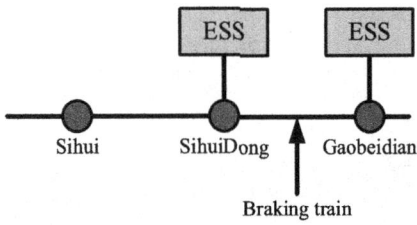

Fig. 15. Case 3

The parameter settings are as follows (Table 8).

Table 8. The parameter settings of case 3

	no-load voltage/V	Charge threshold/V	Disharge threshold/V
Sihui	858.81	–	–
Sihuidong	855.45	898.5	850.45
Gaobeidian	893.5	898.5	888.5

The input waveform of train is as same as case 1. It is shown by the result that the energy storage system has a charging threshold set to 898.5 V, and the train braking resistor has a starting voltage set to 900 V, being very close. When the energy storage can absorb the train's braking power adequately, the braking resistor won't start. In the time intervals of 3 s–4 s, 11 s–15 s, and 21 s–22 s, the train's braking power increases again, the network voltage goes up, and the braking resistor starts (Fig. 16).

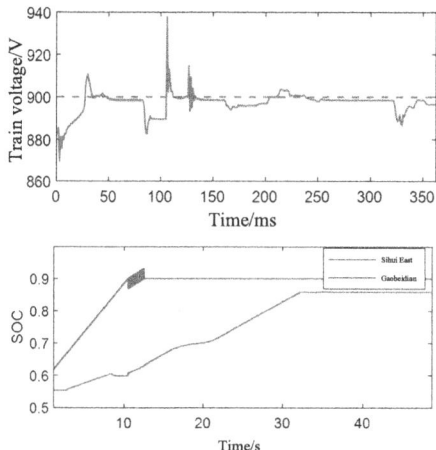

Fig. 16. Simulation results

The Table 9 counts the energy of each part when the no-load voltage of the substation is close to the starting voltage of the braking resistor, and the loss of the braking resistor is 0.676kWh; but when the no-load voltage of the substation takes the value before it is raised, and other conditions remain unchanged, the braking Resistors are basically not activated. Therefore, when the no-load voltage fluctuates to a higher value, the braking resistor is easier to start. The fixed threshold charging and discharging strategy based on the no-load voltage is no longer suitable for the situation where the no-load voltage fluctuates to a higher value, but there is no-load voltage in practice. It fluctuates to the starting voltage of the braking resistor.

Table 9. Energy consumption of each part of the system

Energy	Value/V
Substation output total energy (Gaobeidian)	0.535
Energy storage charge	8.669
Energy storage discharge	0
Train input braking energy	8.988
Braking resistor loss	0.676

5 Summary

This paper mainly analyzes the issue regarding charge and discharge control strategy problem of multi-energy storage system under the condition that the no-load voltage of the substation is different and fluctuates with time in the urban railway distributed power supply system. Firstly, the model of each part of the traction power supply system with stationary hybrid energy storage system is modeled, and then three kinds of work based on the energy circulation between substations and energy storage with different no-load voltages and between multiple energy storages are analyzed. Some working conditions can be avoided by increasing the charging threshold, but at the same time, the braking resistor is easier to start. Therefore, a control strategy needs to be proposed to balance the problem of energy circulation and braking resistor starting.

In view of the above-mentioned problem of energy circulation between multiple energy storage systems, the follow-up plan is to divide the entire line according to the power station and carry out zone management. Charge and discharge distribution.

References

1. WeiI, R., Du, P., Yang, Y., et al.: Analysis on utilization of regenerative braking energy for metro trains and research on timetable optimization method. J. China Railw. Soc. **42**(8), 1–9 (2020)
2. Liu, P.: Research on energy-efficient train timetable optimization based on regenerative braking energy in urban rail transit. Beijing Jiaotong University (2020)
3. Zhang, G., Tian, Z., Tricoli, P., et al.: Inverter operating characteristics optimization for DC traction power supply systems. IEEE Trans. Veh. Technol. **68**(4), 3400–3410 (2019)
4. Ramsey, D., Letrouve, T., Bouscayrol, A., et al.: Comparison of energy recovery solutions on a suburban DC railway system. IEEE Trans. Transp. Electrif. **1**(C), 2332–7782 (2020)
5. Hayashiya, H., Iino, Y., Takahashi, H., et al.: Review of regenerative energy utilization in traction power supply system in Japan: applications of energy storage systems in d.c. traction power supply system. In: IECON 2017 - 43rd Annual Conference of the IEEE Industrial Electronics Society, pp. 3918–3923. IEEE, Beijing (2017)
6. Lin, F., Li, X., Zhao, Y., et al.: Control strategies with dynamic threshold adjustment for supercapacitor energy storage system considering the train and substation characteristics in urban rail transit. Energies **9**(4), 257 (2016)
7. Zhao, Y., Xia, H., Wang, J., et al.: Control strategy of ultracapacitor storage system in urban mass transit system based on dynamic voltage threshold. Trans. China Electrotech. Soc. **30**(14), 427–433 (2015)
8. Zhu, F., Yang, Z., Lin, F., et al.: Dynamic threshold adjustment strategy of supercapacitor energy storage system based on no-load voltage identification in urban rail transit. In: 2019 IEEE Transportation Electrification Conference and Expo, Asia-Pacific, Seogwipo, Korea, pp. 1–6 (2019)
9. Iannuzzi, D., Lauria, D., Tricoli, P.: Optimal design of stationary supercapacitors storage devices for light electrical transportation systems. Optim. Eng. **13**, 689–704 (2012)
10. Ciccarelli, F., Iannuzzi, D., Lauria, D., et al.: Optimal control of stationary lithium-ion capacitor based storage device for light electrical transportation network. IEEE Trans. Transp. Electrif. **3**(3), 618–631 (2017)
11. Wang, J., Yang, Z., Fei. L., et al.: Thresholds modification strategy of wayside supercapacitor storage considering DC substation characteristics. In: IECON 2015 - 41st Annual Conference of the IEEE Industrial Electronics Society. Yokohama, Japan, pp. 002076–002081 (2016)

Intelligent Transportation Systems: Enabling Sustainable Transportation and Efficient Traffic Management—A Review

Roberto D. Rosario[1(✉)], Arjel Alvarez[1], Ronnel C. Quinto[1], and Mark de Guzman[2]

[1] Polytechnic University of the Philippines, Sta. Mesa, Manila, Philippines
robertodrosario@iskolarngbayan.pup.edu.ph
[2] Department of Civil Engineering, Saint Louis University, 2600 Baguio City, Philippines

Abstract. Intelligent Transportation System (ITS) is unquestionably a crucial component of transportation to mitigate the detrimental impacts of traffic congestion. However, despite their tremendous possibilities, ITS cannot be fully beneficial unless it can do so in a sustainable, scalable way and generate support from the public. This literature review evaluated recent studies about the existing principles of traffic management, the development of ITS, and the new generation of ITS for sustainable transportation. This study figured that most of the research is looking towards automated devices that would reduce the effort of drivers and automatically run a self-drive vehicle. It was found that the reality behind vehicular communications requires further and more combined efforts and entails significant public sector investments and engagement from private enterprises. The study determined that ITS cost implications have been neglected in most recent ITS publications. It would generally be more attractive to the public sector to engage in the ITS framework if prompt execution could generate income that could benefit developing countries. Advocating to use the system necessitates solving ITS issues to guarantee that systems are safe, efficient, cost-effective, and user-friendly. It also entails collaboration among governments, businesses, and academic institutions to foster innovation, develop suitable regulations, and implement scalable solutions that might enhance the transportation system.

Keywords: Intelligent Transportation System (ITS) · Sustainable Transportation · Traffic Management · Traffic Congestion · Self-drive Vehicles

1 Introduction

Historically, people everywhere have sought solutions to manage traffic better and reduce congestion [1]. Traffic congestion has increased dramatically in contemporary urban living, and several difficulties related to transportation demand must be addressed, with significant environmental effects such as air and noise pollution, a shortage of space, and traffic congestion [2, 3]. The US traffic congestion cost is expected to be more than 3.7 billion hours of delays each year, or 160 billion US dollars. It corresponds to around

2% of the European Union's gross domestic product (GDP), and it amounts to US$ 80 billion in Brazil [4, 5]. In 2024, the Japan International Cooperation Agency (JICA) estimated the Philippines' loss due to traffic at about 62.4 million US Dollars daily. Given that the traffic issue has not yet been resolved, the losses are expected to grow significantly [6]. Addressing traffic congestion and decreasing logistic productivity has led to the rising utilization of Intelligent Transportation System (ITS) technologies [7]. ITSs are designed to provide safer and more efficient transportation services by increasing productivity, improving safety, reducing travel time, reducing costs, and saving energy. ITS has received a lot of attention in recent years as a solution for the improvement of road safety, rising emissions, and congestion concerns [8–10]. This review reiterates the fundamentals of traffic management, various transportation concerns, and the objectives of transport management to prevent the adverse effects of congestion. Also, the functions of ITS for effective management were presented, with a focus on current research and examples of new-generation vehicle communication and control system technology. Moreover, the development of ITS, a new generation of technologies that will enable the vehicle to infrastructure (V2I), and the introduction of the future's self-driving cars are all covered. Finally, the significance of sustainable mobility is presented in the last chapter, offering a creative and cost-effective approach to improving people's lives while protecting the environment.

2 Methodology

The review sorts of related studies from Scopus Indexes and other academic journals from 2016 to the present are shown in Fig. 1. Using keywords "Traffic congestion" AND "Intelligent Transport System" AND "Traffic Management" AND Limit-to "English," (106) documents were conducted. The researcher performed data analysis and saved and filtered pertinent studies. Cited (37) articles from the Scopus Index and gathered another (49) articles from other journals related to the topic. However, only 24 pertinent journals were utilized upon excluding the duplicated journals. The retrospective study concentrates on publications addressing how this cutting-edge technology reduces traffic jams and accidents, how effective traffic management plans are, and the newly developed ITS and its functions. The methodology's objective is to discover the research gaps within a given outcome so we can develop solutions and figure out how to enhance ITS for sustainable transportation.

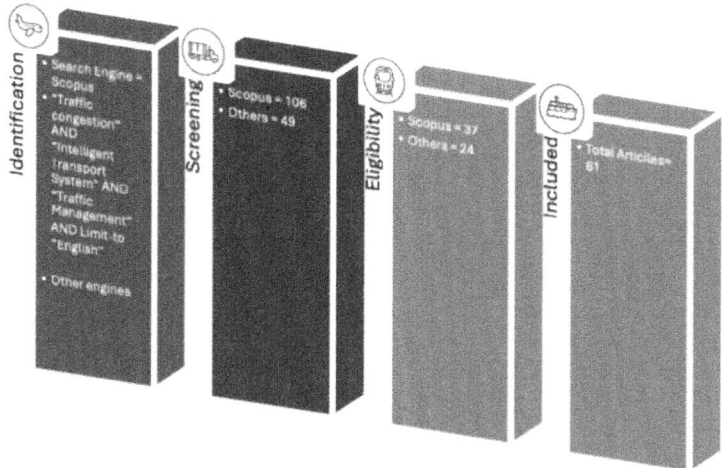

Fig. 1. Research Methodology

3 Principles of Traffic Management

Congestion has a destabilizing effect on quality of life [11, 12]; it costs billions of dollars a year in wasted time and fuel [3, 13, 14]. Traffic management aims to utilize the present road network to its fullest potential while minimizing the detrimental impacts of traffic congestion. Through a variety of programs meant to optimize both the advantages and disadvantages of transportation, it is becoming increasingly important. It impacts the demand for travel and transportation and the supply of transportation systems [15, 16]. The different transportation issues that arise when the transportation system cannot suitably accommodate transport mobility are shown below (see Fig. 2a).

Traffic management is a set of application and management tools that maintain traffic efficiency and capacity [17]. Addressing transportation issues [15] sets a hierarchical goal model with four fundamental higher goals—mobility demands, traffic safety, environmental preservation, and economic efficiency. The goal idea should be followed while managing traffic to improve the road transport system's security, safety, and dependability. These measures impact the effectiveness of the road network (see Fig. 2b), which employs ITS services and projects in daily operations. Numerous novel ideas have been deployed within the framework of cutting-edge operational systems. (see Fig. 2b) shows the elements of traffic management measures to achieve the hierarchical goals [18].

To improve mobility speeds while lowering fuel consumption and cost parameters, introduce desirable safety measures, increase economic yield and other goals, the Intelligent Transportation System (ITS)'s broad scope deals with the effective management of traffic through the fusion of cutting-edge technological frameworks. ITS components work closely together to provide the best results for problems.

Fig. 2. (a) Transportation Challenges; (b) Traffic Management Measures

4 Development of ITS

The use of information technology to provide creative transportation solutions is known as an intelligent transportation system (ITS) [19–21]. It is a comprehensive system that attempts to reduce collisions, traffic jams, emergency management, automated traffic enforcement, and inefficient use of the transportation system [22–25]. Transit navigation, traffic sign control systems, speed cameras, electronic toll collection systems like RFID [26], automatic number plate recognition (ANPR), and security cameras like closed-circuit television (CCTV) systems, as well as more sophisticated applications that integrate real-time data like parking information systems, accident reduction and management systems, and weather information, are all examples of ITS [22, 27]. ITS has developed into a vital tool for controlling and observing moving vehicles on roads and in towns. Additionally, it offers cutting-edge multimodal transport technologies for creating solutions that improve quality of life while conserving time, energy, and resources and preserving the environment [25, 28].

The Comprehensive Automobile Traffic Control System (CACS) project, the first effort at developing ITS applications, was discovered in Japan in the 1970s. The United States made its first systematic move in the late 1980s. The Intelligent Vehicle Highway Systems Act of 1991, passed by the US Department of Transportation, gave rise to the name "ITS" in the early 1990s [10]. The development of intelligent transportation systems (ITS) in the United States has mostly been driven by the need to increase traffic safety and reduce congestion [25, 29, 30]. In Europe, the DRIVE program (Dedicated Road Infrastructure and Vehicle Environment) was established in 1988 after the first formalized program for transportation telematics PROMETHEUS (PROgraM for European Traffic with Highest Efficiency and Unprecedented Safety) was launched in 1986 by European automotive companies (DRIVE, 1990) [16].

Further, China made ITS systems the first area it included in its national planning in 2021. Core issues in ITS research and innovation, such as vision-based innovations, traffic, control, simulation models, information exchange and location-based services, driving safety and assistance, and so forth, were addressed on a high scientific and technological level [16]. ITS America and VERTIS (Vehicle, Road, and Traffic Intelligence Society) work to raise awareness worldwide by performing additional research to improve and revolutionize the transportation system [27]. Despite enormous contributions, most solutions for ITS-based traffic management that have been created are only partly embraced by drivers in actual installations nowadays. Due to the lack of position and speed information for vehicles not linked to the network, centralized control systems cannot provide the benefits they were designed to [29].

5 Functions of Modern ITS for Efficient Traffic Management

Transportation management is an essential area of interest due to the world's population's increasing number of vehicles on the road, and ITS (see Fig. 3) has significantly contributed to a cleaner, safer, and more widespread benefit in the field of enhanced mobility [1, 7, 23]. To support all these intelligent ITS applications, the runtime environment must provide reliable communication between system elements, the procedural chain that generates real-time traffic data illustration shown below (see Fig. 4) [8, 23]. Transportation experts are currently becoming interested in several ITS functions. Indeed, most of those concentrated on enhancing vehicle communications and control systems, such as those mentioned above.

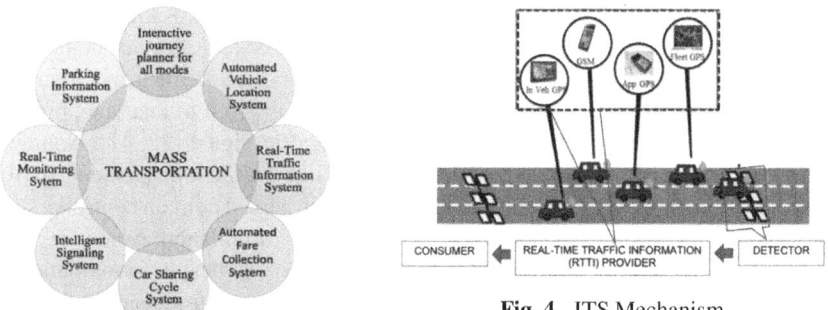

Fig. 3. Existing ITS

Fig. 4. ITS Mechanism

5.1 ITS Installed-on Vehicles

The developed world makes considerable use of the Advanced Traveler Information System (ATIS), which is seen as a crucial part of managing and controlling traffic on roads [31, 32]. ATIS broadcasts travel-related information to commuters through information and communication technology (ICT), assisting them in planning their travels, providing instructions, estimating travel times, and avoiding congestion. Moreover, ATIS may offer information on situations like the exact location of a traffic collision, road construction, traffic conditions, and weather factors that might create traffic congestion [30, 31]. Another example of ITS is floating car data, which uses GPS sensors in moving cars to gather information on the pace of traffic along road segments. Vehicle-mounted GPS units are highly precise in determining geographic locations and speeds. The Automatic Vehicle Location (AVL) has also been used in this approach [33, 34]. A vehicle tracking system receives the locations of cars from AVL on a constant schedule [31, 34]. One of the critical components of ITS is the Vehicular Ad-hoc Network (VANET), which enables communication between vehicles (V2V) and between vehicles and roadside units (RSU). Vehicles may be required automatically to slow their speed in school zones and other accident-prone regions if all of them are VANET-based and have an On-Board Unit (OBU) [22, 35, 36]. The Lane Departure Warning (LDW) is another ITS that often uses a camera positioned on the dashboard or next to the rearview mirror to monitor the area in front of the car while it tracks road markers. The driver receives a warning signal if a vehicle accidentally approaches a lane marker. The most common warning methods are sound, haptics (such as seat or steering wheel vibration), lighting, and display systems [37, 38]. In addition, the GPS (Global Positioning Satellite) is crucial for keeping an eye on, forecasting, and warning drivers of hazardous conditions. As a result, route searches can be performed, and information can be provided based on probe car data about past travel-time patterns. GPS is widely used as sensors in providing instantaneous location information and speed measurements, allowing real-time understanding of changes in traffic conditions [34, 39, 40].

5.2 ITS Used as Control Systems

ITS makes it possible to have more advanced traffic control systems than traditional programmed traffic signals. Intelligent traffic light control systems are one control device that may create the traffic light signal according to the traffic volume in a particular lane. If the traffic is more on a particular lane, then priority is given to that lane, prolonging its green light signal, the system called Adaptive Traffic Signal [22, 41]. Another ITS method is the (Internet of Things) IoT-based Real-Time Traffic Management (IoT-based intelligent traffic strategy). It uses consolidated and dispersed domain controllers to mitigate heavy congestion, allowing for precise planning of collision avoidance amongst linked vehicles. Some IoT components include detectors, internet platforms, data centers, and machine learning (ML) methods [39, 42]. Researchers introduced the Vehicle-to-vehicle (V2V) and Vehicle-to-Everything (V2X), a wireless curbside device placed at an intersection that communicates with a vehicle to provide data to it, ought to make it possible for various vehicles to cooperate automatically on the road [20, 43]. With V2V communication, a vehicle may maintain state convergence with other vehicles cooperatively, allowing for the timely gathering of kinetic data from the preceding [36, 44]. The term "Vehicle-to-Everything"(V2X) refers to how vehicles may communicate with one another to provide useful traffic information about their movements [45]. Examples of these technologies include GPS and IoT [10]. One often used ITS is the license plate recognition system (LPR), a network of cameras that collects license plate numbers, which are then matched to a database of customers who have already paid in advance. A fine is sent to vehicle owners who haven't paid the congestion fee [3, 46]. Similar camera-enforced pricing policies are also applied in Singapore to avoid traffic [37]. The Advanced Traffic Management Systems (ATMS), which is keen to increase road network capacity and traffic flow, is one of the original ITS services. This is made possible by genuine traffic fusion, which only uses 10% of all V2X-enabled traffic as its total population [29].

6 ITS for Sustainable Transport System

When it comes to their functionality, ITS applications may typically be divided into Regarding their functionality, ITS applications may typically be divided into three primary categories: applications for comfort, efficiency, and safety [23, 36]. Real-time, relevant, and trustworthy data are required for these applications to operate effectively. Three key factors contributing to the usage of a more sustainable mode of transportation may be identified: technology, modal shift, and avoidance (see Fig. 5) [2]. Shortening travel distances or eliminating actual travel are two ways to avoid transportation; physical transportation is being replaced by "virtual" transportation due to the Internet of Things (IoT) or using technology [2]. For example, when you choose to deliver your groceries to your home, freight traffic increases, and passenger traffic decreases.

Several nations are adopting the ITS to support sustainable transportation. The Electronic Road Pricing (ERP) concept was implemented in Singapore in 1998 to reduce traffic congestion [47]. ERP generated an economic value of $40 million by reducing traffic congestion by 25,000 cars during peak hours and increasing average speed by 20 kph [48]. Singapore's data-rich intelligent transportation system has reduced traffic there to one of the lowest levels in the world by being able to provide the public with real-time traffic alerts(theaseanpost.com). One of the world's most prominent Automatic Vehicle Location (AVL) systems, the iBus system is now operational in London.

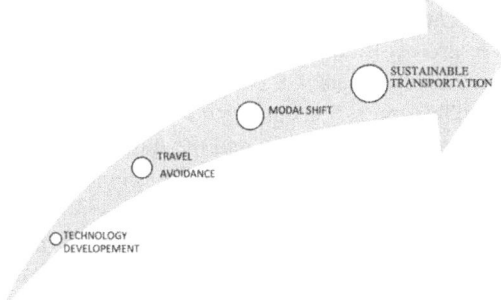

Fig. 5. Components of Sustainable Transport [2]

Each bus' position is sent via iBus to the AVL center every 30 s; the service controllers utilize this information to manage fleets, provide priority at traffic signals and junctions, and the Real Time Passenger Information (RTPI) system [48, 49]. Additionally, the city is improving its infrastructure by setting the standard for 5G operation via The Smart Mobility Living Lab in London (SMLL). By using 5G for road management systems, the economy may save $940 million annually and 370,000 metric tons of CO_2 while saving drivers 10% of their time stuck in traffic [50]. The "Auto-Intellect" system, which offers intelligent traffic management and license plate recognition, was being implemented in Russian Federation territory. In Belgorod, a planned ITS called Singapore's Area Licensing Scheme (ALS) was utilized to prevent cars from entering the restricted area. With this technique, the city decreased CO2 emissions by 25%, fuel consumption by 18%, and traffic congestion by 22% [48]. On the territory of the Russian Federation, the "Auto-Intellect" system—which offers intelligent traffic management and license plate recognition—was in use [51]. Belgorod employed Singapore's Area Licensing Scheme (ALS), a planned version of ITS, to prevent cars from entering the restricted area. With this technique, the city decreased CO_2 emissions by 25%, fuel use by 18%, and traffic congestion by 22% (*seoulsolutions.kr*). ITS provides a viable alternative for reducing traffic congestion and accelerating transportation efficiency. However, numerous critical challenges must be addressed to promote wider adoption and ensure these systems function successfully.

7 ITS Challenges

Most studies have so far focused on the limited techniques, and ensuring compliance with data privacy regulations [2, 15, 30, 33, 55, 56] and finally, the high cost of deployments [4, 44, 60]. The high cost of deploying and maintaining smart infrastructure, such as advanced sensors, communication networks, and control systems, can be a barrier, especially for developing regions. Studies [2, 15, 30, 33, 55, 56] have been addressing the issue of sustainability in the broad transport sector; however, actual data were insufficient or an actual experiment to demonstrate the significant of the research studies. Proposed systems [22, 23, 28, 39, 57–59] have associated fixed parameters and data collections were gathered from heterogeneous sources. ITS components must work seamlessly across different regions, manufacturers, and platforms. Lack of standardization in communication protocols, data formats, and technologies can hinder scalability and efficiency. Future research must advocate developing universal standards and ensure compatibility between ITS components and systems [28, 59]. Several research studies have made significant contributions to ITS and the introduction of automated vehicles [19, 22, 24, 48, 50]. However, studies are valuable for those living in highly urbanized areas, countries using higher 30Mbps faster internet connection and accommodating the cost of calibration investment against internet coverage [50]. Additionally, the reality of vehicle communications demands additional and more coordinated activities. There is virtually little study on the ITS's financial viability [10], and the commercial and broader socioeconomic challenges have not yet been sufficiently examined. Investigating cost-effective capacity of technology [1, 8, 9, 20, 31, 52, 53] some will need to teach the V2I network and convert vehicles to IoT [31], others prerequisite the advance application of intelligent traffic management and Big Data [1] as well as the use of Road Side Unit (RSU) structure to assist the communications between vehicles [5]. Several research challenges on the system implementation [4, 29, 44, 54, 60] because of stochastic phenomena; challenges to predict the drivers' behavior [29], human drivers and pedestrians remain an unpredictable variable. Public trust in automated and intelligent systems is also often low, especially with concerns about job displacement or safety. Moreover, effective communication between vehicles, infrastructure, pedestrians, and other road users V2I architecture (Vehicle-to-Vehicle, Vehicle-to-Infrastructure, etc.) is vital to ITS success [60]. Ensuring low-latency, high-reliability communication is challenging, particularly in urban or densely populated areas. A further ITS challenge is the complexity of the systems in designing secure communication protocols and data encryption solutions, including the use of existing infrastructure, low-cost sensors, and public-private partnerships for funding must take into consideration. Moreover, studies have been concentrating on improving the technology, but no research has been done to assess the downside or adverse effects of ITS machinery operations and disposal. Advocating for the adoption of ITS requires addressing these research challenges (see Table 1) to ensure that systems are secure, efficient, cost-effective, and user-friendly. It also involves collaboration between governments, industry, and academic institutions to drive innovation, create appropriate policies, and implement scalable solutions that might improve the transportation system.

Table 1. ITS Research Challenges

ITS CHALLENGES	REFERENCES
Constraint in technology	[1, 8, 9, 20, 31, 52, 53]
Uncertainty in adaptation level	[4, 44, 54]
Complexity of data	[2, 15, 30, 33, 55, 56]
Inadequate parameters	[22, 28, 39, 57–59]
Limited to highly developed countries	[1, 8, 9, 20, 31, 52, 53]
Cost implications and economic value	[10]

8 Conclusions and Recommendations

The comprehensive literature studies found two primary areas of development: ITS installed in cars and ITS in traffic control activities. ITS is a broad topic, and the classifications of essential technologies are complex. Numerous prototypes are evaluated along with the hundreds of existing systems and equipment such as VANET, V2V, V2I, ITS Architecture, Adaptive Traffic Light, Automatic Vehicle Location (AVL), etc. A growing body of research is looking towards automated devices that would reduce the effort of drivers and automatically run a self-drive vehicle. However, the reality behind vehicular communications requires further combined efforts and entails significant public sector investments and engagement from private enterprises. ITS is unquestionably a crucial component of transportation. However, despite their tremendous possibilities, ITS cannot be entirely beneficial unless it can do so in a sustainable, scalable way and generate support from the public. To encourage the adaptability of ITS systems, comprehensive regulations and legislation governing ITS deployment, legal certainty, and ethical issues in autonomous decision-making must be prioritized. Most ITS publications have so far neglected cost implications; it would generally be more attractive to engage in ITS framework if prompt execution could be income generating. Further research on ITS must concentrate on intelligent traffic enforcement since drivers' behavior is often underestimated yet is one of the significant reasons for transportation problems. The focus of ITS is usually on developed nations. However, a substantial population in developing and underdeveloped nations is susceptible and afflicted by this transportation problem. Further research must focus on the significant parameters of developing countries and their interoperability to promote scalability and efficiency. Finally, future research must focus on promoting the development of universal standards and ensuring compatibility between different ITS components and systems integration to propose an economical new generation of ITS for sustainable transportation.

References

1. Karouani, Y., Elhoussaine, Z.: Toward an intelligent traffic management based on big data for smart city. In: Lecture Notes in Networks and Systems, vol. 37, p. 502–514. Springer (2018)
2. Dura, H., Weil, M.: An approach towards sustainable passenger mobility in urban areas: a life cycle perspective. In: WIT Transactions on the Built Environment, pp. 333–344. WITPress (2014)
3. Rosario, R.D., Padilla, J.A., Bonto, N.G., De Mesa, R.C., Dela Cruz, O.G.: Value engineering on car curbing ownership in metro Manila. In: Transport, Ecology - Sustainable Development: EKOVarna 2022, p. 020017 AIP Publishing (2023)
4. Abraham, A., et al.: Proceedings of the 2015 15th International Conference on Intelligent Systems Design and Applications (ISDA), Marrakesh, Morocco (2015)
5. Institute of Electrical and Electronics Engineers. In: IEEE Symposium on Computers and Communications (ISCC) (2017)
6. Jica: Traffic congestion now costs P3.5 billion a day | Inquirer News. https://newsinfo.inquirer.net/970553/jica-traffic-congestion-now-costs-p3-5-billion-a-day-metro-manila-traffic-jica-cost-of-traffic, Accessed 25 Sept 2024
7. Alam, T., Gupta, R., Nasurudeen Ahamed, N., Ullah, A., Almaghthwi, A.: Smart mobility adoption in sustainable smart cities to establish a growing ecosystem: Challenges and opportunities. MRS Energy Sustain. 1–13 (2024),
8. Khan, T.A., Siddiqui, F.: Role of ACO driven AGV in intelligent transport systems. In: Lecture Notes on Data Engineering and Communications Technologies, vol. 26, pp. 227–232. Springer (2019)
9. Giannoutakis, K.N., Li, F.: Making a business case for intelligent transport systems: a holistic business model framework. Transp. Rev. **32**(6), 781–804 (2012)
10. Rak, J.: Politechnika Gdańska, Polish Association of Telecommunication Engineers, Institute of Electrical and Electronics Engineers, and IEEE Poland Section, Proceedings of 2017 15th International Conference on ITS Telecommunications (ITST): Warsaw, Poland (2017)
11. Mattson, J., et al.: Transportation, community quality of life, and life satisfaction in metro and non-metro areas of the United States. Wellbeing Space Soc. **2**, 100056 (2021)
12. Ghazali, W.N.W.B.W., Atikah B. Zulkifli, C.N., Ponrahono, Z.: The effect of traffic congestion on quality of community life, pp. 759–766 (2019)
13. Fan, Z., Harper, C.D.: Congestion and environmental impacts of short car trip replacement with micromobility modes. Transp. Res. D Transp. Environ. **103**, 103173 (2022)
14. Musa, A.A., Malami, S.I., Alanazi, F., Ounaies, W., Alshammari, M., Haruna, S.I.: Sustainable traffic management for smart cities using internet-of-things-oriented intelligent transportation systems (ITS): challenges and recommendations. Sustainability **15**(13), 9859 (2023)

15. Boltze, M., Tuan, V.A.: Approaches to achieve sustainability in traffic management. In: Procedia Engineering, pp. 205–212, Elsevier Ltd. (2016)
16. Yan, X., Zhang, H., Wu, C.: Research and development of intelligent transportation systems. In: Proceedings - 11th International Symposium on Distributed Computing and Applications to Business, Engineering and Science, DCABES 2012, pp. 321–327 (2012)
17. de Souza, A.M., Brennand, C.A.R.L., Yokoyama, R.S., Donato, E.A., Madeira, E.R.M., Villas, L.A.: Traffic management systems: a classification, review, challenges, and future perspectives. Int. J. Distrib. Sens. Netw. **13**(4) (2017)
18. Advanced International Conference on Telecommunications 12. 2016 Valencia et al., AICT 2016 the Twelfth Advanced International Conference on Telecommunications (2016)
19. Hassanin, I.S.: The role of internet of things on intelligent transport system: a traffic optimization model. In: ICAMS Proceedings of the International Conference on Advanced Materials and Systems, pp. 297–302. Inst. Nat. Cercetare-Dezvoltare Text, Pielarie (2020)
20. Saad Talib, M., Hassan, A.: Converging VANET with vehicular cloud networks to reduce the traffic congestions: a review (2017). http://www.ripublication.com
21. Lv, Z., Shang, W.: Impacts of intelligent transportation systems on energy conservation and emission reduction of transport systems: a comprehensive review. Green Technol. Sustain. **1**(1), 100002 (2023)
22. Shirabur, S., Hunagund, S., Murgd, S.: VANET based embedded traffic control system. In: Proceedings - 5th IEEE International Conference on Recent Trends in Electronics, Information and Communication Technology, RTEICT 2020, pp. 189–192. Institute of Electrical and Electronics Engineers Inc. (2020)
23. IEEE Computer Society. Technical Committee on Parallel Processing, IEEE Computer Society. Technical Committee on Computer Communications, and Institute of Electrical and Electronics Engineers (2016)
24. Shelmakov, S.V., Lobikov, A.V., Grigoreva, T.Y.: Assessment of traffic management measures impact on the population health living near the highway. In: 2021 Systems of Signals Generating and Processing in the Field of on Board Communications, Conference Proceedings. Institute of Electrical and Electronics Engineers Inc. (2021)
25. Waqar, A., Alshehri, A.H., Alanazi, F., Alotaibi, S., Almujibah, H.R.: Evaluation of challenges to the adoption of intelligent transportation system for urban smart mobility. Res. Transp. Bus. Manag. **51**, 101060 (2023)
26. Ravish, R., Swamy, S.R.: Intelligent traffic management: a review of challenges, solutions, and future perspectives. Transp. Telecommun. **22**(2), 163–182 (2021)
27. Mishra, A., Priya, A.: A comprehensive study on intelligent transportation systems. Smart Moves J. Ijosci. **4**(10), 10 (2018)
28. Dalida, J.P.D., Galiza, A.J.N., Godoy, A.G.O., Nakaegawa, M.Q., Vallester, J.L.M., Dela Cruz, A.R.: Development of intelligent transportation system for Philippine license plate recognition. In: IEEE Region 10 Annual International Conference, Proceedings/TENCON, pp. 3762–3766. Institute of Electrical and Electronics Engineers Inc. (2017)
29. Institute of Electrical and Electronics Engineers. In: 2017 13th IEEE International Conference on Control & Automation (ICCA) (2017)
30. Harun, M.: Intelligent transport system. Int. J. Civ. Eng. Technol. **8**(4), 2230–2237 (2017)
31. Ackaah, W.: Exploring the use of advanced traffic information system to manage traffic congestion in developing countries. Sci. Afr. **4**, 79 (2019)
32. Reinolsmann, N., et al.: Dynamic travel information strategies in advance traveler information systems and their effect on route choices along highways. Procedia Comput. Sci. **170**, 289–296 (2020)
33. Bertini, R.L.: Some of the authors of this publication are also working on these related projects: research on mobile systems applied to traffic management and safety, smart vehicles and smart

roads. View project Transportation System Simulation Manual View project. https://www.res earchgate.net/publication/228953732
34. Leurent, F., Sun, D., Xie, X.: On heterogenous sampling rates in origin-destination matrix estimation based on trajectory data and link counts. Transp. Res. Rec. **2677**(1), 1169–1180 (2023)
35. Jeyasheeli, P.G., Deepika, J., Sathya, R.R.: Security for software defined vehicular networks. Smart Innov. Syst. Technol. **370**, 529–545 (2023)
36. Alsaleh, A., Author, C.: How do V2V and V2I messages affect the performance of driving smart vehicles? Comput. Syst. Sci. Eng. **47**(2), 2313–2336 (2023)
37. Bernal, E.A., Wu, W., Loce, R.P., Bernal, E.A., Bala, R.: Computer vision in roadway transportation systems: a survey Computer vision in roadway transportation systems: a survey computer vision in roadway transportation systems: a survey. J. Electron. Imaging (2013)
38. Schnebelen, D., Reynaud, E., Ouimet, M.C., Seguin, P., Navarro, J.: A neuroergonomics approach to driver's cooperation with lane departure warning systems. Behav. Brain Res. **456**, 114699 (2024)
39. Kanamori, R., Takahashi, J., Ito, T.: A study of route assignment strategy based on anticipatory stigmergy. Electron. Commun. Jpn. **99**(3), 1645–1651 (2016)
40. Sivalingam, P., et al.: A review of travel behavioural pattern using GPS dataset: a systematic literature review. Meas. Sens. **32**, 101031 (2024)
41. Palash, M.A., Wijesekera, D.: Benchmarking the performance of 5G CV2X for connected vehicles based adaptive traffic signal, pp. 1853–1858 (2024)
42. Atassi, R., Sharma, A.: Intelligent traffic management using iot and machine learning. J. Intell. Syst. Internet Things **8**(2), 8–19 (2023)
43. Suganthi, K., Kumar, M.A., Harish, N., HariKrishnan, S., Rajesh, G., Reka, S.: Advanced driver assistance system based on IoT V2V and V2I for vision enabled lane changing with futuristic drivability. Sensors **23**(7), 3423 (2023)
44. Jia, D., Ngoduy, D.: Enhanced cooperative car-following traffic model with the combination of V2V and V2I communication. Transp. Res. Part B: Methodol. **90**, 172–191 (2016)
45. Adnan Yusuf, S., Khan, A., Souissi, R.: Vehicle-to-everything (V2X) in the autonomous vehicles domain – a technical review of communication, sensor, and AI technologies for road user safety. Transp. Res. Interdiscip. Perspect. **23**, 100980 (2024)
46. Rao, Z., Yang, D., Chen, N., Liu, J.: License plate recognition system in unconstrained scenes via a new image correction scheme and improved CRNN. Expert Syst. Appl. **243**, 122878 (2024)
47. Li, Z.: Implications of Singapore's congestion pricing policy for China's transportation governance (2024)
48. Hounsell, N.B., Shrestha, B.P., D'souza, C.: Using automatic vehicle location (AVL) data for evaluation of bus priority at traffic signals (2012)
49. Toymentseva, I.A., Chichkina, V.D., Guseva, N.V., Vasetskaya, E.S.: Digitization of transport platforms: an ecosystem approach, pp. 128–135 (2024)
50. Oughton, E.J., Frias, Z.: The cost, coverage and rollout implications of 5G infrastructure in Britain. Telecomm. Policy **42**(8), 636–652 (2018)
51. Kambur, A., Kushchenko, L., Novikov, I.: Improving traffic management through the use of intelligent transport systems. In: MATEC Web of Conferences, vol. 341, p. 00044 (2021)
52. Dimitrov, S.: Optimal control of traffic lights in urban area. In: 2020 International Conference Automatics and Informatics, ICAI 2020 – Proceedings. Institute of Electrical and Electronics Engineers Inc. (2020)
53. SCAD College of Engineering and Technology and Institute of Electrical and Electronics Engineers, Proceedings of the International Conference on Trends in Electronics and Informatics (ICOEI 2019), pp. 23–25 (2019)

54. Telecommunications Society, Univerzitet u Beogradu. School of Electrical Engineering, IEEE Communications Society. Serbia & Montenegro Chapter, and Institute of Electrical and Electronics Engineers. TELFOR 2016: 24th Telecommunications Forum: Belgrade, 22 and 23 November 2016, the SAVA Center = TELFOR (2016)
55. Maaroufi, M.M., Stour, L., Agoumi, A.: A new approach of smart mobility for heavy goods vehicles in Casablanca. In: Intelligent Environments 2021: Workshop Proceedings of the 17th International Conference on Intelligent Environments, vol. 29, pp. 98–107. IOS Press (2021)
56. Hodges, N., Bell, M., Galatioto, F., Hill, G.: Sustainable network management-the integration of intelligent transport systems and 'grid enabled' pervasive sensors cooperative intelligent transport systems (C-ITS) view project customer-led network revolution view project" (2010)
57. Maaroufi, M.M., Stour, L., Agoumi, A.: Striving for smart mobility in Morocco: a case of lanes designated to heavy goods vehicles in Casablanca. Eng. Manage. Prod. Serv. **13**(1), 74–88 (2021)
58. Kolosz, B., Grant-Muller, S.: Comparing smart scheme effects for congested highways. Transp. Res. Part. C Emerg. Technol. **60**, 313–323 (2015)
59. Kim, J., Kurauchi, F., Uno, N., Hagihara, T., Daito, T.: Using electronic toll collection data to understand traffic demand. J. Intell. Transp. Syst.: Technol. Plann. Oper. **18**(2), 190–203 (2014)
60. IEEE Staff, 2016 IEEE Symposium on Computers and Communication (ISCC). IEEE (2016)

A Green Intelligent Transport Model for Urban Mobility

Gerald B. Imbugwa[1(✉)], Tom Gilb[2], and Manuel Mazzara[1]

[1] Innopolis University, 1 Universitetskaya Str., 420500 Innopolis, Russia
g.imbugwa@innopolis.university
[2] Oslo, Norway

Abstract. Rapid population growth, urbanization and increasing mobility demands continue to pose serious challenges for urban transportation systems. Traffic congestion, high costs of transportation, and increased CO2 emission levels contribute to climate change impacts, not only that but also lower user satisfaction.

This paper introduces a mathematical model (TCES). The model leverages Pareto optimization and linear programming to simultaneously minimize two major conflicting objectives: travel time, and cost emissions (Pollution) while maximizing user satisfaction. The model takes into account the factors which influence the transportation sector, such as passenger demand, real-time traffic data and environmental considerations. This makes the model a scalable and resilient solution to modern transportation, that can be applied in carpooling, ridesharing and city planning.

Keywords: Urban transportation · Optimization model · Carpooling · Micro-mobility · Pareto optimization · Traffic congestion · Sustainable mobility · Green · Intelligent · Multi-objective optimization · User satisfaction · Environmental impact · Transportation costs · Urban mobility solutions · Ride-sharing · Public transportation integration

1 Introduction

1.1 Problem Statement

Rapid population growth, urbanization, and rising mobility demands continue to pose serious challenges for urban transport systems. The aforementioned issues have resulted in traffic congestion, high transportation costs, and high levels of CO2 emissions, all of which have affected end-user satisfaction. When we look at current solutions, for example, a taxi app or another transportation infrastructure, we only focus on a single objective. This goal can either be to reduce costs and save time or the other way around. Hence, this infrastructure fails to solve the interconnected issues [2,7].

T. Gilb—Independent Researcher.

Existing approaches often lack an integrated perspective that considers the dynamic interplay between various transportation modes, including carpooling and micromobility. Additionally, these solutions frequently do not adapt to real-time traffic conditions and passenger demands, leading to inefficiencies and suboptimal performance. The absence of a holistic model that optimises multiple objectives-travel time, costs, emissions, and user satisfaction-underscores the necessity for innovative urban transportation planning methods.

1.2 Objective

This paper introduces TCES, a model aimed at the efficiency and sustainability of urban transportation. The model uses carpooling as a case study to show how the model can uniquely integrate carpooling and micromobility. This multimodal goal is to optimize travel time and cost while reducing CO_2 emissions and maximizing user satisfaction. The TCES model is achieved by balancing of the aforementioned variables using Pareto optimisation with linear programming, to be discussed in the model section.

Our proposed model addresses the limitations of the existing model by distinctly incorporating dynamic passenger demands and real-time traffic data while paying attention to environmental pollution. When the factors are incorporated it makes the model dynamic and applicable to optimising carpooling, ridesharing, public transport, or integration with micromobility, to be discussed in the application section.

TCES contributes to the existing literature by providing a robust framework that integrates different data points and real-time optimization.

2 Literature

Carpooling is an effective strategy to alleviate urban congestion, reduce emissions, and lower transportation costs [2,23,24]. This literature review examines 30 papers published between 2014 and 2024, focusing on optimizing carpooling costs and time, especially for last-mile solutions. It highlights key innovations, methodologies, and findings while outlining future research directions [27].

2.1 Algorithmic Integration and Simulation Frameworks

Tamannaei and Irandoost (2019) introduced a branch-and-bound algorithm that improves ride-matching efficiency and user satisfaction by effectively managing multiple carpooling requests [31]. Kumar and Khani (2020) developed a transit-based ridesharing algorithm that enhances cost savings and ride efficiency, though scalability and real-world application pose challenges [17]. Huang et al. (2022) used shared automated vehicle fleets for last-mile deliveries, significantly reducing costs and delivery times, but faced difficulties adapting to various urban environments [16]. Fangxin et al. (2019) created the Car4Pac system for efficient last-mile parcel delivery, showing substantial cost reductions, although scalability remains a concern [32]. Lele and Shah (2023) optimized transportation networks in Washington DC using MST and PERT methods, achieving efficiency gains but facing limitations in regional focus and reliance on existing literature [18]. Li et al. (2020) proposed a real-time peer-to-peer ride-matching algorithm that improves matching efficiency and cost-effectiveness [20]. Wang et al. (2022) introduced advanced optimization techniques for shared mobility, highlighting significant efficiency improvements despite challenges in diverse urban settings [33]. Azimi et al. (2021) explored shared transportation modes, noting benefits like reduced congestion and emissions but faced issues with model application across different urban contexts [4].

2.2 Innovative Approaches to Last-Mile Delivery

Gdowska et al. (2018) applied a stochastic approach to crowd shipping, enhancing the reliability and flexibility of last-mile logistics despite the complexity of coordinating many couriers [14]. Bian et al. (2020) proposed the Mobility-Preference-Based Mechanism (MPMBPC) for first-mile ridesharing, significantly improving service reliability and user satisfaction, although the mechanism's complexity posed implementation challenges [5]. Zhao et al. (2019) optimized vehicle routing for last-mile delivery, achieving cost and time savings but requiring further model refinement [9]. Adnan et al. (2019) studied last-mile bicycle-sharing, highlighting weather's impact on usage and the importance of strategic bike station placement, but faced geographic focus limitations [1]. Mitropoulos et al. (2021) examined factors influencing shared mobility, finding that delivery time, cost, and environmental concerns significantly affect user participation, although potential biases in self-reported data were noted [21]. Mourad et al. (2019) reviewed advancements in shared mobility, identifying research gaps and emphasizing the need for empirical validation of proposed algorithms [22].

2.3 Comparative Analysis of Different Models

Schaller (2021) analyzed ride-hailing services in New York City, noting both their benefits for mobility and drawbacks such as increased congestion and pollution, with limitations including data biases and a single-city focus [25]. Zhang et al. (2015) studied the Feeder system for last-mile delivery, demonstrating cost efficiency improvements but facing scalability issues [35]. Kumar and Khani (2020) enhanced transit efficiency through ridesharing integration, but noted challenges in real-time implementation [17]. Huang et al. (2021) integrated shared automated vehicle fleets with public transit, showing service reliability and cost reductions, yet requiring empirical validation [16]. Wang et al. (2019) developed the Car4Pac system for last-mile parcel delivery, achieving significant efficiency gains but needing further validation in diverse environments [32]. Chen et al. (2020) used machine learning for real-time optimization of dynamic carpooling, improving operational accuracy but requiring empirical validation [8]. Djavadian and Chow (2017) used an agent-based model for day-to-day adjustments in shared mobility systems, showing efficiency improvements but highlighting scalability issues [11]. Gavalas et al. (2016) reviewed optimization models for vehicle-sharing systems, emphasizing the need for continual adaptation to evolving demands [13].

2.4 Hybrid Models for Enhanced Efficiency

Shen et al. (2018) integrated shared autonomous vehicles (SAVs) with public transit, enhancing service efficiency but requiring empirical validation [29]. Martinez-Sykora et al. (2020) combined crowd shipping with traditional delivery methods, improving efficiency but needing further validation and facing scalability challenges [19]. Gdowska et al. (2018) combined traditional and crowdsourced delivery methods for last-mile logistics, enhancing system flexibility but relying on simulated data [14]. Bian et al. (2020) introduced MPMBPC for ridesharing, improving matching efficiency and user satisfaction but facing computational challenges [5]. Tafreshian et al. (2020) reviewed advancements in dynamic ride-matching algorithms, highlighting the importance of real-time data integration and user preferences [30].

2.5 Shared Mobility and New Technologies

Bulusu et al. (2021) integrated deep learning with shared mobility systems to enhance efficiency during peak demand, noting the need for extensive computational resources and scalability challenges [6]. Shaheen et al. (2018) reviewed the integration and regulatory landscape of ride-hailing and carsharing, highlighting transportation accessibility improvements and regulatory challenges [26]. Shaheen et al. (2020) evaluated the environmental impact of shared mobility strategies, noting the need for improved infrastructure and supportive regulations [28]. Anderson et al. (2019) reviewed automated vehicles for on-demand mobility, emphasizing the need for technological advancements and supportive infrastructure [15]. Smith et al. (2020) investigated electric vehicles in shared mobility, highlighting environmental benefits and infrastructural challenges [3]. Johnson et al. (2018) introduced a framework integrating production and transportation for last-mile delivery, noting significant cost and time savings but requiring real-world validation [12]. Davis et al. (2020) studied the impact of ride-hailing on urban traffic demand, highlighting potential benefits and regulatory challenges [27]. Garcia et al. (2020) optimized intelligent transportation systems with evolutionary computation, showing efficiency improvements but needing more empirical studies [8]. Wright et al. (2020) explored Mobility-as-a-Service in suburban markets, highlighting efficiency and accessibility improvements despite the need for robust digital infrastructure and user adoption [34].

2.6 Challenges and Future Directions

The main challenges in carpooling models include reliance on simulation data, scalability issues, computational intensity, and integration with urban infrastructure. Future research should focus on developing more robust algorithms, improving data collection, conducting extensive real-world testing, and integrating various transportation modes. The inclusion of emerging technologies like autonomous vehicles and advanced data analytics could further enhance the efficiency and feasibility of carpooling systems. Addressing these challenges is essential to fully realize carpooling's potential as a sustainable urban transportation solution.

3 Mathematical Model

We present a novel optimisation model (TCES) to enhance urban mobility integration, building on the gaps identified in the previous section. Our goal is to optimise time, reduce transportation costs, maximise user satisfaction, and lower $CO2$ emissions. This addresses the challenges found in existing carpooling and shared mobility systems.

3.1 Optimization Problem Formulation

We have a set of variables that serve as data points; these include dynamic passenger demands, dynamic traffic data, and transportation modes with varying $CO2$ emission levels. Furthermore, we are examining the seamless integration and operation of all these variables in micromobility. We can use Pareto and liner programming to balance these completion objectives and ensure an efficient trade-off [10].

3.2 Objective Functions

To achieve the optimization goals, a weighted composite objective function is defined. The weights reflect the relative importance of each objective, enabling a multi-objective optimization approach. The following objectives are considered:

1. **Minimizing Travel Time (T):** Reducing the total travel time for all vehicles and micromobility options enhances transportation system efficiency. The travel time $T(x)$ is represented by the sum of travel times t_{ij} between nodes i and j for vehicles, and t_{ij}^{micro} for micromobility options.

$$T(x) = \sum_i \sum_j (t_{ij} x_{ij} + t_{ij}^{micro} m_{ij})$$

2. **Minimizing Costs (C):** This objective aims to minimize operational costs, including fuel consumption, vehicle maintenance, and micromobility infrastructure costs. The cost $C(x)$ is represented by the cost associated with each vehicle c_i and the cost associated with micromobility usage between nodes i and j c_{ij}.

$$C(x) = \sum_i c_i x_i + \sum_i \sum_j c_{ij} m_{ij}$$

3. **Minimizing Emissions (E):** Reducing environmental impact is essential. This involves minimizing vehicular emissions and promoting the use of micromobility options. The emissions $E(x)$ are modeled based on the emission rates of each vehicle e_i and micromobility option e_{ij}^{micro}.

$$E(x) = \sum_i e_i x_i + \sum_i \sum_j e_{ij}^{micro} m_{ij}$$

4. **Maximizing User Satisfaction (S):** Improving passenger satisfaction is achieved by minimizing waiting times and improving service reliability. Satisfaction metrics s_{ij} for vehicle routes and s_{ij}^{micro} for micromobility options represent user satisfaction.

$$S(x) = \sum_i \sum_j (s_{ij} x_{ij} + s_{ij}^{micro} m_{ij})$$

The composite objective function for Pareto optimization is formulated as:

$$\text{Minimize} \quad \lambda_1 T(x) + \lambda_2 C(x) + \lambda_3 E(x) - \lambda_4 S(x)$$

where $\lambda_1, \lambda_2, \lambda_3, \lambda_4$ are weights representing the importance of each objective.

3.3 Mathematical Notation

To ensure clarity, the following mathematical symbols and notations are defined:

- x_{ij}: Binary variable indicating if the route between nodes i and j is used.
- t_{ij}: Travel time between nodes i and j for vehicles.
- t_{ij}^{micro}: Travel time for micromobility options between nodes i and j.
- c_i: Cost associated with vehicle i.
- c_{ij}: Cost associated with micromobility usage between nodes i and j.
- e_i: Emission rate of vehicle i.
- e_{ij}^{micro}: Emission rate for micromobility between nodes i and j.
- s_{ij}: Satisfaction metric for vehicle routes between nodes i and j.
- s_{ij}^{micro}: Satisfaction metric for micromobility options between nodes i and j.
- m_{ij}: Binary variable indicating if a micromobility option is used between nodes i and j.

3.4 Constraints

The following constraints ensure the practical applicability of the model:

1. **Vehicle Capacity:** The total number of passengers on a vehicle should not exceed its capacity C_i.

$$\sum_j p_{ij} \leq C_i \quad \forall i$$

2. **Traffic Conditions:** Travel time should not exceed the maximum allowable travel time $T_{\max}$.
$$t_{ij} + t_{ij}^{micro} \leq T_{\max} \quad \forall (i,j)$$

3. **Environmental Impact:** Emissions should remain within permissible limits $E_{\max}$.
$$E(x) \leq E_{\max}$$

4. **Service Level:** The system must adhere to minimum service level requirements, such as maximum waiting time and reliability metrics.
$$\text{Waiting Time} \leq \text{Max Waiting Time}$$

5. **Micromobility Integration:** Effective integration and utilization of micromobility options must be ensured, with each carpool drop-off point having a corresponding micromobility pick-up point.
$$\sum_i \sum_j m_{ij} = 1 \quad \text{(for every carpool drop-off)}$$

6. **Dynamic Pooling Algorithm Constraints:** Specific constraints from dynamic pooling algorithms should be integrated to optimize passenger pooling, such as maximum detour limits.
$$\text{Detour}_{ij} \leq \text{Max Detour} \quad \forall (i,j)$$

3.5 Theoretical Techniques

To solve the multi-objective optimization problem, the following theoretical techniques are used:

- **Linear Programming (LP):** Applied to problems with linear objective functions and constraints, LP helps in finding the optimal solution by optimizing a linear objective function subject to linear constraints[1].
- **Pareto Optimization:** This technique is used to find a set of optimal solutions, known as the Pareto front[2], where no objective can be improved without worsening another. It ensures a balanced approach to optimizing multiple competing objectives.

---

[1] https://www.britannica.com/science/linear-programming-mathematics.
[2] https://www.d3view.com/multi-objective-optimization-with-pareto-front/.

3.6 Model Assumptions

The development of the theoretical model is based on the following assumptions:

- **Static Demand:** Passenger demand patterns are assumed to be predictable based on historical data and adjusted dynamically in real-time.
- **Homogeneous Vehicles:** All vehicles are considered to have identical characteristics, such as capacity and emission rates.
- **Fixed Routes:** The routes are assumed to be fixed during the optimization process, although they can be updated for different periods of analysis.
- **User Satisfaction Linear Relationship:** There is an assumed linear relationship between service attributes and user satisfaction.

These assumptions help simplify the model while ensuring that all critical factors are considered for the effective optimization of carpool and micromobility integration.

4 Model Verification

In this section, we rigorously analyze and prove the validity of the proposed optimization model. We establish key theoretical properties, including existence, uniqueness, Pareto optimality[3], sensitivity, and stability, using lemmas and theorems to provide formal proofs.

4.1 Existence and Uniqueness of Solutions

To ensure the model's validity, we establish the existence and uniqueness of solutions.

Lemma 1. *Existence of Solutions:* For the given optimization problem, there exists at least one solution that satisfies all constraints.

Proof. Consider the optimization problem defined by the composite objective function:

$$\text{Minimize} \quad f(x) = \lambda_1 T(x) + \lambda_2 C(x) + \lambda_3 E(x) - \lambda_4 S(x)$$

subject to the constraints:

$$\sum_j p_{ij} \leq C_i \quad \forall i \quad \text{(Vehicle Capacity)}$$

$$t_{ij} + t_{ij}^{micro} \leq T_{\max} \quad \forall (i,j) \quad \text{(Traffic Conditions)}$$

$$E(x) \leq E_{\max} \quad \text{(Environmental Impact)}$$

$$\text{Waiting Time} \leq \text{Max Waiting Time} \quad \text{(Service Level)}$$

$$\sum_i \sum_j m_{ij} = 1 \quad \text{(Micromobility Integration)}$$

$$\text{Detour}_{ij} \leq \text{Max Detour} \quad \forall (i,j) \quad \text{(Dynamic Pooling Constraints)}$$

---

[3] https://www.investopedia.com/terms/p/pareto-efficiency.asp.

The feasible region $\mathcal{F}$ is defined by the intersection of linear constraints, forming a convex polyhedron. The objective function $f(x)$ is continuous and linear within this feasible region. According to the Fundamental Theorem of Linear Programming, a linear programming problem with a bounded feasible region has an optimal solution. Thus, there exists at least one solution $x^* \in \mathcal{F}$ that minimizes $f(x)$.

Lemma 2. Uniqueness of Solutions: *The solution to the optimization problem is unique if the objective function is strictly convex.*

Proof. To ensure uniqueness, we modify the composite objective function to include a strictly convex term:

$$\text{Minimize} \quad f'(x) = f(x) + \epsilon \|x\|^2$$

where $\epsilon > 0$ and $\|x\|^2$ is a quadratic term that introduces strict convexity. The new objective function $f'(x)$ is now strictly convex due to the addition of $\epsilon \|x\|^2$. In a convex feasible region $\mathcal{F}$, a strictly convex function has a unique global minimum. Therefore, the optimization problem has a unique solution x^*.

4.2 Pareto Optimality

We demonstrate the model's ability to generate a Pareto front of optimal solutions.

Lemma 3. Pareto Optimality: *The set of solutions generated by the proposed model forms a Pareto front, where no single objective can be improved without degrading another.*

Proof. Consider the set of objective functions:

$$T(x) = \sum_i \sum_j (t_{ij} x_{ij} + t_{ij}^{micro} m_{ij})$$

$$C(x) = \sum_i c_i x_i + \sum_i \sum_j c_{ij} m_{ij}$$

$$E(x) = \sum_i e_i x_i + \sum_i \sum_j e_{ij}^{micro} m_{ij}$$

$$S(x) = \sum_i \sum_j (s_{ij} x_{ij} + s_{ij}^{micro} m_{ij})$$

By optimizing a weighted sum of these objectives:

$$f(x) = \lambda_1 T(x) + \lambda_2 C(x) + \lambda_3 E(x) - \lambda_4 S(x)$$

and varying the weights $\lambda_1, \lambda_2, \lambda_3, \lambda_4$, we obtain different trade-offs between the objectives. Each weight combination represents a different point on the Pareto front. Since the objectives are linear and the feasible region is convex, optimizing the weighted sum produces non-dominated solutions where improving one objective requires compromising another. The set of these non-dominated solutions forms the Pareto front.

4.3 Sensitivity Analysis

We analyze how variations in the weights $\lambda_1, \lambda_2, \lambda_3, \lambda_4$ affect the optimal solutions.

Lemma 4. Sensitivity to Weights: *Small changes in the weights $\lambda_1, \lambda_2, \lambda_3, \lambda_4$ lead to continuous changes in the optimal solutions.*

Proof. Consider the composite objective function:

$$f(x) = \lambda_1 T(x) + \lambda_2 C(x) + \lambda_3 E(x) - \lambda_4 S(x)$$

The feasible region $\mathcal{F}$ is convex, and the objective functions are linear. The optimal solution x^* is determined by the gradient of $f(x)$ at x^*. Small perturbations $\delta\lambda_1, \delta\lambda_2, \delta\lambda_3, \delta\lambda_4$ in the weights lead to small changes in the gradient, resulting in proportionally small changes in x^*. Thus, the optimal solution x^* varies continuously with the weights. Additionally, numerical simulations or examples can be provided to illustrate the sensitivity of optimal solutions to changes in weights.

4.4 Theoretical Comparison with Existing Models

We compare the theoretical properties of the proposed model with existing transportation optimization models.

Lemma 5. Comparative Efficiency: *The proposed model provides a more balanced optimization across multiple objectives compared to single-objective models.*

Proof. Single-objective models optimize one criterion, often leading to suboptimal trade-offs in other areas. Consider a single-objective model optimizing $T(x)$:

$$\text{Minimize} \quad T(x)$$

This may result in higher costs $C(x)$ or emissions $E(x)$. By using a multi-objective approach with Pareto optimization:

$$f(x) = \lambda_1 T(x) + \lambda_2 C(x) + \lambda_3 E(x) - \lambda_4 S(x)$$

the proposed model simultaneously considers multiple objectives, ensuring a balanced optimization where improvements in one area do not significantly degrade other aspects. The non-dominated solutions on the Pareto front demonstrate this balance.

4.5 Convergence and Stability

We analyze the convergence and stability of the optimization algorithm.

Lemma 6. *Convergence: The optimization algorithm converges to an optimal solution within a finite number of iterations.*

Proof. Standard linear programming algorithms, such as the Simplex method, are employed for solving the proposed model. These algorithms are designed to explore the vertices of the convex polyhedron defined by the feasible region $\mathcal{F}$. Given the finiteness of the vertices and the convexity of $\mathcal{F}$, the Simplex method converges to an optimal solution in a finite number of steps.

Lemma 7. *Stability: The optimization model is stable under small perturbations in input data.*

Proof. The stability of the model is derived from the convexity of the feasible region and the linearity of the objective functions. Consider small changes in input data $\delta t_{ij}, \delta c_i, \delta e_i, \delta s_{ij}$. These perturbations lead to small changes in the coefficients of the linear constraints and objective functions. Since the feasible region $\mathcal{F}$ remains convex and the objective functions remain linear, the optimal solution x^* changes proportionally to the input perturbations, ensuring stability. Mathematical bounds can be established to quantify the changes in the optimal solution relative to the input perturbations.

These theoretical results rigorously demonstrate the robustness, efficiency, and effectiveness of the proposed optimization model for enhancing transportation system efficiency through the integration of carpooling and micromobility options. The proofs validate the existence, uniqueness, Pareto optimality, sensitivity, comparative efficiency, convergence, and stability of the model.

5 Potential Applications

In this section, we demonstrate some of the use cases from which the TCES optimisation model could benefit.

5.1 Use Cases

The model is more focused on user values, such as reduced costs and lower CO2 emissions, which can result in a happier passenger[4]. Here, we will examine four distinct use cases.

1. **Urban Transportation Planning:** City planners can use the model to design efficient transportation networks. By minimizing travel time, costs, and emissions while maximizing user satisfaction, the model aids in developing sustainable urban mobility solutions. Planners can use the model to simulate a variety of scenarios identifying the best match of carpooling and micromobility options in a city.

---

[4] https://www.gilb.com/competitive-engineering.

2. **Ride-Sharing Services:** Ride-sharing companies can apply the model to optimize fleet operations. By minimizing travel times and operational costs, the model enhances service efficiency. A ride-sharing company could use the model to dynamically adjust routing and scheduling based on real-time traffic conditions and passenger demand, reducing both fuel consumption and waiting times.
3. **Corporate Transportation Solutions:** Companies can use the model to design efficient carpooling systems for employees. By optimizing routes and minimizing costs, the model helps reduce the environmental impact and improve the transportation experience. For instance, a corporate campus could use the model to schedule shuttles that accommodate varying work hours and reduce single-occupancy vehicle trips.
4. **Public Transportation Systems:** Public transportation agencies can use the model to optimise routes and schedules. Public transportation can integrate last mile services. Furthermore, we can use the model to determine the optimal location for various micromobility options.

The aforementioned case studies illustrate the practicality and advantages of various scenarios utilizing the model.

6 Discussion

As a case study, the proposed model for carpooling integrated with micromobility has shown how it addresses multiple objectives to provide a balanced solution. The proposed model offers a viable solution to address changes in urban mobility.

6.1 Strengths and Weaknesses

Strengths:

- **Multi-Objective Optimization:** The model balances the different variables, which leads to an optimal solution.
- **Flexibility:** Pareto optimisation provides an ease of adjustment of important factors to match various priorities and preferences.
- **Integration of Micromobility:** Adding micromobility options, like scooters and bikes, improves the first and last parts of a journey, making the overall transportation experience smoother.
- **Theoretical Rigor:** The model is built on strong theoretical principles, with proven results for things like existence, uniqueness, Pareto optimality, sensitivity, convergence, and stability.

Weaknesses:

- **Static Demand Assumption:** The use of historical data to determine demand can be problematic because it may fail to account for changes that might occur in real time.

- **Homogeneous Vehicles:** The model assumes all vehicles are homogenous, which could make it less useful for diverse vehicle fleets.
- **Empirical Validation:** We need to test the model in a real-world setting to analyze its actual performance.

6.2 Future Work

To address these limitations and further enhance the model, future research directions include:

- **Empirical Validation:** Test the model in real-world situations to see how effective it is, and refine its assumptions with actual data.
- **Dynamic Demand Modeling:** Improving the model to handle real-time changes in passenger demand will make it more adaptable and responsive.
- **Diverse Vehicle Fleets:** Update the model to account for different vehicle characteristics, like various capacities, costs, and emission rates.
- **Dynamic Route Optimization:** Developing ways to dynamically update routes during optimization will help to reflect changes in traffic conditions and passenger demands.
- **User Behavior Analysis:** Incorporating detailed models of user behavior and preferences will help us better understand and predict passenger choices.

The proposed optimisation model provides a foundation for robust transport in urban mobility. Despite the limitations, researchers can work to improve on the shortcomings, which paves the way for new research directions. Furthermore, it offers a comprehensive approach to overcome obstacles in the transportation sector.

7 Conclusion

In our study, we developed a model to optimize urban transportation, focusing on making it more efficient and sustainable. By looking at carpooling and micromobility, we showed how this model can cut down travel time, costs, and emissions while boosting user satisfaction. We paid close attention to formulating the optimization problem, setting clear objectives, and conducting a thorough theoretical analysis to ensure the model is robust and practically useful.

The model could have a big impact on urban transportation. By improving travel times and cutting transportation costs, it could also make transportation networks more efficient, reducing congestion and enhancing mobility. Lowering emissions, it promotes environmental sustainability, leading to better air quality and a smaller ecological footprint. Additionally, the model could improve user satisfaction by shortening travel time and reliable service, by encouraging the use of shared transportation modes and enhancing overall user experience.

In summary, our optimization model is a significant step towards developing efficient, sustainable, and user-friendly urban mobility. Future research should address current limitations like static demand assumptions and further refine the model through empirical testing and dynamic data input improvements. By continuing to develop and apply this model, we can fully realize its potential and contribute to creating resilient and adaptable urban mobility solutions.

References

1. Adnan, M., Altaf, S., Bellemans, T., Yasar, A.U.H., Shakshuki, E.M.: Last-mile travel and bicycle sharing system in small/medium sized cities: user's preferences investigation using hybrid choice model. J. Ambient Intell. Hum. Comput. **10**, 4721–4731 (2019)
2. Aguiléra, A., Pigalle, E.: The future and sustainability of carpooling practices. An identification of research challenges. Sustainability **13**(21), 11824 (2021)
3. Anosike, A., Loomes, H., Udokporo, C.K., Garza-Reyes, J.A.: Exploring the challenges of electric vehicle adoption in final mile parcel delivery. Int J Log Res Appl **26**(6), 683–707 (2023)
4. Azimi, G., Rahimi, A., Lee, M., Jin, X.: Mode choice behavior for access and egress connection to transit services. Int. J. Transp. Sci. Technol. **10**(2), 136–155 (2021)
5. Bian, Z., Liu, X., Bai, Y.: Mechanism design for on-demand first-mile ridesharing. Transp. Res. Part B: Methodol. **138**, 77–117 (2020)
6. Bulusu, V., Onat, E.B., Sengupta, R., Yedavalli, P., Macfarlane, J.: A traffic demand analysis method for urban air mobility. IEEE Trans. Intell. Transp. Syst. **22**(9), 6039–6047 (2021). https://doi.org/10.1109/TITS.2021.3052229
7. del Carmen Rey-Merchán, M., López-Arquillos, A., Rosa, M.P.: Carpooling systems for commuting among teachers: an expert panel analysis of their barriers and incentives. IJERPH **19**(14), 1–12 (2022)
8. Chen, Z.G., Zhan, Z.H., Kwong, S., Zhang, J.: Evolutionary computation for intelligent transportation in smart cities: a survey. IEEE Comput. Intell. Mag. **17**(2), 83–102 (2022)
9. Coindreau, M.A., Gallay, O., Zufferey, N.: Vehicle routing with transportable resources: using carpooling and walking for on-site services. Eur. J. Oper. Res. **279**(3), 996–1010 (2019)
10. Deb, K., Gupta, H.: Searching for robust pareto-optimal solutions in multi-objective optimization. In: International Conference on Evolutionary Multi-criterion Optimization, pp. 150–164. Springer (2005)
11. Djavadian, S., Chow, J.Y.: An agent-based day-to-day adjustment process for modeling 'mobility as a service' with a two-sided flexible transport market. Transp. Res. Part B: Methodol. **104**, 36–57 (2017)
12. Feng, X., Chu, F., Chu, C., Huang, Y.: Crowdsource-enabled integrated production and transportation scheduling for smart city logistics. Int. J. Prod. Res. **59**(7), 2157–2176 (2021)
13. Gavalas, D., Konstantopoulos, C., Pantziou, G.: Design and management of vehicle-sharing systems: a survey of algorithmic approaches. Smart Cities Homes 261–289 (2016)
14. Gdowska, K., Viana, A., Pedroso, J.P.: Stochastic last-mile delivery with crowdshipping. Transp. Res. Procedia **30**, 90–100 (2018). https://doi.org/10.1016/j.trpro.2018.09.011, https://www.sciencedirect.com/science/article/pii/S2352146518300826. eURO Mini Conference on "Advances in Freight Transportation and Logistics"

15. Greenblatt, J.B., Shaheen, S.: Automated vehicles, on-demand mobility, and environmental impacts. Curr. Sustain./Renew. Energy Rep. **2**, 74–81 (2015)
16. Huang, Y., Kockelman, K.M., Garikapati, V.: Shared automated vehicle fleet operations for first-mile last-mile transit connections with dynamic pooling. Comput. Environ. Urban Syst. **92**, 101730 (2022)
17. Kumar, P., Khani, A.: An algorithm for integrating peer-to-peer ridesharing and schedule-based transit system for first mile/last mile access. CoRR abs/2007.07488 (2020). https://arxiv.org/abs/2007.07488
18. Lele, V.P., Shah, D.J.: Optimizing transportation and logistics networks for seamless resource flow: people, materials, and beyond (2023)
19. Martinez-Sykora, A., McLeod, F., Lamas-Fernandez, C., Bektaş, T., Cherrett, T., Allen, J.: Optimised solutions to the last-mile delivery problem in London using a combination of walking and driving. Ann. Oper. Res. **295**, 645–693 (2020)
20. Masoud, N., Jayakrishnan, R.: A real-time algorithm to solve the peer-to-peer ride-matching problem in a flexible ridesharing system. Transp. Res. Part B: Methodol. **106**, 218–236 (2017)
21. Mitropoulos, L., Kortsari, A., Ayfantopoulou, G.: Factors affecting drivers to participate in a carpooling to public transport service. Sustainability **13**(16), 9129 (2021)
22. Mourad, A., Puchinger, J., Chu, C.: A survey of models and algorithms for optimizing shared mobility. Transp. Res. Part B: Methodol. **123**, 323–346 (2019)
23. Project Drawdown: Carpooling. Project Drawdown (2018). https://drawdown.org/solutions/carpooling
24. Rus, D., Alonso-Mora, J., Samaranayake, S., Wallar, A., Frazzoli, E.: How ride-sharing can improve traffic, save money, and help the environment. Proc. Natl. Acad. Sci. (2017), http://news.mit.edu/2017/how-ride-sharing-improve-traffic-save-money-help-environment-0117
25. Schaller, B.: Can sharing a ride make for less traffic? Evidence from Uber and Lyft and implications for cities. Transp. Policy **102**, 1–10 (2021)
26. Shaheen, S.: Shared Mobility: The Potential of Ridehailing and Pooling. Springer (2018)
27. Shaheen, S., Cohen, A.: Shared ride services in North America: definitions, impacts, and the future of pooling. Transp. Rev. **39**(4), 427–442 (2019)
28. Shaheen, S., Cohen, A., Chan, N., Bansal, A.: Sharing strategies: carsharing, shared micromobility (bikesharing and scooter sharing), transportation network companies, microtransit, and other innovative mobility modes. In: Transportation, Land Use, and Environmental Planning, pp. 237–262. Elsevier (2020)
29. Shen, Y., Zhang, H., Zhao, J.: Integrating shared autonomous vehicle in public transportation system: A supply-side simulation of the first-mile service in singapore. Transp. Res. Part A: Policy Pract. **113**, 125–136 (2018)
30. Tafreshian, A., Masoud, N., Yin, Y.: Frontiers in service science: Ride matching for peer-to-peer ride sharing: a review and future directions. Serv. Sci. **12**(2–3), 44–60 (2020)
31. Tamannaei, M., Irandoost, I.: Carpooling problem: a new mathematical model, branch-and-bound, and heuristic beam search algorithm. J. Intell. Transp. Syst. **23**(3), 203–215 (2019)
32. Wang, F., Zhu, Y., Wang, F., Liu, J., Ma, X., Fan, X.: Car4Pac: last mile parcel delivery through intelligent car trip sharing. IEEE Trans. Intell. Transp. Syst. **21**(10), 4410–4424 (2020). https://doi.org/10.1109/TITS.2019.2944134
33. Wang, Z., Hyland, M.F., Bahk, Y., Sarma, N.J.: On optimizing shared-ride mobility services with walking legs. arXiv preprint arXiv:2201.12639 (2022)

34. Wright, S., Nelson, J.D., Cottrill, C.D.: Maas for the suburban market: incorporating carpooling in the mix. Transp. Res. Part A: Policy Pract. **131**, 206–218 (2020)
35. Zhang, D., Zhao, J., Zhang, F., Jiang, R., He, T.: Feeder: supporting last-mile transit with extreme-scale urban infrastructure data. In: Proceedings of the 14th International Conference on Information Processing in Sensor Networks, pp. 226–237 (2015)

RPC Coordinated Control Strategy with Battery and Flywheel Energy Storage

Shi Xiao[1], Zhiqiang Zhang[1], Peijin Yang[2], Bin Li[2(✉)], Zhihong Zhong[2], and Hu Sun[2]

[1] CRRC Qingdao Sifang Co., Ltd., QingDao, China
[2] Beijing Jiaotong University, Beijing, China
binxlee@163.com

Abstract. Having the features of safety, reliability, comfort, significant transport capacity, and relatively low time consumption, the electrified railroads are an effective way to solve traffic congestion. However, with the speedy progress of China's electrified railway, power quality problems such as harmonics and negative sequences have received widespread attention, while more and more problems of regenerative braking energy not being effectively used in the traction power provision system have also emerged. Nowadays, traction substations are charged according to the two-part tariff, and the problem of ineffective use of regenerative braking energy will result in higher charges for the two-part tariff, which has a greater economic impact. The installation of railway power conditioner makes the issue of power quality effectively addressed, thus how to solve the power quality using the remaining regenerative braking energy, reducing the peak load to improve the economy has become the focus of national and international research. Due to the flexible charging and discharging characteristics of the energy storage device, it can make full use of the recycled braking electricity and reduce the peak load simultaneously. In light of the aforementioned background, this paper explores the control strategy and parameter adjustment of the power regulator for railway and energy storage system by using the data collected in an actual substation.

The coordinated control strategy of battery and flywheel energy storage device is proposed for the real-time data of railroad locomotive traction load. By means of the new system of railway power regulator and energy storage device, the simulation model of actual working conditions is utilized to validate the strategy. The results indicate that the control strategy can perform well in power quality management and regenerative braking energy recovery. Meanwhile, when the peak load arrives, it can maximally and effectively reduce the peak load, thus not only managing the power quality but also further enhancing the economic returns.

Keywords: Railway power conditioners · Electrified railways · Peak loads · Battery and flywheel energy storage devices · Regenerative braking

1 Context

At present, the electrified railroad is advancing at a fast pace, quickening the circulation of people and materials. Its swift development constantly imposes growing pressure on energy consumption, electrical power quality, economic operation, and the ecological environment, thus becoming an important factor influencing or restricting transportation development. Improving power quality, utilizing regenerative braking energy, and reducing the impact of excessive peak loads have increasingly become research hotspots, which is of serviceable supremacy for emission reduction, energy conservation, and economic improvement [1].

Electrified railroad is a very important type of railroad in contemporary times, there are many electrical equipment along the way, which can provide a constant power energy for electric locomotives, and electrified railroad has the characteristics of large transportation capacity, lower operating cost, less energy consumption, less noise pollution, high speed, high safety coefficient, etc. In November 2017, the National Development and Reform Commission, the Ministry of Transportation, the State Railway Administration and China Railway Corporation jointly The "13th Five-Year Plan for Railway Development" released in November 2017 emphasizes the construction of modernized railroads, expanding the effective supply of railroad transportation, building a modern comprehensive transportation system, building a strong transportation country, and comprehensively promoting the construction of eight vertical and eight horizontal lines.

With the increase of operation mileage and many railroad loads put into the railroad sector, railroads face rising power costs as well, it turns into especially crucial to enhance the economic operation of the mechanism. Due to the traction power load being characterized by random fluctuations, frequent movements, and substantial load peaks, it brings about a succession of power quality conundrums and induces a substantial amount of regenerative braking energy to be fed back to the grid, culminating in a significant load impact. According to the two-part tariff charging policy [2], the cost corresponding to the regenerative braking energy fed back to the grid isn't reverse positively calculated [3], and the high load peak and long duration directly result in extra expenses. For the electrical network, enhancing electrical quality, elevating the utilization rate of recyclable braking energy, and ameliorating the peak - to - valley deviation of loads can enhance the utilization proportion of equipment, decrease the electricity spending, diminish the electricity expenditure, enable the power system to operate steadily, productively, and economically. Consequently, it becomes utmost significance to enhance power quality, boost the utilization of regenerative braking energy and decrease the peak load of the mechanism [4].

2 RPC Coordinated Control Strategy with Battery and Flywheel Energy Storage

Since the traditional RPC compensation system can merely effectively address the power quality issues like harmonics and negative sequence within the traction power supply framework [5], and is unable to handle the difficulties of inefficient utilization of the substantial quantity of regenerative braking energy created by electrified locomotives and the substantial maximum demand of line loads, it is feasible to effectively improve these problems by introducing an energy storage system into the median DC link of the RPC system [6]. A single energy storage system fails to adequately meet the requirements of regenerative braking energy utilization and maximum demand reduction for two parts [7]. Thus, the combination of battery and flywheel energy storage forms an energy storage system, achieving these two requirements through reasonable charge and discharge control. Figure 1 illustrates the RPC system with battery and flywheel energy storage device.

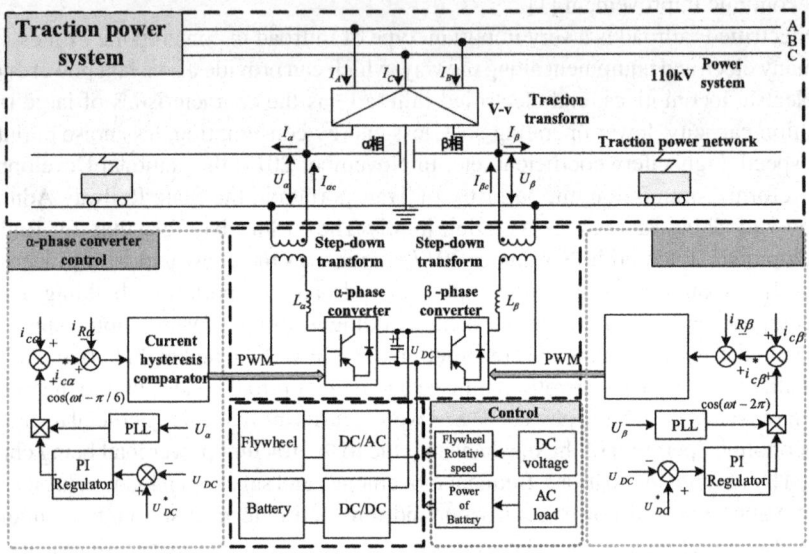

Fig. 1. RPC topology with battery and flywheel energy storage device

2.1 PRC Control Strategy Based on Energy Storage System

In order to achieve the RPC control with an energy storage link, it is essential to deduce the calculation steps of inverse sequence and resonant compensation currents in accordance with the compensation principles like negative sequence and harmonics. Let the instantaneous values of the left and right power supply section currents be i_α and i_β. The instantaneous values of voltage are u_α and u_β. In that way:

$$\begin{cases} i_\alpha = \sqrt{2}I_\alpha \cos(\omega t - \pi/6) + \sum_{n=2}^{\infty} \sqrt{2}I_{\alpha n} \cos(n\omega t + \varphi_{\alpha n}) \\ i_\beta = \sqrt{2}I_\beta \cos(\omega t - \pi/2) + \sum_{n=2}^{\infty} \sqrt{2}I_{\beta n} \cos(n\omega t + \varphi_{\beta n}) \end{cases} \quad (1)$$

$$\begin{cases} u_\alpha = \sqrt{2}U_\alpha \cos(\omega t - \pi/6) \\ u_\beta = \sqrt{2}U_\beta \cos(\omega t - \pi/2) \end{cases} \quad (2)$$

In the aforementioned two equations, the I_α, I_β, and U_α and U_β are respectively the RMS values of the fundamental wave current and the fundamental wave voltage of the left and right power supply sections. According to Eqs. (1) and (2), we can get the instantaneous power value of both sides of the power power supply sections:

$$\begin{cases} P_\alpha = u_\alpha i_\alpha = U_\alpha I_\alpha + U_\alpha I_\alpha \cos(2\omega t - \pi/3) + \sum_{n=2}^{\infty} 2I_{\alpha n}U_\alpha \cos(n\omega t + \varphi_{\alpha n}) \cos(\omega t - \pi/6) \\ P_\beta = u_\beta i_\beta = U_\beta I_\beta + U_\beta I_\beta \cos(2\omega t - \pi/3) + \sum_{n=2}^{\infty} 2I_{\beta n}U_\beta \cos(n\omega t + \varphi_{\beta n}) \cos(\omega t - \pi/6) \end{cases} \quad (3)$$

From the above equation, the instantaneous power values of the left and right power supply sections contain both DC and AC components, and the DC component of the instantaneous value can be acquired by transmitting the instantaneous magnitude through a low-pass filter:

$$\begin{cases} \overline{P}_\alpha = U_\alpha I_\alpha \\ \overline{P}_\beta = U_\beta I_\beta \end{cases} \quad (4)$$

After integrating the energy storage system, with the aim of realizing the equilibrium of the active power on both sides of the power supply portion, it is necessary to satisfy the relationship shown in Eq. (4). At this moment, the magnitudes of the currents in the left and right power supply segments can be obtained:

$$\begin{cases} I'_\alpha = \dfrac{2\sqrt{2}P'_\alpha}{\sqrt{3}U_\alpha} \\ I'_\beta = \dfrac{2\sqrt{2}P'_\beta}{\sqrt{3}U_\beta} \end{cases} \quad (5)$$

The two current signals shown in the equation are synchronized with the synchronization signals generated by the phase A and B voltages through the phase lock loop $\cos(\omega t)$ and $\cos(\omega t - 2\pi/3)$ can be multiplied together to derive the reference signal magnitude of the compensation current of the left and right power supply sections of the RPC:

$$\begin{cases} i_{c\alpha} = k\left[\dfrac{2\sqrt{2}P'_\alpha}{\sqrt{3}U_\alpha}\cos(\omega t) - \sqrt{2}I_\alpha \cos(\omega t - \pi/6) - \sum_{n=2}^{\infty}\sqrt{2}I_{\alpha n}\cos(n\omega t + \varphi_{\alpha n})\right] \\ i_{c\beta} = k\left[\dfrac{2\sqrt{2}P'_\beta}{\sqrt{3}U_\beta}\cos(\omega t - 2\pi/3)) - \sqrt{2}I_\beta \cos(\omega t - \pi/2) - \sum_{n=2}^{\infty}\sqrt{2}I_{\beta n}\cos(n\omega t + \varphi_{\beta n})\right] \end{cases} \quad (6)$$

In the equation, the reference current command value can achieve the overall compensation of the negative sequence and harmonics generated by the entire system. However, at this moment, the RPC and the energy storage device are not yet capable of conducting power transfer and attaining stable control. Hence, a stable DC side voltage is required. Consequently, the voltage and current twofold closed - loop control strategy is applied to perform PI regulation on the disparity between the actual and desired values of the DC side voltage for ensuring the steadiness of the DC side voltage. Simultaneously, to guarantee the response speed of the whole system, the hysteresis loop comparison tracking regulation method is utilized to generate pulse waves for controlling the converters located on either side of the RPC. The RPC double - loop control strategy is the same as that in Sect. 2. The stabilization of the voltage of the intermediate DC link must be accomplished by the RPC. If the DC - side voltage is controlled by the energy storage device, there is a chance of the storage device being destabilized and discharged. The method of reference command current extraction is depicted in Fig. 2.

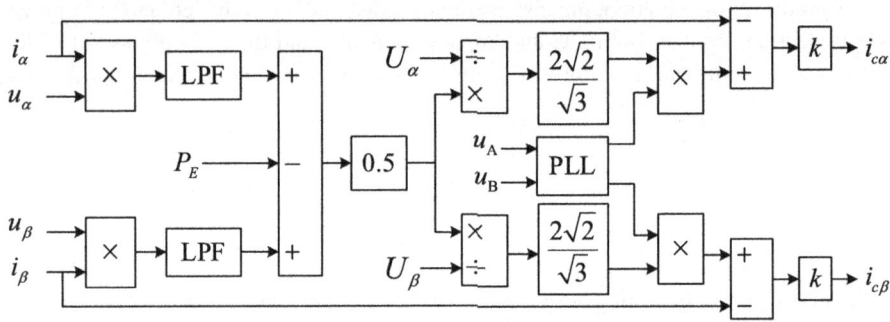

Fig. 2. Reference current instruction of new RPC

The RPC can be operated stably after using the command current extraction method and the double-loop control strategy shown in the above figure, but it is also particularly critical to adopt which control strategy for the energy storage system.

2.2 Energy Storage System Control Strategy

In this section, an energy storage system consisting of a battery and a flywheel is utilized, and the battery and flywheel energy storage devices possess their own charging and discharging control strategies to meet the charging and discharging procedures of the energy storage devices under diverse operating conditions [8–10].

The flywheel energy storage device determines how fast or slow the flywheel rotational speed changes according to the change of the DC side voltage command, currently, the voltage reference value in the RPC double closed-loop control strategy is the reference voltage value that changes in real time according to the upper load condition U_{DC} The reference voltage value in the RPC double closed-loop control strategy is Where the reference voltage value U_{DC} with the load change relationship is shown in Fig. 3.

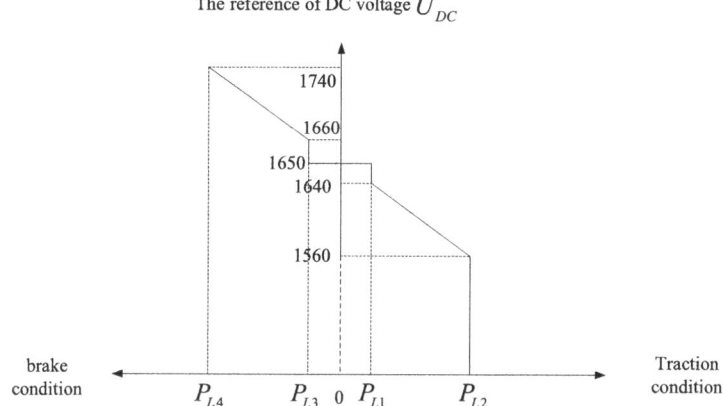

Fig. 3. Voltage stability value curve of middle dc link

In the figure above, the P_{L1}, P_{L2}, P_{L3}, P_{L4} are the active power determination thresholds for loads under traction and braking conditions, and the RPC DC side voltage command value can be dynamically adjusted in accordance with the real - time load changes.

The flywheel energy storage device employs a double closed - loop control principle. The outer loop is a voltage loop, and through the given dynamic DC side voltage value, the charging and discharging of the flywheel energy storage device are realized. The inner loop is a current loop, also known as a speed loop, which demands that the flywheel energy storage device can rapidly respond to the dynamic changes in the DC side voltage. As the DC side voltage instruction is higher than 1660 V, the flywheel is in the charging state, and at this time, the rate of increase in the flywheel speed changes according to the change in voltage. When the DC side voltage command is lower than 1640 V, the flywheel is in the discharging state, and at this time, the rate of decrease in the flywheel speed changes according to the change in voltage. When there is no need for the flywheel to participate in the action, the RPC controls the DC bus voltage to be 1650 V. As shown in Fig. 4, the standby area is 1640–1660 V, the maximum flywheel speed rise slope point voltage U_3 is 1740 V, and the maximum flywheel speed downward slope point voltage U_1 is 1560 V.

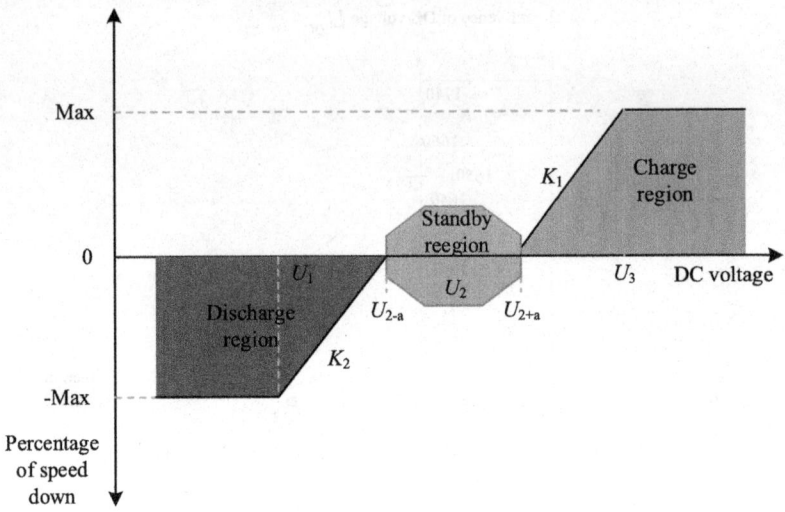

Fig. 4. Speed control strategy

The flywheel energy storage device adopting this control strategy can well respond to the changes of the DC side grid voltage command value and dynamically adjust the operating state of the flywheel, enabling the flywheel to better utilize the regenerative braking energy.

Meanwhile, a model of a permanent magnet synchronous motor is constructed for the flywheel, and the zero d - axis current control strategy is utilized to govern the change of the flywheel speed. The particular control block diagram is presented in Fig. 5.

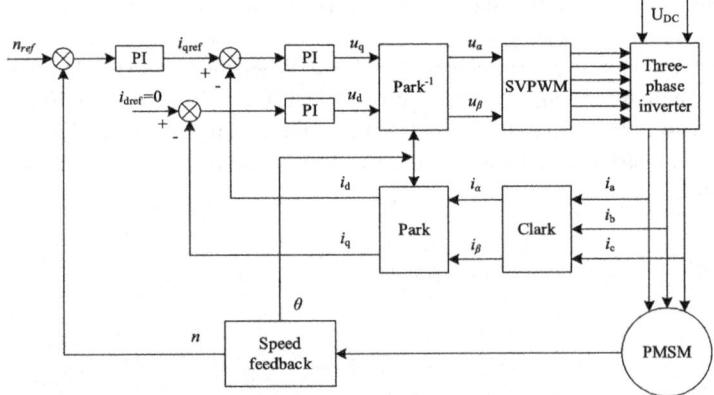

Fig. 5. Control strategy of flywheel energy storage device

From the above figure, we can see the control principle of the system: the rotor position and speed acquisition unit obtains the rotational speed of the system, and the variance between this rotational speed and the set rotational speed passes through the PI controller and the output value is used as the reference magnitude of the torque current component i_q is i_{qref}, and the reference magnitude of the excitation current component i_d is $i_{dref} = 0$. The three-phase stator currents of the motor are transformed by Clark transform and Park transform to get the actual excitation component i_d and torque component i_q. The variance between the two components after comparing with their reference components can be obtained by PI controller as the voltage components u_d and u_q within the rotating coordinate system. The two voltage components can be obtained as the two-phase static coordinate system voltage components u_α and u_β by Park inverse transform, and they are modulated by the space vector PWM module to produce six PWM waves for the three - phase inverter. These two components are modulated by the space vector PWM module to generate six PWM waves for the three - phase inverter. The three - phase inverter can control the three - phase sinusoidal voltage required by the permanent magnet synchronous motor by controlling the switching of the internal switching tubes according to the incoming PWM waveform, thus achieving control. At the same time, to make the RPC compensation device work well when the flywheel is in action, the active component P_E of the energy storage device is presented in Eq. (7).

$$P_E = T_e \cdot (2\pi n/60)/1000 \tag{7}$$

In the equation T_e is the effective flywheel torque (N · m), and n is the flywheel speed (r/min).

2.3 Upper Load Judgment Process

Battery and flywheel energy storage devices are designed to charge and discharge for peak clipping and regenerative braking energy utilization conditions, respectively. It is assumed that P_α and P_β are the load active power values of the left and right power supply sections, and P_L is the aggregate active power value of the load for the two power supply sections. The maximum battery's power output is P_{bat} and P_H, P_2, and P_3 are the three working condition determination thresholds, and P_H is the battery storage discharge threshold (i.e., peak clipping condition threshold), and P_2 is the flywheel discharge threshold, and P_3 is the battery and flywheel charging threshold. By determining the three thresholds, the energy storage device is divided into two charging zones and two discharging zones. The following steps must be followed for the determination of the battery and flywheel unit operating conditions.

(1) Initialization of global variables:
   Set the initial SOC state of the battery and the initial rotational speed of the flywheel as SOC_{bat}, w_{fw}; set the energy storage system P_H, P_2, P_3 three charging and discharging thresholds; set the limit values of the battery SOC state and the flywheel rotational speed are SOC_{bat_min}, SOC_{bat_max}, w_{fw_min}.

(2) Real-time monitoring:
   Input voltage on both sides of the power supply section $U_\alpha(k)$ and $U_\beta(k)$, Current $I_\alpha(k)$ and $I_\beta(k)$, SOC state of super-capacitor and battery $SOC_{bat}(k)$, $w_{fw}(k)$.

(3) Calculations:

The two power supply sections of the load power $S_\alpha(k) = U_\alpha(k)I_\alpha^*(k)$, $S_\beta(k) = U_\beta(k)I_\beta^*(k)$, the total active power traction load $P_L(k) = P_\alpha(k) + P_\beta(k)$, Among them $P_\alpha(k)$ and $P_\beta(k)$ are the value of power of load on both power supply sections.

(4) Determine whether the active power of the electric traction load exceeds or equals to the threshold of the peak clipping condition based on the peak clipping condition:

If yes, the battery storage system enters the discharging state to determine $SOC_{bat}(k)$ whether it is working within the normal range. If it is within the normal range, the battery energy storage device discharges at constant power with maximum discharge power.

If not, skip to step 5.

(5) Determine whether the overall active power value of the traction load is larger than or equal to the flywheel discharge threshold value P_2:

If yes, the flywheel unit enters the discharge state to determine $w_{fw}(k)$ whether it is operating within the normal range. If within the normal range, the flywheel speed change is determined from the DC side voltage command.

If not, skip to step 6.

(6) Determine whether the overall active power of the traction load is less than or equal to the braking condition judgment threshold value P_3:

If yes, the battery storage system enters the charging state to determine $SOC_{bat}(k)$ whether it is working within the normal range. If it is within the normal range, the battery storage system is charged with the peak set charging power.

If not, skip to step 7.

(7) Determine the flywheel speed condition:

Judgment $w_{fw}(k)$ whether it is operating within the normal range. If so, the flywheel unit determines the flywheel speed change based on the DC side voltage command.

Otherwise, the entire energy storage system is in a pending state.

(8) Judge whether it is finished, if it is finished, stop the operation; if it is not finished, jump to the next decision point of step 2. The general step flow chart is presented in Fig. 6.

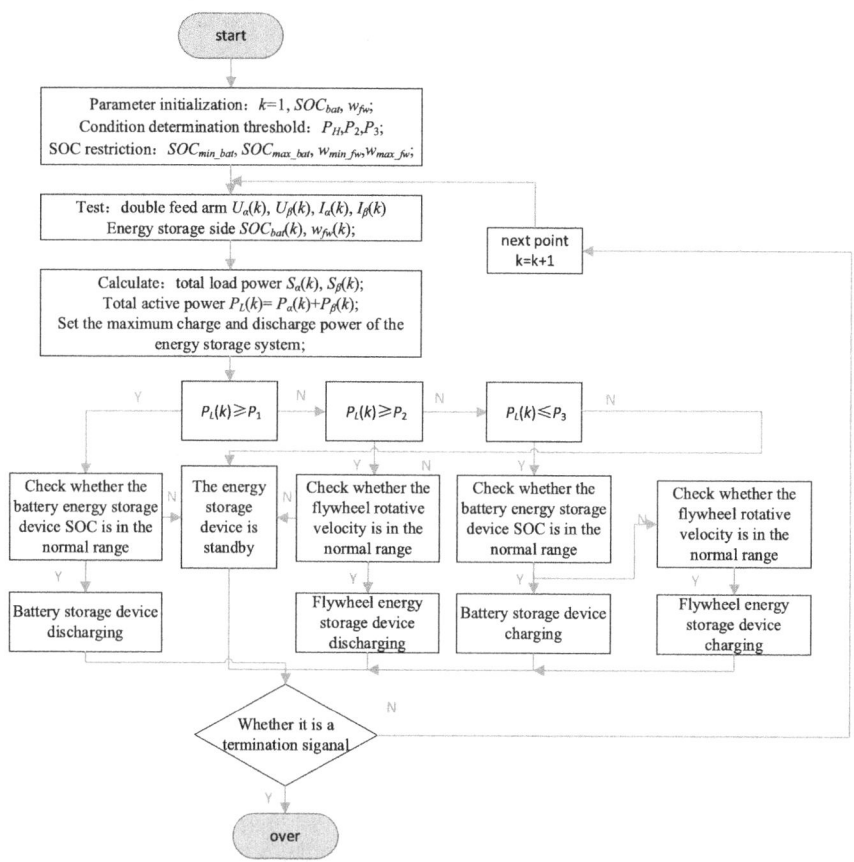

Fig. 6. Upper load detection condition determination flow II

3 Simulation Outcomes and Analysis

To verify the RPC coordinated control strategy with battery and flywheel energy storage, the feasibility of each control strategy and the efficacy of the corresponding energy management strategy, the simulation uses the data of a certain substation in Hebei Province on the peak day of a certain month and employs MATLAB to conduct the RPC simulation of the entire traction power provision system and RPC simulation of the energy storage system with batteries and flywheel. The main characteristics of the battery energy storage device are presented in Table 1:

Table 1. Battery energy storage device parameters table

Parameters used	Batteries
Energy storage capacity/MWh	0.989
Discharge power/MW	2
Charging power/MW	0.2
SOC range/%	20–80
Initial SOC status/%	80

The principal parameters of the flywheel energy storage device are presented in Table 2:

Table 2. Flywheel energy storage device parameters

Parameters used	Batteries
Stator winding/Ω	18.7
Inductance of stator d-phase winding/H	0.02682
Inductance of stator q-phase winding/H	0.02682
Moment of inertia/kg-m2	4.5e−5
Rotor flux linkage/Wb	0.1717
Number of pole pairs	2
Range of speeds/(rad/s)	200–800

The remaining traction network system parameters are presented in Table 3:

Table 3. Traction network system parameters table

Placement	Parameters	Measure value
Traction network	voltage	110 kV
	total capacity	1000 MVA
Traction transformer	capacity	50 MVA
	transformation ratio	110/27.5
Step-down transformer	capacity	10MVA
	transformation ratio	27.5/1

(*continued*)

Table 3. (*continued*)

Placement	Parameters	Measure value
Intermediate direct current link	voltage	1560–1740 V
Operating condition threshold	PH	19.545 MW
	P2	200k W
	P3	−200 kW
Flywheel response load variation direct current side voltage threshold	PL1	200 kW
	PL2	2 MW
	PL3	−200 kW
	PL4	−2 MW

Through MATLAB simulation modeling and analysis, after adding battery and flywheel energy storage system in the intermediate DC link of the RPC system, the battery charging and discharging intervals, flywheel charging and discharging intervals, and the three-phase current governance during a certain period are analyzed.

(1) Battery charge and discharge

The operation of the battery energy storage unit is presented in Fig. 7.

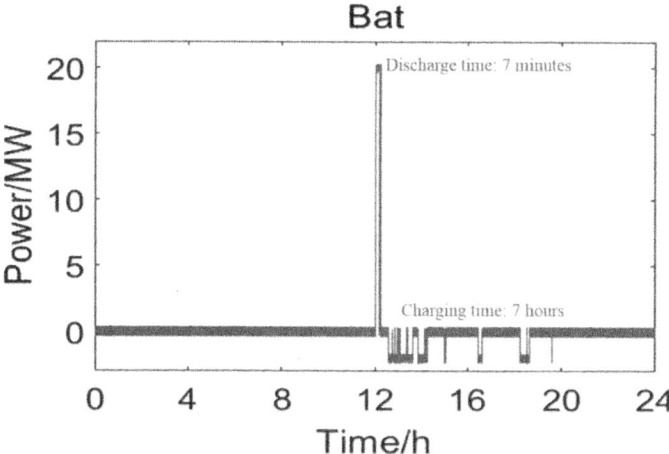

Fig. 7. The operating status of the battery energy storage device in the peak-clipping area

As shown in the figure above, the battery energy storage device is simply discharged for a total of 7 min during the peak load period, the battery is charged at a low power of 200 kW, and the subsequent charging interval takes about 7 h to replenish the power, so the battery device is only operated in a small interval each month, and the small rate of charging doesn't affect the flywheel regenerative braking energy utilization too much. The variation of the rotational speed of the flywheel unit during the process of battery charge and discharge, relevant data or a particular phenomenon is presented in Fig. 8 and Fig. 9.

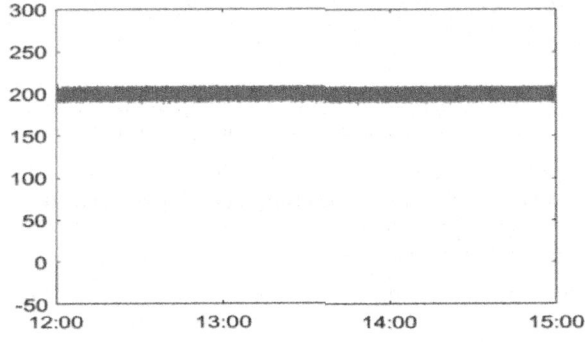

Fig. 8. Flywheel working condition I

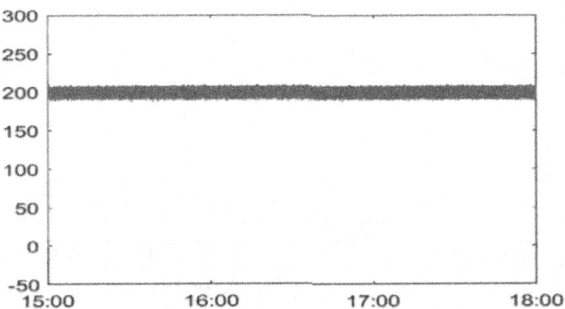

Fig. 9. Flywheel working condition II

The flywheel has been reduced to its lowest speed before the battery is discharged, and the flywheel maintains its speed unchanged throughout the battery charging and discharging intervals, during which time the flywheel does not perform regenerative braking energy utilization. The peak clipping effect of the battery energy storage device is presented in Fig. 10 and Fig. 11.

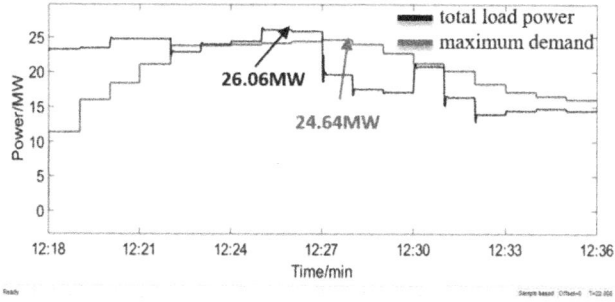

Fig. 10. Overall active power and maximum demand before load of energy storage device

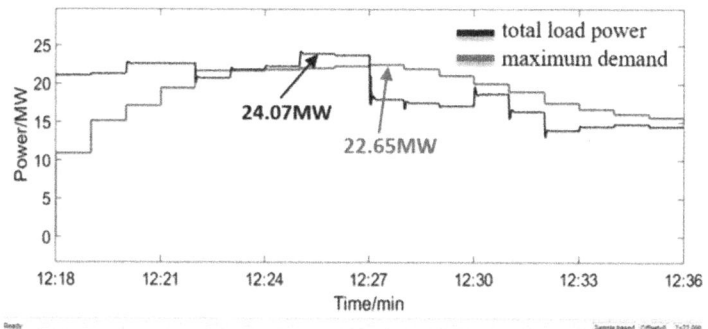

Fig. 11. Overall active power and maximum demand after load of energy storage device

(2) Flywheel charging and discharging

The remaining periods of flywheel operation are shown in Figs. 12, 13, 14, 15, 16, 17, 18 and 19.

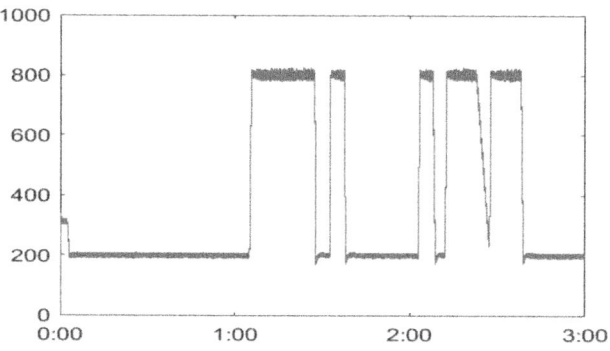

Fig. 12. Flywheel working condition III

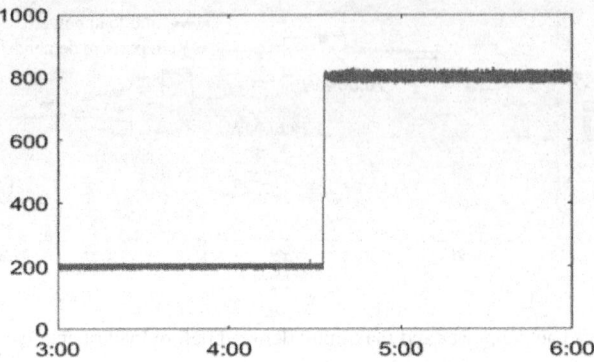

Fig. 13. Flywheel working condition IV

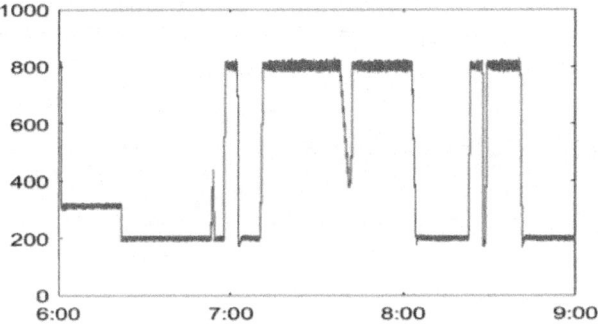

Fig. 14. Flywheel working condition V

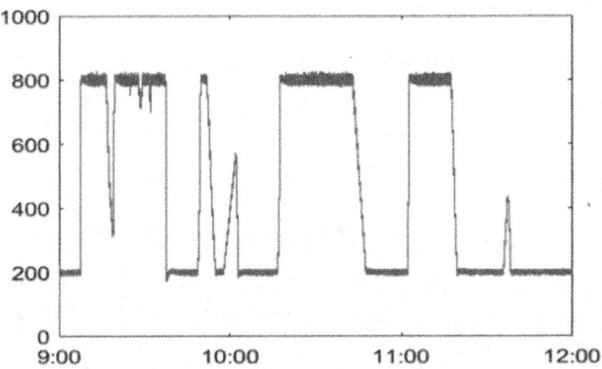

Fig. 15. Flywheel working condition VI

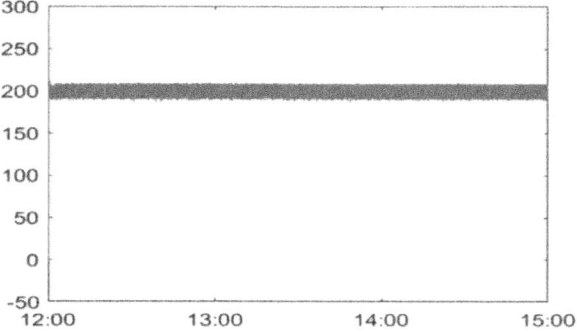

Fig. 16. Flywheel working condition VII

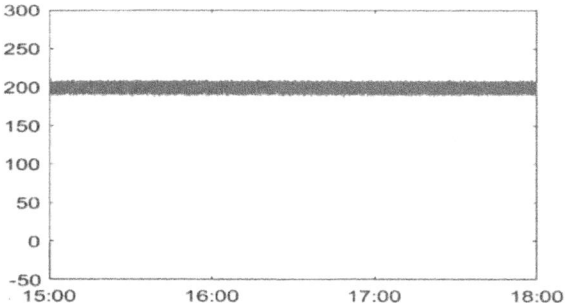

Fig. 17. Flywheel working condition VIII

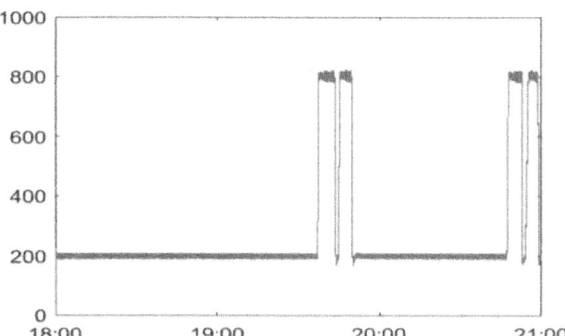

Fig. 18. Flywheel working condition IX

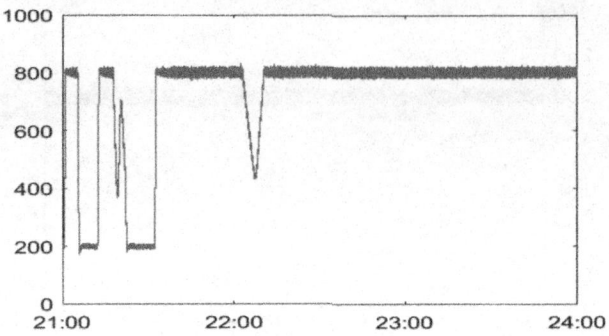

Fig. 19. Flywheel working condition X

As shown in the figure above, the flywheel energy storage device is charged and discharged a total of 31 times, which allows for effective regenerative braking energy utilization.

(3) Improvement of three-phase current imbalance

Between 12 h 18 min and 12 h 37 min in the morning of the same day, the RPC system with the energy storage link was involved in the power quality management before the situation is shown in Fig. 20 and Fig. 21.

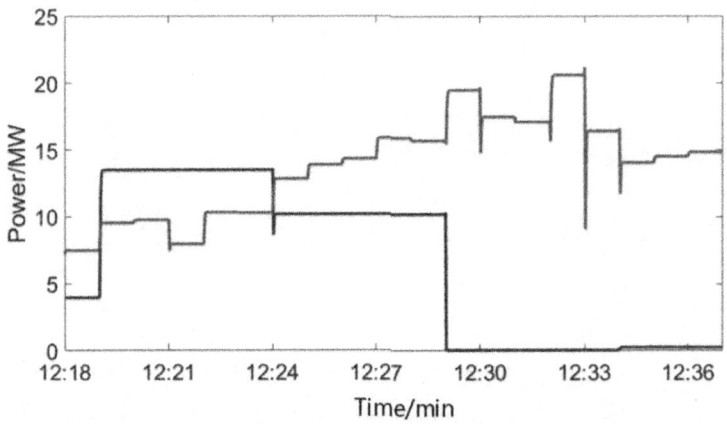

Fig. 20. Both sides of the load before power quality control

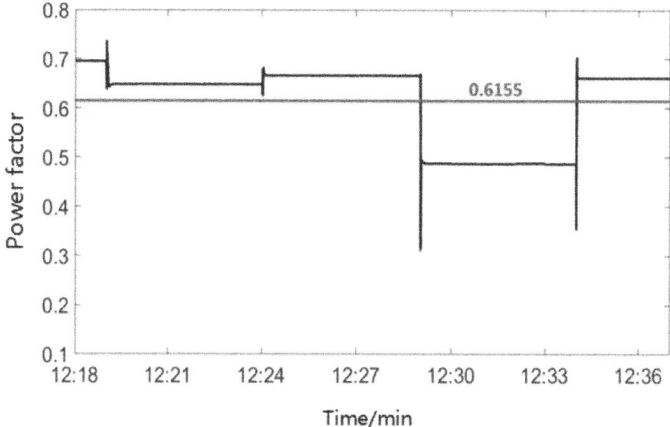

Fig. 21. Power factor before power quality control

From Fig. 20, at this moment, the active power values of the bridge arm loads on both sides are large, is the maximum demand value appears in the period. As depicted in Fig. 21, for the power factor at this time, it fluctuates between 0.7 and 0.5, the average power factor is only 0.6155. Moreover, the load changes on two flanks of the bridge arm are large at 12:29, and the power factor fluctuates significantly. The situation after the RPC system with an energy storage link participates in power quality management is presented in Fig. 22 and Fig. 23.

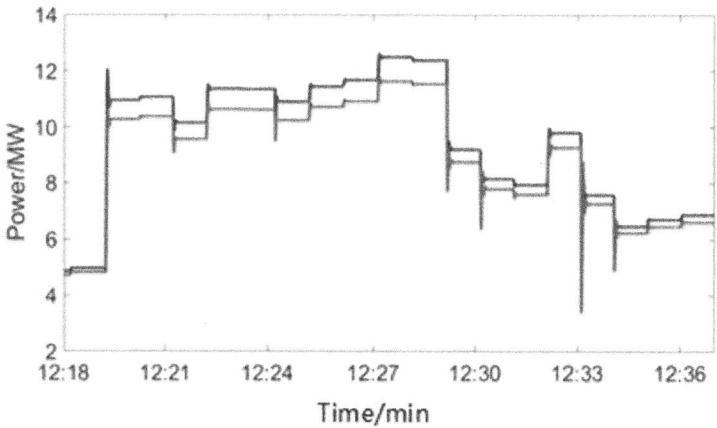

Fig. 22. Both sides of the load before power quality control

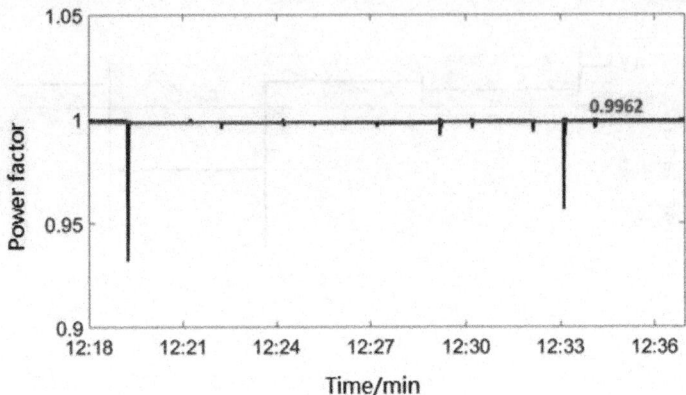

Fig. 23. Power factor before power quality control

From Fig. 22, the battery storage device has been in the discharging state during this period. At this time, in addition to balancing the active energy, the dual flanks of bridge arm also used regenerative braking energy stored in the battery for peak clipping. Meanwhile, from Fig. 23, we can see that compared with the power factor before adding the new RPC compensation device is basically 1, the average power factor reaches 0.9962, which is a great improvement, so the addition of the battery and flywheel storage system in the RPC intermediate DC link can be a very good power quality management.

4 Summary and Outlook

Swift speed, great capacity and supreme safety are the advantages of electrified railways. Nevertheless, the traction substation generates a substantial quantity of negative sequence, reactive power and harmonics, and simultaneously, a copious amount of regenerative braking energy due to the constant change of electric locomotive load power. The employment of energy storage systems and RPC compensation devices not only addresses various power quality issues in substations, but also diminishes peak loads, alleviates shocks, stabilizes the DC-side network voltage and improves the application proportion of regenerative braking energy.

In this article, based on the actual situation of the traction substation and under the precondition of ascertaining the secure and stable functioning of the overall system, appropriate control strategies are proposed for the RPC compensation device, the battery energy storage device and the flywheel energy storage device. Eventually, an RPC compensation system with battery and flywheel energy storage is formed, and the results of the coordinated work of each device are analyzed, which verifies the feasibility of the system for power quality, the employment of regenerative braking energy of the storage part, peak clipping control and the employment of RPC power transfer regenerative braking energy. The feasibility of the system for power quality, employment of recyclable braking energy in aspect of energy storage, peak clipping control and employment of regenerative braking energy in RPC power transfer is verified.

References

1. Khayyam, S., Ponci, F., Goikoetxea, J., et al.: Railway energy management system: centralized–decentralized automation architecture. IEEE Transactions on Smart Grid **7**(2), 1164–1175 (2016)
2. Huang, Y.: Study on two systems electricity price. Water Resour. Power **017**(003), 57–60, 72 (1999)
3. Wei, W., Hu, H., Wang, K., et al.: Energy storage scheme and control strategies of high-speed railway based on railway power conditioner. Trans. China Electrotech. Soc. **034**(006), 1290–1299 (2019)
4. Jiang, S., Shu, N., Liang, T., et al.: Control strategy and capacity optimization of energy-storage-based railway power conditioner. Sci. Technol. Energy Transit. **79**, 46 (2024)
5. Barros, L.A.M., Martins, A.P., Pinto, J.G.: Balancing the active power of a railway traction power substation with an sp-RPC. Energies **16**(7), 3074 (2023)
6. Qin, Y., Yuan, B., Pi, S.: Research on framework and key technologies of urban rail intelligent transportation system. In: Proceedings of the 2015 International Conference on Electrical and Information Technologies for Rail Transportation: Transportation, pp. 729–736. Springer, Heidelberg (2016)
7. Chen, H., Che, Y., Fu, R., et al.: Study on regenerative braking energy utilization and power quality control in electrified railways. In: IEEE International Power Electronics & Application Conference & Exposition (2018)
8. Cai, J., Xu, Q., Ye, J., et al.: Optimal configuration of battery energy storage system considering comprehensive benefits in power systems. In: 2016 IEEE 8th International Power Electronics and Motion Control Conference (IPEMC 2016 - ECCE Asia) (2016)
9. Choi, M., Lee, J., Seo, S.: Real-time optimization for power management systems of a battery/supercapacitor hybrid energy storage system in electric vehicles. IEEE Trans. Veh. Technol. **63**(8), 3600–3611 (2014)
10. Song, Z., Li, J., Han, X., et al.: Multi-objective optimization of a semi-active battery/supercapacitor energy storage system for electric vehicles. Appl. Energy **135**(dec.15), 212–224 (2014)

Experimental Virtual-Reality Assessment of a Cycling Environment Using a One-Boundary Drift-Diffusion Model

Kaori Nakamura[1(✉)], Shun Su[2], Yusak Susilo[2], and Daisuke Fukuda[1]

[1] Department of Civil Engineering, The University of Tokyo, Tokyo, Japan
310kaorinne712@g.ecc.u-tokyo.ac.jp
[2] University of Natural Resources and Life Sciences, Vienna, Austria

Abstract. This study investigates the effects of changes in the cycling environment on cyclists' dynamic behavioral responses and explores how the environment contributes to the cyclists' overall subjective rating of safety and comfort. Twenty-four participants cycling through virtual-reality simulations of environments of a local town road in Japan were exposed to six infrastructure design scenarios having different widths, bike lane colors, and lane separation types in the virtual reality. The estimation results of one-boundary drift-diffusion decision models across the different scenarios showed that designs with clear separations from car lanes facilitated faster reactions, aligning with higher subjective safety ratings. Moreover, a changepoint analysis was conducted to assess the cycling performance in terms of braking and steering over the different designs. The main conclusion of this study is that a drift-diffusion decision model can be applied to the analysis of cycling behaviors and such analysis might provide insights into safety infrastructure evaluation.

Keywords: Active transport modes · Cognitive and decision-making modeling · Road environment designs · Virtual-reality experiment

1 Introduction

Active modes of transportation, such as walking and cycling, are crucial in promoting healthy and sustainable urban environments. Encouraging the use of these modes requires a well-designed road environment that ensures travelers' safety and comfort. Understanding what constitutes a safe environment for these users is essential.

Many experimental studies have been conducted to address these fundamental issues. For example, Gössling and McRae (2022) [2] conducted a large-scale online survey in Berlin and pointed out the importance of wide bicycle tracks separating cyclists from motorized and pedestrian traffic. In addition, bicycle simulators with virtual reality (VR) environments are popular in experimental studies. Nazemi et al. (2022) [4] analyzed the perceived level of safety from stated preference data. Guo et al. (2023) [3] studied not only the stated preference but also the gaze and heart rate, and they found that in unsafe bicycle environments, there was a shorter fixation duration relating to a higher hazard estimation by cyclists.

Previous research has mainly focused on subjective safety or the likelihood of collisions, and there needs to be more analysis of the real-time behaviors of travelers and the dynamic transitions of their discomfort perceptions. The evaluation methods typically adopted in past studies on post-ride or pre-ride evaluations by cyclists may not accurately capture the immediate reactions and discomfort of cyclists, with the evaluations possibly being biased or retrospective. In contrast, dynamic analysis captures immediate, observable reactions and perceptions and thus provides objective insight into how travelers interact with their environment in real time and enables a more accurate assessment of safety and risk factors. Although some past experimental studies have used physiological and behavioral data to evaluate road designs, they only assessed road designs by comparing time-average physiological and behavioral data across different designs. Dynamic analysis has the potential to identify specific moments and conditions that trigger road users to perceive the road as dangerous or uncomfortable. Such analysis is thus expected to provide infrastructure designs that mitigate these triggers, enhancing overall road safety and traveler comfort.

Drift-diffusion models (DDMs) are cognitive models used to initially analyze two-choice decision-making tasks on the basis of evidence accumulation. The models assume that decisions are made when accumulated evidence reaches one of two thresholds, at which point the corresponding decision is made. DDMs have been used in numerous psychological and cognitive neuroscience studies. Currently, an increasing number of traffic engineering studies are using DDMs. For example, Pekkanen et al. (2022) [6] and Theisen et al. (2024) [10] used a two-choice DDM to study pedestrian road-crossing decisions. These studies built models that explain how pedestrians decide when to cross a road, considering how they use their senses, such as seeing gaps in the traffic or a car slowing down to signal that it will yield its right of way.

DDMs have three main parameters, namely the drift rate, decision boundary/threshold, and non-decision time. The drift rate reflects how quickly evidence for or against crossing builds up. The response threshold is the level of evidence needed for the person to make a decision. DDMs divide the reaction time (RT) into two segments, namely the non-decision time and decision time. The decision time is the time required for evidence accumulation. The non-decision time excludes such thinking and focuses on the automatic tasks the brain does before a person even considers crossing. This includes seeing the situation and preparing the body to move. DDMs thus capture the inherent trade-off between speed and accuracy in decision making.

Ratcliff and Van Dongen (2011) [9] presented a one-boundary diffusion-decision model for performance in one-choice tasks. Ratcliff and Strayer (2014) [8] adopted a one-boundary DDM for a driving task. Within the model framework, noisy evidence accumulates progressively following stimulus presentation. This accumulation continues until a predefined decision threshold, or boundary, is reached. At this point, a response is initiated. This one-boundary DDM has the three parameters given above. These parameters capture the differences in reaction between models, making it possible to analyze the road infrastructure design from the perspective of the ease or speed of the response.

By considering the potential of DDMs in the safety and comfort analysis of an active transport mode and its driving environment, this study investigates the effects of riding environmental changes on cyclists' dynamic physiological and behavioral responses and explores how the conditions contribute to the cyclists' overall subjective safety and comfort. Specifically, this study examines the single-choice DDM in fitting RT data for a cycling task. The study creates a VR environment in which participants cycle six road designs and explores how the parameters of the single-choice DDM change under the various environmental conditions.

2 Methodology

2.1 Experiment Design

This study investigated the cognitive effects of road infrastructure using a bicycle simulator. The bike simulator was developed to represent the scenarios and to record data of cycling behaviors such as speed, braking, and lane position data of the bicycle riders. The bicycle was equipped with rotation sensors on the handlebar and a pedal and bending sensors on the brakes to translate the bicycle's movements in VR. The platform had a projector that provided a front view and a sound system. Participants were asked to ride the bike as they would in the real world in terms of using the brakes, handles, and pedals.

The immersive virtual environments in VR were deployed in the Unity 3D game engine. During the experiment, participants were exposed to six road infrastructure design scenarios in VR. These scenarios were modeled after Numazu Sun-Sun Street Road in Numazu, Shizuoka Prefecture, Japan, using a free three-dimensional city model. This road has two lanes for traffic and a dedicated pedestrian lane, as frequently found in the centers of mid-sized cities in Japan. The bike lane was developed in VR following existing guidelines published by the Japanese Ministry of Land, Infrastructure, Transport and Tourism. The cycling course was designed to be straight and approximately 560 m long

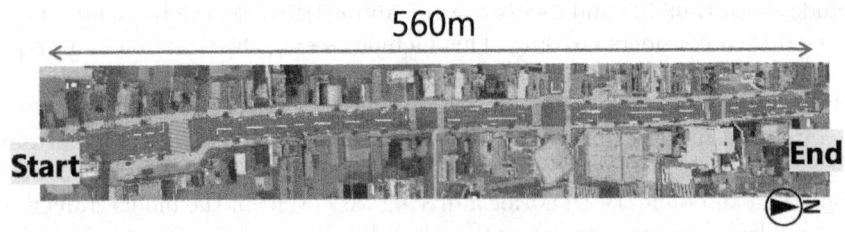

Fig. 1. Course configuration

Fig. 2. Bicycle simulator

Table 1. Designs of the bike lane

	width	color	separation type
Design 1	1 m	Asphalt/Arrow	Line (Visual)
Design 2	1 m	Blue/Pictogram	Block (Physical)
Design 3	1.5 m	Asphalt/Arrow	Block (Physical)
Design 4	1.5 m	Blue/Pictogram	Line (Visual)
Design 5	2 m	Asphalt/Arrow	Line (Visual)
Design 6	2 m	Blue/Pictogram	Block (Physical)

to limit the ride time and lessen motion sickness. The course configuration is shown in Fig. 1 (Fig. 2).

Six designs having different road widths, road separation materials, and road colors were used for the cycling environments, as shown in Table 1 and Fig. 3. Each immersive virtual street scenario represented an alternative bicycle infrastructure design, namely 1) a narrow 1.0-m wide bike lane with minimal road signage and a line indicating a separation from the car lane, 2) a narrow 1.0-m-wide bike lane with a differentiating blue color and a physical separation from the car lane by blocks, 3) a 1.5-m-wide bike lane with minimal road signage and a physical separation from the car lane by blocks, 4) a 1.5-m-wide bike lane with a differentiating blue color and a line indicating separation from the car lane, 5) a wide 2.0-m-wide bike lane with minimal road signage and a line indicating separation from the car lane, and 6) a 2.0-m-wide bike lane with a differentiating blue color and a physical separation from the car lane by blocks. All other factors, such as the traffic volume, vehicle speeds, and car lane width, were constant across the scenarios. Each participant was instructed to cycle through all six environments in the bicycle simulator. The order in which participants cycled in these design scenarios was randomized to avoid possible ordering biases in the response. Moreover, the participants completed a questionnaire before and after each cycling each road design.

Fig. 3. Designs of the cyclist viewpoint in VR

2.2 Participants

Participants were recruited directly by the research group and through word of mouth. The participants were required to be 18 years old or older without any health condition preventing them from riding a bike or receiving visual stimuli information. The 24 tested participants comprised 9 women and 15 men with a mean age of 35.8 years, standard deviation in age of 12.9 years, maximum age of 60 years, and minimum age of 20 years. No participant dropped out due to motion sickness or feeling unwell. All tests were conducted at the Institute for Transport Studies at the University of Natural Resources and Life Sciences (BOKU), Vienna, Austria, in December 2023. The experiment was carried out with the approval of the BOKU ethical board and complied with data privacy laws of the European Union (Table 2).

Table 2. Socio-demographic characteristics

Variable	Sample (N = 24)
Gender: Male	62.5%
Gender: Female	37.5%
Age: 18–29 years old	41.7%
Age: 30–49 years old	41.7%
Age: 50 + years old	16.7%
Nationality in Europe Ratio	66.7%
Nationality in Asia Ratio	33.3%
Biking Attitude: Expert Ratio	8.3%
Biking Attitude: Confident Ratio	12.5%
Biking Attitude: Occasional Ratio	70.8%
Biking Attitude: Beginner Ratio	8.3%

2.3 Data Collection

The collected data were sociodemographic data, stated preference data, physiological data comprising heart rate measurements, and skin conductance and cycling performance data.

Sociodemographic Data. Sociodemographic data on gender, age, nationality, educational level, and biking attitude were collected in pretest surveys. The participants' bicycle use habits and confidence in riding a bike were measured to assess biking attitudes. The participants were asked to answer 16 questions. Following this, the participants completed screening questions, including questions on their frequency of drinking alcohol and caffeine and their length of sleep.

Stated Preference Data. Before each riding task, the participants were asked to give their first impressions on biking infrastructure, and after each riding task, the participants evaluated their assigned safety and comfort levels on a seven-point Likert scale across the three dimensions of safety, stress, and comfortability. The safety section focused on the perceived accident risk. The stress section addressed obstruction perception, fear of proximity, number and speed of other road users, prediction of movements, and the mental and physical effort required of the participants. For comfortability, there were two questions on the participants' comfort level and willingness to ride a bike regularly in the environment.

The participant-reported ratings of perceived safety for each infrastructure scenario were susceptible to systematic and random biases. The underlying assumption that all participants use identical rating scales could not be guaranteed. Therefore, the participants were also asked to select the most and least preferred design scenarios according to their perceived safety at the end of the experiment.

Heart Rate Measurement. Smart wearable devices connected to a smartphone app via Bluetooth collected heart rate (HR) data at a frequency of 1 Hz on the participants' chests. A researcher turned on the data collection function before the experiment, and no further action was required.

Skin Conductance. Smartwatches connected to a smartphone app via Bluetooth collected electrodermal activity (EDA) data expressed as microsiemens (μS) at a frequency of 4 Hz. The researcher turned on the data collection function before the experiment, and no further action was required.

Cycling Performance. Coordination, steering, and braking data were extracted from Unity environments. The performance data were automatically recorded at a frequency of 12.5 Hz.

3 Results of Fundamental Data Analysis

3.1 Stated Preference Data

On average, participants rated Design 6 as the most comfortable on a seven-point Likert scale (2.0 m wide, colored, physically divided) (= 6.12). Design 6 was followed by Design 5 (2.0 m wide, not colored, visually divided) (= 5.29), Design 4 (1.5 m wide, not colored, physically divided), Design 3 (1.05 m wide, colored, visually divided), Design 2 (1.0 m wide, colored, physically divided), and Design 1 (1.0 m wide, not colored, visually divided), which had the lowest subjective comfort rating (= 3.21), as shown in Fig. 4. Other stated preference results, such as the perceived accident risk and fear of proximity of cars, had the same trends across the six designs. Design 6 had significantly higher comfort ratings than Designs 1 ($\beta = 2.1875, s.e. = 0.2644, p = 2.22 \times 10^{-16}$), 2 ($\beta = 0.7917, s.e. = 0.1477, p = 8.25 \times 10^{-8}$), 3 ($\beta = 0.6992, s.e. = 0.1067, p = 5.71 \times 10^{-11}$), and 4 ($\beta = 0.2604, s.e. = 0.0711, p = 0.00025$). However, there was no significant difference in comfort ratings between Designs 5 and 6 ($\beta = 0.0375, s.e. = 0.0674, p = 0.5779$). Participants indicated a preference toward Design 6 (with 67% of participants ranking the design as the most comfortable) and least preferred Design 1 (with 54% of participants ranking the design as the least comfortable). No age or gender difference was found in the stated preferences (Fig. 5).

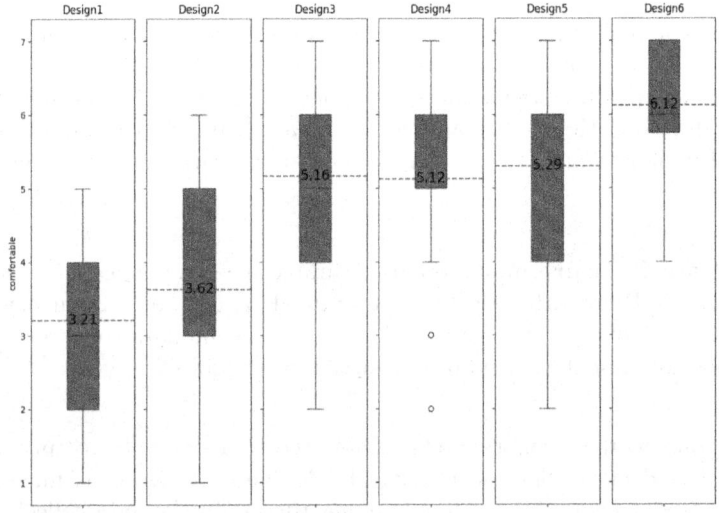

Fig. 4. Comfort level of each design

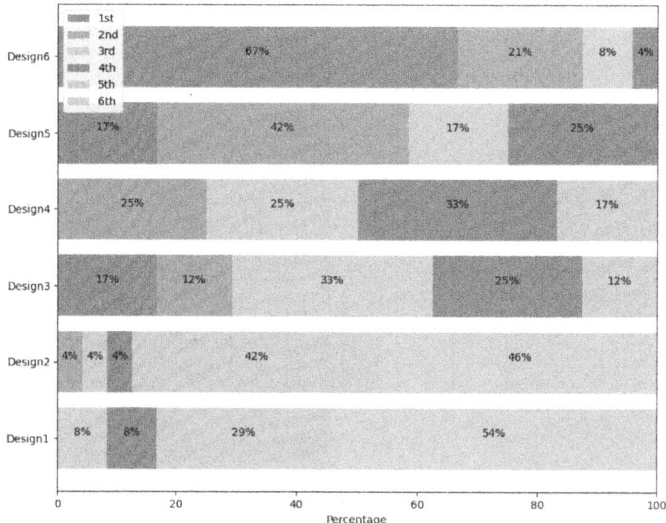

Fig. 5. Percentages of participant rankings for each design

3.2 Physiological Data

A linear mixed model was used to assess the effect of the road design on the HR and EDA. The road design had a significant effect on the mean HR during each riding task ($\beta = -0.281, s.e. = 0.127, p = 0.027$), indicating that the different road designs affected the participants' HR levels differently. In addition, the task had a significant effect on the difference in the mean HR between the rest time before a ride and each riding task ($\beta = -0.437, s.e. = 0.214, p = 0.042$), suggesting that the task type affected the change in HR. The road design had no significant effect on the mean EDA ($\beta = -0.021, s.e. = 0.036, p = 0.567$) or the difference in the EDA between resting and riding ($\beta = -0.017, s.e. = 0.022, p = 0.432$), suggesting that EDA level was not affected by the design of the road or the type of task. Moreover, the abrupt changes in the participants' HR and EDA were analyzed using the binary segmentation algorithm with a radial basis function model. The findings suggest that the road design affected physiological responses, specifically the HR, during cycling.

3.3 Cycling Performance

A changepoint analysis was conducted to assess the cycling performance, especially that relating to braking and steering, over the different designs. No significant difference was found between designs, as shown in Fig. 7 (Fig. 6).

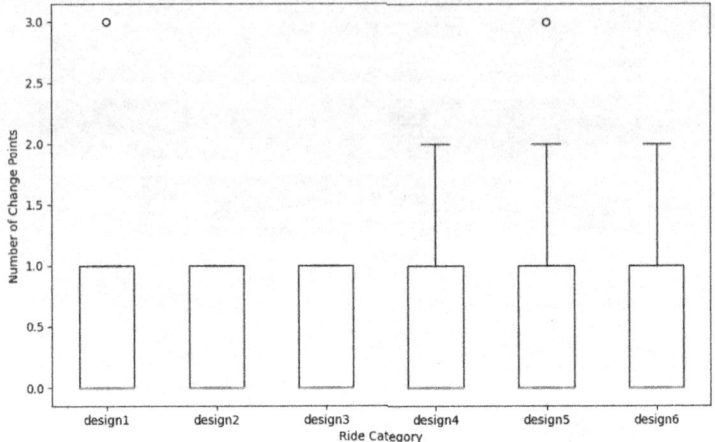

Fig. 6. Change points of HR

4 Estimation Results of One-Boundary DDMs

4.1 Estimation Procedure

In the experiment, participants rode a bicycle straight to the end of the street. As stimuli, cars traveled in the lane next to the bike lane. Some participants used their brakes as they may have felt that they were too close to a vehicle, and this reaction was analyzed. Specifically, the braking response was extracted from the cycling performance data immediately after a car passed by, and the RT was taken as the time between a vehicle appearing on the display and the participant applying the brakes.

A DDM was fitted to the modeling set for each design, and the parameters were independently estimated. The one-boundary DDM was adopted in this study. The experiment had only one choice, which was whether to use the brakes. In this model, the observation to be fit is a distribution of RTs for hitting the response key. For the one-boundary DDM, there is an explicit solution for the distribution of RTs for a single positive drift rate (i.e., the inverse Gaussian or Wald distribution) (Ratcliff, Citation2015) [7]. The current model faces a challenge in the form of negative drift rates. These values, originating from the left tail of the empirical distribution of drift rates across trials, lack an analytical mathematical solution within the framework of RT distributions. The model was therefore replicated with a simulation using a random walk approximation to the diffusion process with 20,000 iterations and a step size of 0.5 ms (Tuerlinckx et al., 2001) [11].

In this study, the 0.05, 0.10, ..., and 0.95 quantiles of the RT distribution were calculated from the observed RT data to fit the model to the data. In addition, as in the work of Ratcliff and Strayer (2014) [8], the first two quantiles were grouped to produce much better patterns of correlations among model parameters. The quantiles were used to find the proportions of responses in the

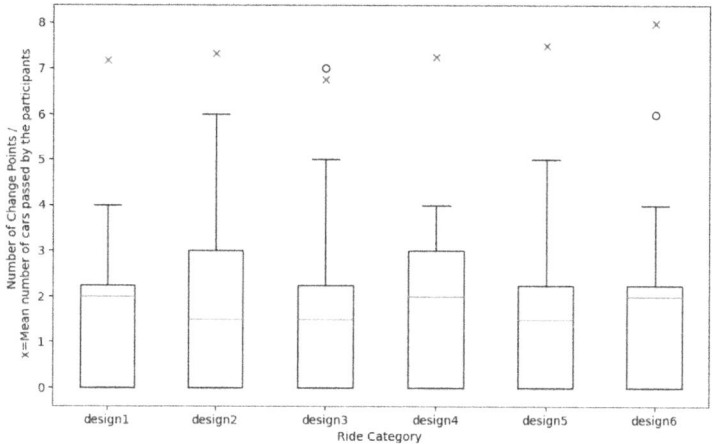

Fig. 7. Change points of braking

estimated RT distribution between the observed data quantiles, and they were multiplied by the number of observations to give the expected values (E). The proportions of responses between the observed data quantiles fixed to 0.05 were multiplied by the number of observations to provide the observed values (O). A chi-square statistic was computed 1, and the model parameters were adjusted by a genetic algorithm (GA) to find a unique minimum chi-squared value.

$$\chi^2 = \Sigma \frac{(O - E)^2}{E} \qquad (1)$$

The GA parameters were configured as follows. The initial population size was set at 100, where 100 sets of model parameters, called genes, were considered at each step. The cost function, defined as the chi-square statistic, was computed for each set within this population. In each generation, 50% of the new population was derived by selecting the best chromosomes, which minimized the cost function. Another 30% of the new population was created using the two-point crossover method, whereas the remaining 20% underwent mutation. As generations advanced, the population evolved toward an optimal solution. Consequently, the GA generated a population of chromosomes at each iteration, with the best chromosome converging toward the optimal solution. The GA process included 100 restarts, where each cycle involved the generation of 20,000 simulated RTs per model evaluation. This modeling followed past studies [1,5,7].

4.2 Estimation Results

As mentioned previously, the changepoint of braking was considered to be a reaction of each respondent, and the RT was calculated as the time from when a car appeared on the display to when the braking changepoint was detected. This

RT distribution was used to estimate parameters in the one-boundary DDM. Figure 8 presents the box plot of RTs for each design.

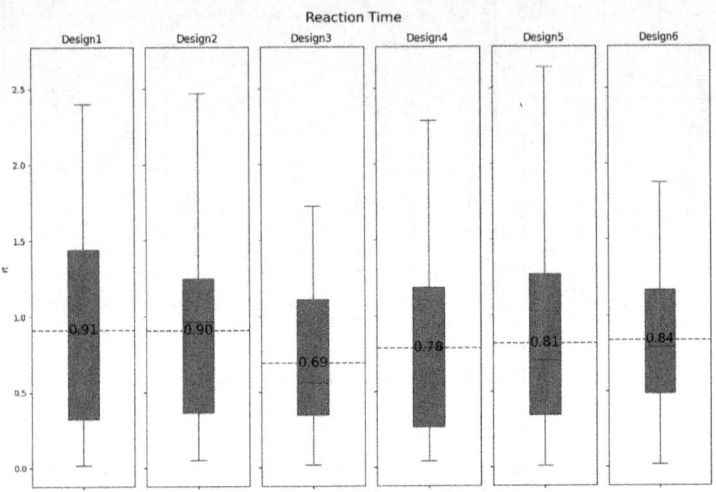

Fig. 8. Observed reaction time for six designs

Five parameters of the decision process, namely the decision boundary, mean non-decision time, standard deviation in the drift rate across trials, range of the distribution of the non-decision time, and mean drift rate, were estimated to fit the model to the data. However, Ratcliff (2015) [7] pointed out that there is an issue of identifiability in model parameters in one-boundary DDMs, such that only two parameters out of the drift rate, across-trial standard deviation in the drift rate, and boundary setting can be uniquely determined. Thus, two ratios of the parameters are identified. In this model fitting, the standard deviation in the drift rate across trials was fixed at 0.6, such that only four parameters were estimated.

Table 3 gives the estimated parameters of DDMs for each design setting. There were 18 degrees of freedom in the data, identical to the number of quantiles. As there were four parameters, there were 13 degrees of freedom, and the critical value of the chi-square value was 22.36 for 13 degrees of freedom. In the table, a is the boundary separation, T_{er} is the non-decision component of the response time, η is the standard deviation of the drift rate, S_t is the range of the distribution of non-decision times, and v is the mean of the drift rate.

Cumulative distributions for the model fits and data for each design are shown in Fig. 9. The figure shows the good correspondence between the observation data and theoretical data predicted using the estimated parameters. A two-way analysis of variance on the threshold values (a) and mean drift rate (v) showed no significant interaction between the effects of the designs.

Table 3. Mean RTs and estimated parameters of the one-boundary DDMs

	RTs	a	T_{er}	η	S_t	v	χ^2
Design1	0.962	0.542	0.659	0.600	1.935	2.181	14.655
Design2	0.850	0.694	0.323	0.600	1.470	1.625	33.155
Design3	0.722	0.399	0.511	0.600	1.489	2.402	18.949
Design4	0.893	0.297	0.620	0.600	1.981	2.081	15.962
Design5	0.890	0.165	0.722	0.600	1.944	2.295	39.367
Design6	0.842	0.526	0.430	0.600	1.596	1.657	26.816

5 Discussion

This study investigated the effects of different designs on parameters of the one-boundary DDM, especially on the drift rate and threshold, and examined how these effects relate to stated preference data, which previous studies have mainly focused on.

The analysis of the stated preference data found that Design 6 was clearly the most preferred and safest design whereas Design 1 was the least preferred design and considered unsafe.

Estimation results of the DDMs provide insights into the temporal dynamics of the reaction process. The drift rate represents the quality of evidence accumulation, with a higher drift rate indicating more sufficient evidence accumulation (Daneshi et al., 2020) [1]. Ratcliff & Strayer (2014) [8] concluded that distraction alters performance by affecting the rate of evidence accumulation (drift rate) and increasing the boundary settings. In our modeling, Design 1 had the longest mean RT, the highest threshold, and a lower mean drift rate across Designs 1, 3, and 4, with a low chi-square value. Among these three designs, Design 3 had the best parameters, meaning the participants reacted more quickly when they received stimuli from cars.

A comparison of parameters from DDM analysis with stated preference survey data revealed that subjectively comfortable designs had higher drift rates, which means that participants accumulated evidence more efficiently and made quicker decisions for these designs. This relationship suggests that design elements perceived as more comfortable facilitate faster cognitive processing and reduce decision-making times. Consequently, such designs can enhance the overall safety and usability of shared spaces by eliciting quicker and more confident responses from road users.

It was demonstrated that road infrastructure evaluation is feasible using a one-boundary DDM. Notably, the safety evaluation results of the one-boundary DDM were consistent with the results obtained from the stated preference survey, and they aligned with the results of Gössling and McRae (2022) [2], who found that lane widths of 1.5 m or more were preferred in online surveys. To the best of the authors' knowledge, the present study is the first study to apply the one-boundary DDM to cyclists. The findings indicate that this model, for-

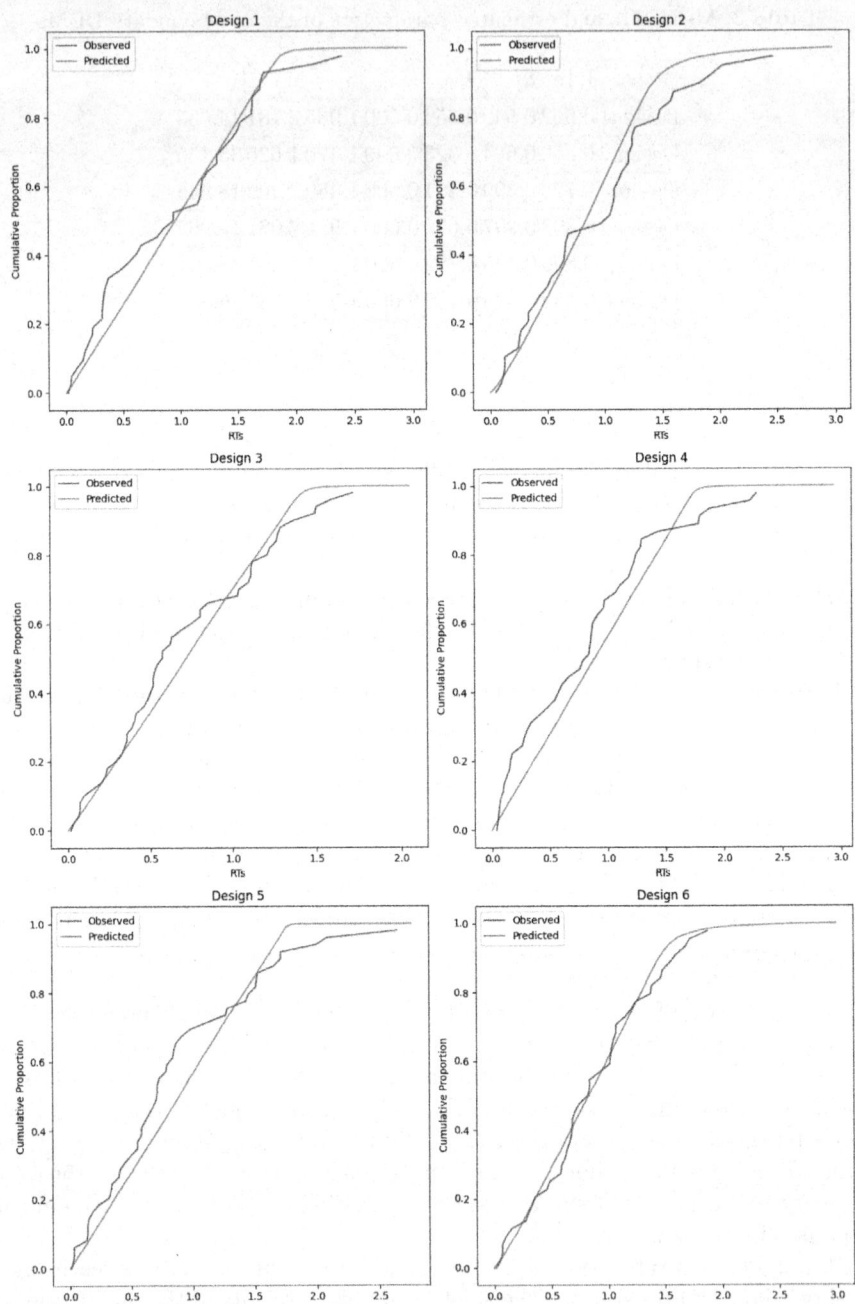

Fig. 9. Cumulative RT distributions for the six designs

merly used in evaluating driver decisions, can be effectively applied to cyclists. The results are expected to facilitate detailed evaluations of road infrastructure using cognitive models, contributing to the development of safe and comfortable cycling environments.

The main conclusion of this study is that one-boundary drift-diffusion modeling can be applied to the analysis of cyclist behavior, and such analysis will potentially provide insights into safety infrastructure evaluation. However, the study had limitations. First, there were only a limited number of participants. Future studies should include a broader range of participant attributes. Second, our analysis focused exclusively on braking reactions when participants encountered nearby passing cars and did not consider alternative reaction scenarios. This methodology may not be optimal for infrastructure safety analysis as it does not encompass the normal cycling behaviors of participants. Our approach only assessed safety according to the rapid responses of riders to dangerous situations. Moreover, future research could adopt two-choice DDMs to assess safety by analyzing reactions in typical cycling behaviors. Finally, validating RTs using VR is essential to enhancing the real-world applicability of experiments.

Disclosure of Interests. The authors declare that they have no relevant or material financial interests that relate to the research described in this paper.

References

1. Daneshi, A., Azarnoush, H., Towhidkhah, F.: A one-boundary drift-diffusion model for time to collision estimation in a simple driving task. J. Cogn. Psychol. **32**(1), 67–81 (2019). https://doi.org/10.1080/20445911.2019.1688336
2. Gössling, S., McRae, S.: Subjectively safe cycling infrastructure: new insights for urban designs. J. Transp. Geogr. **101**(103340) (2022). https://doi.org/10.1016/j.jtrangeo.2022.103340
3. Guo, X., Tavakoli, A., Angulo, A., Robartes, E., Chen, T.D., Heydarian, A.: Psycho-physiological measures on a bicycle simulator in immersive virtual environments: how protected/curbside bike lanes may improve perceived safety. Transp. Res. F: Traffic Psychol. Behav. **92**, 317–336 (2023). https://doi.org/10.1016/j.trf.2022.11.015
4. Nazemi M, Van Eggermond MAB, Erath A, Schaffner D, Joos M, Axhausen KW.: Studying bicyclists' perceived level of safety using a bicycle simulator combined with immersive virtual reality. Accid Anal. Prevent. **151**(105943) (2022). https://doi.org/10.1016/j.aap.2020.105943
5. Patanaik, A., Zagorodnov, V., Kwoh, C.K.: Parameter estimation and simulation for one-choice Ratcliff diffusion model. In: Symposium on Applied Computing-Association Special Interest Group on Applied Computing (SIGAPP) (2014). https://doi.org/10.1145/2554850.2554872
6. Pekkanen, J., et al.: Variable-drift diffusion models of pedestrian road-crossing decisions. Comput. Brain Behav. **5**(1), 60–80 (2021). https://doi.org/10.1007/s42113-021-00116-z
7. Ratcliff, R.: Modeling one-choice and two-choice driving tasks. Attent. Percept. Psychophys. **77**(6), 2134–2144 (2015)

8. Ratcliff, R., Strayer, D.: Modeling simple driving tasks with a one-boundary diffusion model. Psychon. Bull. Rev. **21**(3), 577–589 (2013). https://doi.org/10.3758/s13423-013-0541-x
9. Ratcliff, R., Van Dongen, H.P.A.: Diffusion model for one-choice reaction-time tasks and the cognitive effects of sleep deprivation. Proc. Natl. Acad. Sci. U.S.A. **108**(27), 11285–11290 (2011). https://doi.org/10.1073/pnas.1100483108
10. Theisen, M., Schießl, C., Einhäuser, W., Markkula, G.: Pedestrians' road-crossing decisions: Comparing different drift-diffusion models. Int. J. Hum.-Comput. Stud. **183**(103200) (2024). https://doi.org/10.1007/s42113-021-00116-z
11. Tuerlinckx, F., Maris, E., Ratcliff, R., De Boeck, P.: A comparison of four methods for simulating the diffusion process. Behav. Res. Methods Instrum. Comput./Behav. Res. Methods Instrum. Comput. **33**(4), 443–456 (2001). https://doi.org/10.3758/BF03195402

Reinforcement Learning Based Smart Charging for Electric Vehicle Fleet

Biao Xu[1,2], Aivars Rubenis[3,4], and Chao Long[1(✉)]

[1] School of Electrical Engineering, Electronics and CS, University of Liverpool, Liverpool, UK
xubiao@tju.edu.cn, Chao.Long@liverpool.ac.uk
[2] School of Electrical and Information Engineering, Tianjin University, Tianjin, China
[3] Lesla Ltd., Weybridge, UK
aivars.rubenis@lesla.eu
[4] Latvia University of Life Sciences and Technologies, Jelgava, Latvia

Abstract. The aggregated flexibility of electric vehicle (EV) charging presents significant potential for mitigating the adverse impacts that EV mass adoption imposes on the power grid. A three-stage framework, consisting of aggregation, bidding, and disaggregation, is proposed to enable smart charging for multiple EVs. To address modeling errors and uncertainties, a data-driven reinforcement learning (RL) algorithm is employed to optimize the bidding decisions for the aggregated EVs and to coordinate the bidding and disaggregation processes. The effectiveness of the proposed method is validated using both simulated and real-world datasets. With RL-based smart EV charging, the daily costs for each season are reduced, and the overload on distribution transformers is alleviated. Using simulated EV charging data for charging at home, the average daily cost is reduced by approximately 11%. When the proposed algorithm is applied to the real-life EV chargers in a commercial hub in England, UK (where EVs predominantly charge during the day), the average daily cost is reduced by approximately 3.83%, and transformer capacity violations are eliminated for all seasons except Autumn. The cost reduction with the commercial hub example is lower than that of the home charging case, because most cars park in the commercial hub for a short period of time during daytime and provide little flexibility for rescheduling.

Keywords: Aggregator · Electric Vehicle · Reinforcement Learning · Smart Charging

1 Introduction

The electrification of transportation provides an important opportunity for energy efficiency and sustainability [1]. Even though the number of electric vehicles (EVs) is growing rapidly both in the UK and Europe, reaching 931 thousand in the UK [2] and 4.4 million in EU at the end of 2023 [3], it still represents less than 3% of all vehicles in most countries and less than 2% of energy used in transportation [4]. However, taking into account the aspirations of UK and EU to gradually phase out internal combustion engine vehicles from 2035 onwards [5], the share of the energy used for EV charging will increase drastically and may present a tremendous challenge to the operation of the distribution networks, such as the transformer capacity violation [6]. At the same time, EVs offer flexibility as battery storage systems. Exploiting and optimizing the aggregate flexibility of these EVs fleet offer significant potential for mitigating the negative impacts caused by EVs [7].

There are many studies on the EV charging optimization problem, which can be divided into individual optimization methods [8] and aggregate optimization methods [9]. The individual optimization method directly optimizes the charging power of individual EVs. This ensures an optimal solution of the optimization; however, it is suffering from scalability problem. When managing a large number of EVs' charging activities simultaneously, the computation burdens for the decision-maker are becoming unacceptable. In contrast, the aggregate optimization is based on the aggregate modeling of the EV cluster. The large decision space is greatly reduced to one-dimension. However, the aggregate flexibility modeling brings about modeling errors and uncertainty problems [10], which hinders the utilization of the aggregate optimization methods.

Inspired by the current works, a model-free data-driven framework is proposed for achieving the smart charging of aggregate EVs. The main contributions include:

- A three-stage framework including aggregation, bidding and disaggregation is proposed for the smart charging of clustered EVs.
- To handle modeling errors and uncertainties, a data-driven reinforcement learning (RL) algorithm is employed to optimize the bidding decisions of aggregate EVs.
- The effectiveness of the proposed method is validated in both simulated dataset and real-world dataset. Under both datasets, the total EV charging cost is reduced and the transformer capacity violation is alleviated or mitigated.

2 Problem Formulation

2.1 Overall Framework

This paper investigates the participation of aggregate EVs in the intraday energy market, while considering limited transformer capacity. Figure 1 shows an overview of the proposed framework, which includes three stages: 1) *aggregation stage*; 2) *bidding stage*; and 3) *disaggregation stage*. An EV aggregator interacts with individual EVs during *aggregation stage* and *disaggregation stage*, to perform aggregate flexibility evaluation and charging power allocation, respectively. According to the aggregate flexibility, the EV aggregator optimizes the hourly energy bidding and submits it to the intraday energy market during the *bidding stage*.

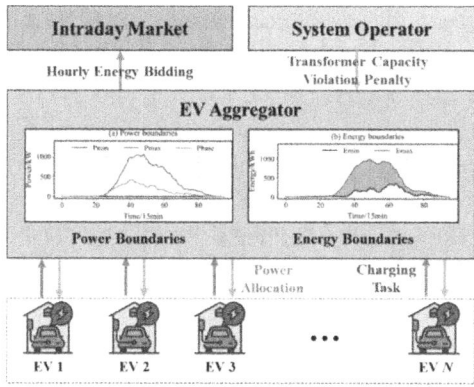

Fig. 1. Overall framework.

The detail of each stage is as follows:

Aggregation Stage. The EV aggregator collects the charging task information, including the departure time and charging demand of each newly arrived EV. Based on the charging task parameters, the EV aggregator calculates the boundary parameters of the aggregate flexibility model, consisting of aggregate energy boundaries and aggregate power boundaries.

Bidding Stage. According to the aggregate flexibility model, the EV aggregator runs a RL-based algorithm for optimizing the hourly energy bidding. The proposed method optimizes decisions in a data-driven and model-free manner, thus handling uncertainties and modeling errors.

Disaggregation Stage. During real-time delivery, the EV aggregator conducts power disaggregation to allocate charging power among EVs using a rule-based method. The power allocation aims to track the bidding quantity and fulfill the charging demand of each EV. The EV aggregator gets penalties if the aggregate response power of EVs violates the transformer capacity.

2.2 Markov Decision Process

To solve the aggregate EV charging scheduling problem in a data-driven and model-free manner, at first the problem is formulated as a Markov Decision Process (MDP). In MDP, an EV aggregator is represented by an agent and selects the action a_t at time t based on the environmental state observation s_t. Then, at time $t+1$, the agent receives a reward signal r_t from the environment, and the environment state transits to s_{t+1}. In this process, the action is optimized by maximizing its long-term cumulative discounted reward $R = \sum_{t=0}^{T} \gamma^t r_t$, where γ is the discount factor and r_t is the reward signal. Details of MDP are clarified as following:

State Space. For hour t, the environment states observed by the EV aggregator is denoted as $s_t = [\lambda_t^e, E_t, P_t^{non}]$, including the intraday energy prices λ_t^e, the total energy stored in all EVs E_t and the power of non-flexible load P_t^{non}.

Action Space. For hour t, the agent selects action $a_t = P_{t+1}^{\text{bid}}$ from the policy based on the state observation s_t, where P_{t+1}^{bid} denotes the energy bidding volume for next hour $t+1$.

State Transition. The state transition of the environment describes the process of transitioning from the current state s_t to the next state s_{t+1}. The transition of λ_t^e and P_t^{non} are influenced by exogenous uncertainty factors. In contrast, the transition of E_t is impacted by the endogenous bidding action. During real-time disaggregation, the energy bidding volume is allocated among EVs, thus controlling the energy transition of individual EVs. The power allocation is performed by a proportional allocation method:

$$\varpi_{\tau,t} = \frac{P_{\tau,t}^{\text{bid}} - \sum_{i \in \Phi} p_{i,\tau,t}^{\vee}}{\sum_{i \in \Phi} p_{i,\tau,t}^{\wedge} - \sum_{i \in \Phi} p_{i,\tau,t}^{\vee}} \tag{1}$$

$$p_{i,\tau,t} = (1 - \varpi_{\tau,t}) p_{i,\tau,t}^{\vee} + \varpi_{\tau,t} p_{i,\tau,t}^{\wedge} \tag{2}$$

where $p_{i,\tau,t}^{\vee}$ and $p_{i,\tau,t}^{\wedge}$ denote the real-time upper and lower power limits of EV $i \in \Phi$ at the τ th sub-interval of hour t. $p_{i,\tau,t}$ is the allocated power of EV $i \in \Phi$ at the τ th sub-interval of hour t. The energy transition of individual EVs is then represented as:

$$e_{i,\tau+1,t} = e_{i,\tau,t} + p_{i,\tau,t} \cdot \Delta\tau, \forall i \in \Phi \tag{3}$$

where $e_{i,\tau,t}$ denotes the stored energy of EV $i \in \Phi$ at the τ th sub-interval of hour t.

Reward. The reward evaluates the immediate influence of the action a_t on the EV aggregator at hour t. It consists of two terms: 1) the energy trading cost c_t^e of hour t; and 2) the penalty cost c_t^{pen} for transformer capacity violation.

$$r_t = -\left(c_t^e + c_t^{\text{pen}}\right) \tag{4}$$

$$c_t^e = \lambda_t^e \sum_{\tau=1}^{H} \sum_{i \in \Phi} p_{i,\tau,t}/H \tag{5}$$

$$c_t^{\text{pen}} = \lambda^{\text{pen}} \sum_{\tau=1}^{H} \left| \min\left\{P_{t,\tau}^{\text{total}}, P^{\max}\right\} - P_t^{\text{total}} \right|/H \tag{6}$$

$$P_t^{\text{total}} = \sum_{i \in \Phi} p_{i,\tau,t} + P_t^{\text{non}} \tag{7}$$

where H denotes the total number of sub-intervals within one hour, λ^{pen} is the penalty price for transformer capacity violation, $P^{\max}$ is the maximum active power capacity of the transformer.

3 Proximal Policy Optimization

3.1 Actor-Critic Architecture

In proximal policy optimization (PPO), two neural networks are introduced to learn from historical interactions between the agent and the environment. The first one is the *actor network*, which is used for approximating the optimal policy $\pi(a_h|s_h)$. It maps the observation s_h to means $\mu_h \in \mathbb{R}^{M \times 1}$ and standard deviations $\sigma_h \in \mathbb{R}^{M \times M}$ of a_h, where M denotes the dimension of $a_{g,h}$. During training, a_h is randomly sampled from the Gaussian distribution:

$$a_h \sim \pi(a_h|s_h) = \frac{1}{\sqrt{(2\pi)^M |\sigma_h|}} \exp\left(-\frac{1}{2}(a_h - \mu_h)^T \sigma_h^{-1}(a_h - \mu_h)\right) \quad (8)$$

The second one is the *critic network*, which is adopted to predict the average return $V(s_h)$ (cumulative future rewards) of all possible actions given state s_h. Based on $V(s_h)$, we can estimate the relative benefit of taking the chosen action a_h under a given state s_h, compared with the average return.

3.2 Parameter Update

The actor network and critic network are updated by maximizing a clipped surrogate objective and minimizing a squared-error loss function, respectively.

Actor Network Update. For the $(k+1)$th iteration of the training, the actor network parameter set θ is updated by maximizing a clipped surrogate objective:

$$L_k^{\text{CLIP}}(\theta) = \mathbb{E}_t\left[\min\left\{\varpi_k \hat{A}_t, \varpi_k^{\text{CLIP}} \hat{A}_t\right\}\right] \quad (9)$$

$$\varpi_k = \frac{\pi(a_t|s_t)}{\pi^k(a_t|s_t)}, \quad \varpi_k^{\text{CLIP}} = \text{clip}(\varpi_k, 1 \pm \epsilon) \quad (10)$$

where $\text{clip}(\varpi_k, 1 \pm \epsilon)$ refers to enforcing ϖ_k in $[1-\epsilon, 1+\epsilon]$, $\pi^k(a_t|s_t)$ denotes the old policy obtained at the kth iteration, $\hat{A}_t$ is estimated advantage value, which is calculated by the generalized advantage estimation method.

Critic Network Update. The parameter set ϕ of the critic network is updated by minimizing a squared-error loss function:

$$L(\varphi) = \frac{1}{|\mathcal{D}|} \sum_{i \in \mathcal{D}} \left(V_\varphi(o_i) - \left(\hat{A}_i + V_{\overline{\varphi}}(o_i)\right)\right)^2 \quad (11)$$

where $\overline{\varphi}$ denotes the parameter set of the critic network before this update, $\mathcal{D}$ is a minibatch sampled of historical transitions.

4 Case Study

4.1 Experiment Setup

To validate the effectiveness of the proposed method in optimizing the charging cost of EVs and dealing with transformer capacity violation, both simulated dataset and real-world dataset were utilized for validation. The historical intraday energy prices and non-flexible load data were collected from UK, ranging from 2021 to 2022. The simulated data about EV charging behavior was generated from truncated normal distributions and uniform distributions for EV charging at home. The real-world data about EV charging behavior was collected from 8 chargers in a commercial hub in England.

4.2 Results Based on Simulated Data

To simulate different levels of transformer violation, four cases with different numbers of EVs were designed as shown in Table 1.

Figure 2 shows the total demand of EVs and residents under four cases. The shaded area of each curve represents the volatility of the total demand within one year. The total demand increases with the number of EVs. Two load peaks are observed during 9:00–10:00 and 19:00–21:00, respectively. It is shown in Fig. 2 that transformer capacity overload happens in case 3 and case 4 during the two peak load periods. The transformer capacity overload demonstrates that the electricity distribution network is not capable of supply the energy required to satisfy charging demand of all EVs without scheduling.

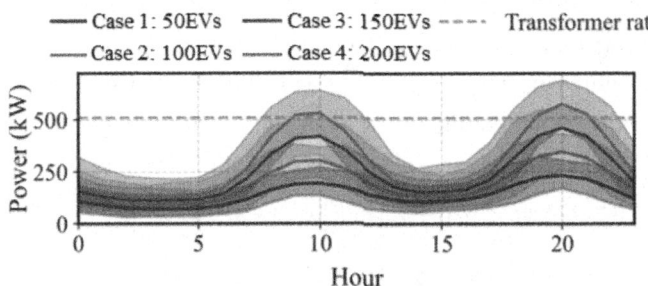

Fig. 2. Total demand of EVs and residents under four cases.

Table 2 shows the economic and technical performances of the proposed method across four scenarios. For all four scenarios, the average daily charging costs per EV under the proposed method (w/ RL) are reduced compared to the un-scheduled situation (w/o RL). The problem of transformer capacity overload is eliminated with RL for case 3. The maximum power overload under case 4 is reduced by 15%. Compared with the case without RL, the average daily cost per EV is reduced by about 11%.

Table 1. Case setup.

Case	Number of residents	Transformer capacity (kVA)	Number of EVs
Case 1	350	500	50
Case 2	350	500	100
Case 3	350	500	150
Case 4	350	500	200

Table 2. Evaluation metrics for different cases.

Case	Daily cost on average (GBP)		Average daily cost per EV (GBP)		Maximum power violation volume (kW)	
	w/o RL	w/ RL	w/o RL	w/ RL	w/o RL	w/ RL
Case 1	266.79	236.59	5.34	4.73	0.0	0.0
Case 2	529.13	463.48	5.29	4.63	0.0	0.0
Case 3	796.80	706.09	5.31	4.71	7.3	0.0
Case 4	1089.95	951.26	5.45	4.76	119.5	101.2

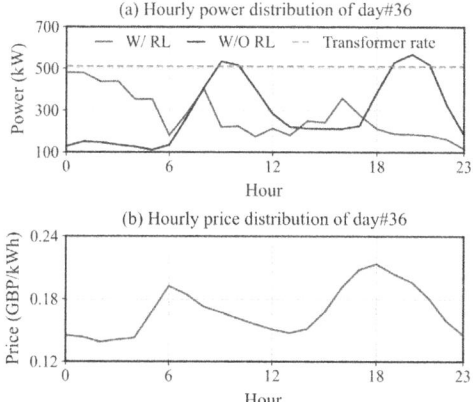

Fig. 3. Hourly power and price distributions of day #36.

To demonstrate how the proposed method reduces the charging cost of EVs, we visualize the hourly power and price distributions of a typical day in Fig. 3. It can be seen from Fig. 3 that a load shift from high-price or peak-load periods to low-price or valley-load periods can be observed. Specifically, the majority of charging demand is shifted to 00:00–6:00 when the intraday energy prices reach the valley.

4.3 Results Based on Real-World Data

The real-world EV charging dataset was collected from 8 chargers located in a commercial hub in England. The community system comprises 350 households and 400 electric vehicles. These EVs park at commercial hubs or retail center. The parking time of these EVs is concentrated between 7:30 a.m. and 5:30 p.m. Based on the dataset, we investigate the impacts of seasonal characteristics on the charging scheduling results. The evaluation metrics and average daily power for different seasons are given in Table 3 and Fig. 4, respectively.

It can be concluded from Table 3 that the daily cost for each season is reduced, and transformer capacity overload is alleviated. Specifically, the average daily cost is reduced by 3.83%; the transformer capacity overload is eliminated for all seasons except Autumn;

Table 3. Evaluation metrics for the real-world dataset

Season	Daily cost on average (GBP)		Average daily cost per EV (GBP)		Maximum power violation volume (kW)		Cost saving
	w/o RL	w/ RL	w/o RL	w/ RL	w/o RL	w/ RL	
Spring	475.44	468.46	1.19	1.17	45.52	0.00	1.7%
Summer	859.64	842.79	2.15	2.11	29.68	0.00	1.9%
Autumn	701.64	621.76	1.75	1.55	177.41	47.64	11%
Winter	581.61	574.74	1.45	1.44	60.75	0.00	0.7%
Average	654.58	626.93	1.64	1.57	78.34	11.91	3.8%

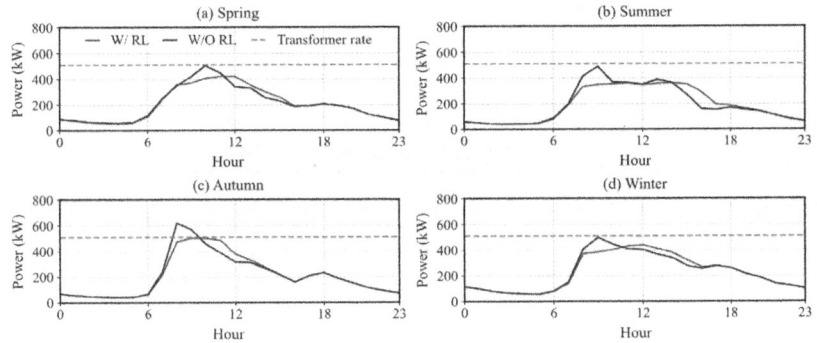

Fig. 4. Average daily power for each season.

The maximum power violation volume of Autumn is reduced by 73.1%. The charging cost reduction is benefited from the adoption of the proposed RL-based method. With RL-based smart charging, a load shift from peak-load periods to valley-load periods can be observed in Fig. 4. Due to the reason that most EV charging events in the commercial hub happen on daytime, the rescheduling for EV charging has less flexibility than that in the simulated dataset. Therefore, the proposed method exhibits lower cost reduction in the real-world commercial hub dataset than the simulated dataset for charging at home.

5 Conclusions

This paper develops a reinforcement learning-based smart charging method for EV fleet. The effectiveness of the proposed method is validated in both simulated dataset and real-world dataset. With RL-based smart EV charging, the daily cost for each season reduces and transformer capacity violations are alleviated. With simulated EV charging data for charging at home, the average daily cost was reduced by approximately 11%. When applying the reinforcement learning algorithm in a real-life example of EV chargers in a commercial hub in England with mainly daytime EV charging, the average daily charging cost was reduced by approximately 3.83%, and the transformer capacity overloads were eliminated for all seasons except Autumn. The cost reduction with the real-life charging hub example is lower than that of the simulated data of EV charging at home, because most EVs in the commercial hub park for a short period of time and thus have little flexibility of rescheduling.

Acknowledgement. This work was supported in part by the Innovate UK project (Ref 10078489 and 10123670), "Using AI to optimize energy flow management in urban electric vehicle charging", 2023–25. The projects funders were not directly involved in the writing of this article.

References

1. Chen, N., Kurniawan, C., Nakahira, Y., Chen, L., Low, S.H.: Smoothed least-laxity-first algorithm for electric vehicle charging: online decision and performance analysis with resource augmentation. IEEE Trans. Smart Grid **13**(3), 2209–2217 (2022). https://doi.org/10.1109/TSG.2021.3138615
2. Tyers, R., Hutton, G., Walker, A., Stewart, I.: Electric Vehicles and Infrastructure, UK House of Commons, CBP-7480 (2024). https://researchbriefings.files.parliament.uk/documents/CBP-7480/CBP-7480.pdf. Accessed 24 Sep 2024
3. Eurostat, "Passenger cars, by type of motor energy" (2024). https://ec.europa.eu/eurostat/databrowser/view/road_eqs_carpda/default/table?lang=en&category=road.road_eqs. Accessed 24 Sep 2024
4. Eurostat, "Final energy consumption in transport by type of fuel" (2024). https://ec.europa.eu/eurostat/databrowser/view/ten00126/default/table?lang=en&category=t_nrg.t_nrg_indic. Accessed 24 Sep 2024
5. European Commission and Directorate-General for Climate Action, "Fit for 55: EU reaches new milestone to make all new cars and vans zero-emission from 2035." https://climate.ec.europa.eu/news-your-voice/news/fit-55-eu-reaches-new-milestone-make-all-new-cars-and-vans-zero-emission-2035-2023-03-28_en
6. Li, S., et al.: A multiagent deep reinforcement learning based approach for the optimization of transformer life using coordinated electric vehicles. IEEE Trans. Ind. Inf. **18**(11), 7639–7652 (2022). https://doi.org/10.1109/TII.2021.3139650
7. Zhang, J., Guan, Y., Che, L., Shahidehpour, M.: EV charging command fast allocation approach based on deep reinforcement learning with safety modules IEEE Trans. Smart Grid, 1 (2023). https://doi.org/10.1109/TSG.2023.3281782
8. Lyu, R., Guo, H., Zheng, K., Sun, M., Chen, Q.: Co-optimizing bidding and power allocation of an ev aggregator providing real-time frequency regulation service. IEEE Trans. Smart Grid (2023). https://doi.org/10.1109/TSG.2023.3252664
9. Yan, D., Huang, S., Chen, Y.: Real-time feedback based online aggregate EV power flexibility characterization (2023). *arXiv*: arXiv:2301.03342. http://arxiv.org/abs/2301.03342. Accessed 13 Oct 13
10. Wen, Y., Hu, Z., You, S., Duan, X.: Aggregate feasible region of DERs: exact formulation and approximate models. IEEE Trans. Smart Grid **13**(6), 4405–4423 (2022)

Materials Recycling and Transportation Infrastructure

Christian Paglia(✉)

Institute of Materials and Construction, Supsi, V. Flora Ruchat 15,
CH-6850 Mendrisio, Switzerland
christian.paglia@supsi.ch
https://www.researchgate.net/profile/C-Paglia

Abstract. The sustainability of building materials and infrastructure became a main concern in recent years. The CO_2 emissions, energy requirements, durability, natural resource depletion, lack of landfill space and recycling are some of the main issues to be faced. Railways, roadways, bridges, tunnels, supporting walls and pipelines are the main transportation infrastructures that require a special care with respect to the future use of materials, especially cementitious and bituminous compounds. This work addresses the main features related to these materials, that are widely used in the transportation infrastructure and the main issues for the development of more sustainable transportation infrastructures. The durability, the restoration and the recycling are main topics that requires a deep knowledge of materials technology coupled with the environmental conditions and the traffic loads to which the infrastructures are subjected. In this concern, the performance of recycled cement-based and bituminous materials or a combination of both, should not be underestimated. The strength is not always directly correlated with the durability, while the recycling potential must be carefully evaluated with adequate dosages and material designs. In this manner, it is possible to achieve the required properties and reduce the environmental pollutions. Planning, construction, maintenance and restoration or replacement of infrastructures may be chosen accordingly.

Keywords: Materials · recycling · infrastructure

1 Introduction

The transportation infrastructure is widely present worldwide and connects geographical regions, people and countries for millions of kilometers in length. The enormous horizontal dimension of such infrastructures is less visible, if compared to the vertical dimension of buildings and skyscrapers, that try to constantly beat the world record in height. Nonetheless, impressive long tunnels through mountains or below the sea water level, as well as long bridges across narrow and deep valleys, emphasize the magnificence of transportation infrastructure. Tall buildings are composed of a multitude of different materials with a high degree of complexity of the construction systems. Therefore, the recycling of the composite materials within buildings, requires an adequate separation of the various components to facilitate the future recycling. On the other hand, transportation infrastructure seems easier from the point of view of the variety of the materials, but the durability requirements are higher as compared to buildings. Both type of structures, e. g. buildings and infrastructures, are subjected to similar environmental actions. Nonetheless, the transportation infrastructures or at least some parts of them, are particularly exposed to more frequent and cyclic detrimental effects, such as chloride-rich waters splashing from the circulating traffic, the frost action in the presence of deicing salts, and a higher frequency of fatigue solicitations. Thus, the durability is a main requirement for infrastructure. Recently, service life of 50 years up to 100 years are required. This may be more easily attained by single materials, but rarely in composite construction systems, where the weaker point of the construction system, may accelerate the degradation of the whole structure. In this regard, water not adequately managed, a too rapid strength and stiffness development of cementitious materials, and the presence of multiple material's interface may largely promote the deterioration.

The high durability requirements of transportation infrastructure largely hinder the use of recycled materials, such as recycled concrete or reclaimed asphalt. However, this behaviour still belongs to an old and insufficient knowledge about the potential of recycled materials to be used in the construction field. The aim of this work is to highlight the application potential of recycled materials in infrastructures.

2 Roadways

The stone cubes paved roadways of the past (Fig. 1 left) switched to concrete roads, that are still used in many parts of the world and generally exhibit a long-lasting behaviour (Fig. 1 right). A general high compressive strength [1] above 50 MPa and a modulus of elasticity [2] above 28'000 MPa up to 40'000 MPa, can be reached after more than half a century of service life. This fact is observed in both highly trafficked roads at 200 m above sea level and in altitude from 2000 to 2400 m above sea level. The porosity and water permeability [3] tend to be low, although with variable results. The carbonation of concrete slabs [4] may exhibit low values (< 4 mm). The frost resistance may vary from 30 g/m^2 to 6'000 g/m^2. The variation is also correlated to the type of accelerated tests [3], while half of the chloride content measurements [5] may exhibit chloride concentrations above 0.200% by mass of concrete in a depth of 30 mm from the surface. Nonetheless, the durability of the steel mesh reinforced concrete slabs is high [6]. However, this largely depends on the chloride spreading during the winter, or from the chloride enrichments in marine environments.

Fig. 1. Granitoids stone cubes-based road across a mountain pass up to 2000 m above sea level (left) and concrete slabs roadway more than 50 years old (right).

A main issue is represented by the junctions. A good water sealing with bituminous-based material may reduce the entrance of chloride-rich solutions during time and limits the induced localized corrosion of the dowels and rebars. Furthermore, a constant maintenance of the junctions, particularly along transverse joints, may prevent a high degradation of the steel dowels, even after 60 years of service life in a roadway located between 1200 and 1500 m above sea level. This is observed on cored specimens taken 2 cm away from the sealed bituminous joints center line. Only a limited corrosion of dowels (Fig. 2 left) is seen. Most of the dowels embedded in a depth of 15 cm within the concrete slabs do not show relevant corrosion (Fig. 2 right). The upper steel mesh layer placed 5 cm below the concrete surface still exhibits a satisfactory conservation state, with only occasional corrosion.

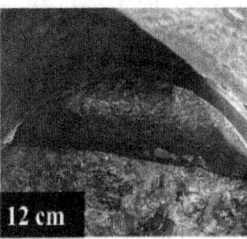 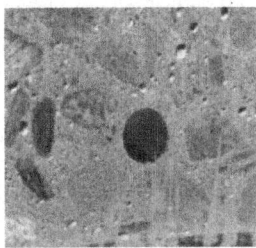

Fig. 2. Conservation state of a dowel bar between concrete slabs in a 15 cm depth after 60 years of service (left) and clean dowel embedded in concrete close to a joint center line (right).

Considering the age of the concrete pavement slabs, reinforced with double steel mesh layers, a visual inspection of the surface does not indicate frequent longitudinal or transverse cracking, or D-cracking due to the frost action. Consequently, a replacement of the old bituminous joints, with a special care if asbestos is present, a cleaning of the dowels, a renewed joint sealing and a replacement of the seriously damaged concrete slabs, would be sufficient for a long-lasting restoration of the pavement. In some cases, epoxy-coated dowels with plastic caps at the edges, replace the corroded dowels. In addition, an asphalt overlay of about 4 cm with a macro-roughness surface texture and a stress adsorption membrane interlayer (SAMI) is placed. The asphalt surface course is cut along the below concrete transverse joints, to avoid reflective cracking. Nonetheless, reflective cracking is difficult to mitigate on a long-term with such a thin asphalt overlay. Moreover, in harsh weather conditions, such restoration procedure may exhibit a reduced durability, no longer than 20 years.

Thus, prior to any decision concerning the partial or complete replacement of old materials and infrastructures and the subsequent demolition and crushing of the old material to be recycled, a careful evaluation of the residual life of existing infrastructures is necessary. The shorter the durability of the restored systems, the lower the sustainability and the higher the frequency of the maintenance works along the roadways.

The use of recycled concrete crushed aggregates for pavement applications may produce high quality recycled concrete aggregates (RCA). These latter exhibit an angular shape, sharp edges and rough surface and usually require a high amount of water that ranges from 3% up to 10%. High quality RCA should not exhibit water adsorption after

24 h above 7%. The natural aggregates show water adsorption below 1% by weight. The replacement of natural aggregates (NA) with coarse RCA with a maximum dosage of 50% allows to attain a satisfactory performance in terms of mechanical and durability of the recycled concrete as well as in the unbound status for the base and the subbase layers [7]. The aggregates susceptible to chemical or physical reactions, such as frost D-cracking can be recycled in concrete with no additional cracking after 35 years of service [8]. The quality of the source concrete is essential to produce RCA with good quality [9]. The old concrete slabs cast several decades ago tend to exhibit satisfactory performance, in spite of some cracking and surface scaling, due to ageing. Therefore, old concrete pavements are a good source for RCA in pavement applications, especially for recycled concrete. The use of unbound RCA as a subbase or base material may also be thought. Nonetheless, the high energy consumption and CO_2 emission to produce the Ordinary Portland cement used in the past, rather require a high level of recycling of the cementitious materials to produce recycled concrete and not unbound material. Furthermore, the widespread of roadway nets requires a high amount of cementitious waste material. This is not very sustainable.

On the other hand, quality control of the RCA quality, chemical reactions occured in the past, such as the alkali-silica reaction, chloride and other pollutants contaminations, need to be carefully evaluated for the recycling in an unbound or bound status. However, a washing procedure and a partial dilution in new concrete, largely reduce the concentration of reactive species. In addition, some pollutants, such as heavy metals, present in a relative low concentration, are bond into the cementitious matrix of recycled concrete and are difficult to be leached out into the ground [10].

Hydraulic road binders are used for soils and unbound materials to be stabilized. Portland cement clinker, granulated blast furnace slags, pozzolanic materials, siliceous and calcareous fly ashes, burnt shales, limestone, silica fume, recycled concrete fines with some limitations, circulating fluidized bed fly ashes, paper sludge ashes, crystallized basic oxygen furnace aggregates and lime are recycled materials that may be used with adequate chemical, physical and performance controls [11]. Cement-based materials can be used in soils that are difficult to stabilize, such as for clay-rich materials, as long as the cost of the soil removal and the replacement with adequate compounds is not sustainable, due to long distances or quarry operations. However, the use of cement-based materials in soils should be limited.

Roller compacted concrete (RCC) is a concrete with no workability, that allows a relative rapid placing with asphalt high density pavers [12, 13]. High quality RCA can be used to produce such pavements, although the dosage with respect to the natural aggregates needs to be evaluated. On going studies are clarifying the addition of 100% RCA and up to 30% reclaimed asphalt pavement for the NA replacement [14].

Mixed granulates mainly containing rocks, ceramic, and concrete tend to be used to produce recycled concrete for low grade applications and as filling material for pipelines basements. This type of concrete exhibits a relative high inhomogeneity, porous coarse aggregates, such as ceramics, and low durability [15]. Some attempts were done to use the waste aggregates in an unbound status as a base or subbase layer. Nevertheless, the general variable performance of such material, makes it inadequate for pavements.

Supporting walls are widely used along roadways, especially where a lack of space exists. The contact with earth, humidity, water and the pressure of the landscape, causes a relevant degradation of the structures. The water infiltration causes salt deposition within cracks. The frost action coupled with the previously mentioned conditions, cause cracking and scaling of the surface. Not rarely, white salts, mainly carbonates, are located on the upper parts of the walls (Fig. 3 left).

Fig. 3. Supporting wall along a roadway located at 1500 m above sea level with sprayed concrete in the upper part and an old steel net fixed with anchorages to improve safety (left). Unsorted RAP and excavation material from a roadway with maximum aggregate size of 90 mm (right).

Recycled concrete may be an option for such structures, especially when the lowered stiffness may allow the walls to partially adapt to the service conditions and lower the cracking susceptibility. On the other hand, the water drainage and filling with drainage material behind the walls, also with recycled minerals materials, is relevant for the long-term behaviour.

Asphalt pavements are widely used, due to their riding comfort and tyre adhesion to the street surface. On the other hand, the increasing traffic loads, frequency and the more extreme weather conditions of the last decades cause a premature damage. This is especially observed in the surface course of highly trafficked roadways or highways, that are also more exposed to the oxygen, the humidity and solar radiation. A permanent plastic deformation, e. g. rutting, at high temperatures and cracking during cold temperatures are the main damage types [16]. Restoration works cause traffic jams and the large amount of milled asphalt creates stockpiles with reclaimed asphalt pavements (RAP). In full-depth reclaimed asphalt pavements, RAP are often mixed with excavation stone materials, that need to be separated (Fig. 3 right). The RAP can be more easily recycled in countries, where a high number of new roadways are under construction and the RAP can be added especially in the lower layers. In regions where restoration works mainly take place, the use of RAP is less abundant and stockpiles form quickly, in spite of the higher percentage allowed to be used within the asphalt layers. The export in the neighboring countries is not always a sustainable way. Rejuvenators or soft bituminous material may be solutions to be adopted to increase the use of RAP and decrease the rigidity of the bituminous material, although the durability and the interaction with the polymer-modified bituminous material still remains to be clarified. In this concern, the mixed use of RAP within concrete merits a further clarification for pavement applications [17].

3 Railways

The trains increased their speed with time and the railway materials slightly changed. Wood sleepers are partially replaced by steel and concrete sleepers. Concrete sleepers crushed and recycled to produce recycled concrete aggregates for self-compacting concrete may exhibit a variable water adsorption after 24 h up to 10% [18]. This is considered a too high value for medium or high quality recycled concrete aggregate, that should remain below 7%. The low quality of the recycled aggregates may not be necessarily caused by the crushing technique or by fatigue cracking, due to the service life. The original mix design may be a reason of the low RCA quality and a careful quality investigation of concrete sleepers is required. Recycled concrete may not be excluded for high grade applications, if an adequate mix design and stiffness development of the cementitious material is reached.

Stone ballast remains a main supporting material for sleepers and railway lines. The ballast provides the resistance of the sleepers to the vertical, lateral and longitudinal displacements, transfers the load to the subgrade, maintains the elasticity of the whole track system, enhances the drainage, mitigates the noise and the vibration, exhibits insulating properties and avoids vegetation growth. The type of rocks depends on geography and climate conditions. Magmatic and metamorphic rocks are preferred, while limestone may show a slow degradation in rainy regions [19]. Hard rocks with an angular shape and a maximum diameter varying from 54 mm up to 90 mm, guarantee the necessary resistance to the train load by the interlocking of the aggregates (Fig. 4 left).

Fig. 4. Crushed hard rock ballast and concrete sleepers (left) and a steep 100 years old funicular railway line under repair (right).

A constant cleaning of the fine particles or partial replacement of material promote the appropriate functionality of the ballast. In fact, an enrichment of fines along steel sleepers, especially in steep railway lines (Fig. 4 right), tends to collect earth material and dirt, thus promoting the permanence of humidity and steel corrosion [20].

Steel slags may exhibit high mechanical properties, high shear strength and modulus of elasticity and less permanent deformation as compared to conventional crushed rocks [21]. Nonetheless, electrical conductivity concerns [22], stray-current issues [23] and metal leaching [24], should not be underestimated with the use of such waste material. Crushed concrete debris mixed with bottom ashes tend to show low aggregate crushing values [25]. Rubber chips were also introduced to the ballast material to reduce ballast abrasion, breakage and degradation [26]. In this concern, as for most waste materials, the addition of rocky aggregates, concrete or asphalt systems requires a limit on the dosage, in order to achieve a satisfactory performance. Asphalt binders were also used to bind the ballast [27], although asphalt materials are more likely to be used as a sub-ballast layer. Asphalt layer with limited porosity of a maximum of 2.5% by volume with 22 mm maximum aggregate size, may be placed below the ballast as a water sealant layer, that conveys the water on the side towards the soil [28]. Polyethylene fibres may reduce the settlement and lateral expansion [29]. Nonetheless, the addition of polymer materials in mineral aggregates arises some questions about the sustainability. Ballast waste usually account for 30% of the material to be periodically managed and can be reused, if sieved in grain size above 30 mm, while finer particles are not to be reused. However, ballast waste can also be mixed with asphalt to produce asphalt pavements [30]. A direct relationship does not always exist between the Los Angeles abrasion rate and the durability. Complex service life actions contribute to the degradation and some further clarification is needed between ballast aggregate shape and performance. The use of recycled materials for ballast still needs to be discussed also on a standard level worldwide [31]. In this concern, recycled materials may be an opportunity to investigate. This also depends on local materials availability and environmental conditions.

Excavation materials washed in special plants to eliminate the pollutants at a maximum extent and sorting of the gravel may be an option. Crushing of the stone component may be partially required to attain the load bearing capability. However, the energy requirements and the costs should not be too high. This by considering the enormous amount of excavation materials brought to disposal.

Recycled materials may also be used for sleepers. The fibres or filler reinforced recycled plastic is an investigation topic [32], although high temperature behavior, creep and the high energy consumption for the production of plastics, makes this solution only provisional. Recycled concrete sleepers appear a more reliable solution, since it is possible to obtain satisfactory performance. The addition of crumb rubber into sleepers arises some questions on the strength, while some durability aspects can be improved, depending on the rubber dosage [33]. The recovered glass and recycled crushed aggregates can be used in the sub-ballast layer, although an optimal dosage needs to be found [34]. However, the recyclability of glass should be better exploited outside the infrastructure field.

4 Bridges

The increased dimensions of long span bridges represented a significant development in engineering achievements, that were able to speed up the transportation and the connection between the countries across narrow valleys and mountains. From coupled twisted railway tunnels, towards curved roadways that followed the landscape, to more straight highway long span bridges, the transportation systems exhibited a constant evolution in shape and materials (Fig. 5 left).

Fig. 5. Three generation of transportation ways and materials across the European Alps (left). Crushed high quality recycled concrete aggregates 0–32 mm from a demolition of a 60 years old highway bridge (centre) and bridge curb with some initial coating detachments (right).

The long-term service life of bridges in a main target. The chloride attack of the reinforced steels [35], the resistance to the frost and combined environmental actions [36] and the differentiated chloride penetration and carbonation in the various parts of a bridge [37], together with the mechanical actions, are the main degradation factors. The bridges, like other infrastructures, show an asymmetrical deterioration [38]. The absence of the direct contact with chloride-rich waters in railway bridges provides a longer lasting service life. The bridge dilatation junctions do not suffer from chloride-rich water infiltrating towards the sub-structure. Steel bridges can be recycled, since the material's scrap can be reused to produce recycled steel, which exhibits similar or even better properties than the steel originated from the ore deposits. The reuse of steel elements may be possible, as for pre-cast concrete elements, although a careful investigation of the extent of the damage, such as fatigue microcracks and environmental degradation processes, are required. Demolished and crushed infrastructure concrete, such as the concrete deriving from bridges, tends to exhibit high quality recycled concrete aggregates (Fig. 5 centre). They exhibit sharper edges, a more angular shape and a rough surface as compared to round-smooth glacial-alluvial natural aggregates. The RCA water adsorption ranges between 4% and 5%, while the NA adsorption is < 1%. Nonetheless, the water adsorption of such RCA is satisfying. The material comes from an original concrete with a compressive strength beyond 68 MPa and with an adequate mix design, a high quality recycled concrete can be prepared. The NA replacement below 40% allows to produce a durable concrete, which achieves the standard durability requirements [39]. Not rarely, bridge curbes are particularly subjected to degradation, in spite of the coatings, that may only partially increase the durability (Fig. 6 right). Restoration cement-based mortars may contain recycled aggregates. The decrease in the modulus of elasticity of recycled cementitious systems is even desirable, due to the good adhesion to the existing more rigid concrete substrate, which exhibits a higher modulus of elasticity. Nevertheless, the use of recycled cement-based materials in the infrastructure sector is still limited and a higher level of confidence must be created with the time.

5 Tunnels

In tunnels several type of materials are also used and a significant part of them being concrete (Fig. 6 left). Usually, it should exhibit a water proofing function to limit the water infiltration from the rocks. The supplementary cementitious materials are used as recycled materials in most tunnel concrete linings and shotcrete applications, due to the performance improvement [40]. High quality recycled concrete aggregate may be used for such infrastructures, with a dosage ranging from 25–50% for the NA replacement, in order to achieve the durability specifications. Water proofing is a main concern to clarify for recycled concrete, while tunnel basement layers are more prone to be prepared with recycled concrete.

Tunnel rock break materials can be recycled [41] to produce concrete or shotcrete [42]. Explosion excavation advancement methods provide more cubic aggregates, while TBM techniques creates chipped, angular and platy shaped particle (Fig. 6 centre). These latter may be crushed and rounded with mobile plants to produce aggregates for concrete. Hardness and mineralogical composition are the main parameters for adequate recycled rocky materials for concrete. In the infrastructure sector, a large quantity of material often needs to be managed during long tunnel excavations (Fig. 6 right) or demolition of old structures. An adequate sorting and quality control of the material, allows a more targeted and increased recycling potential. In this regard, a high amount of space is required for the materials to be provisionally deposited, treated and recycled.

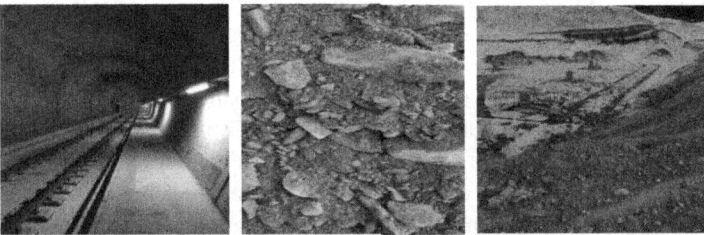

Fig. 6. A concrete railway tunnel > 10 km (left), TBM chipped, and platy broken rocks with a maximum dimension of 18 cm (centre) and provisional deposition of excavation tunnel debris (right).

6 Maintenance and Monitoring

The use of recycled materials in the infrastructure sector may require an increased constant monitoring during an initial stage. Basically, smart sensors can be inserted as in conventional materials to predict damage or timing for the repairs along the structures. On the other hand, a careful monitoring may reduce the frequency of repair interventions. Sensors need to be adapted and placed in the most critical parts of the infrastructures and the data need an adequate interpretation and a correlation with regular visual inspections. From the one side, an increase in the smart monitoring of infrastructures made by using a higher percentage of recycled materials may add some costs in an initial phase. On the other side, it lowers the overall costs during the service life of the structures. In this manner, maintenance will take place only if strictly necessary and along part of the infrastructure, that exhibit a degradation level beyond acceptable limits and prevent an invasive restoration of the whole infrastructure.

7 Conclusions

The use of recycled materials, such as recycled concrete and RAP asphalt, is possible in the field of infrastructure. A special attention must be given to the quality of the source materials, to an adequate sorting, separation and crushing of the components, to the limitation of the dosage of the recycled components and the use of rejuvenators, in order to achieve the desired performance. Thus, durable infrastructures can also be attained with recycled materials. Sufficient provisional space availability for the recycled material production and management, local materials knowledge and a higher confidence in today's technological developments, will constantly increase the use of recycled materials.

Acknowledgments. The author would like to thank the collaborators of the Institute of materials and construction, Supsi, Switzerland.

References

1. Standard EN 12390–3, Testing hardened concrete, Part 3: Compressive strength of test specimens (2019)
2. Standard EN 12390–13, Testing hardened concrete, Part 13: Determination of secant modulus of elasticity in compression (2013)
3. Standard SIA 262/1 Concrete construction, complimentary specifications (2016)
4. Standard SN EN 14630, determination of carbonation depth in hardened concrete by the phenolpthalein method (2006)
5. Standard SN EN 14629, products and systems for the protection and reparation of concrete structures, measurements of the concrete chloride content (2007)
6. Paglia, C., Mosca, C.: The concrete road pavement durability: a study of over 50 years old concrete structures in a South alpine space. In: The 14th International Symposium on Concrete Roads 25–29 June 2023 - Krakow, Poland (2023)
7. Panda, B., Imran, N.T., Samal, K.: A study on replacement of coarse aggregate with recycled concrete aggregate (RCA) in road construction. In: Recent Developments in Sustainable Infrastructure: Select Proceedings of ICRDSI 2019 LNCE, vol. 75, pp. 1097–1106. Springer, Cham (2021).https://doi.org/10.1007/978-981-15-4577-1_91
8. Zeller, M.: Performance history of recycling D-Cracking susceptible concrete into TH 59 in Minnesota. In: Workshop 3: Recycled Concrete Aggregate. 11th International Conference on Concrete Pavements. August 28–September 1, San Antonio, TX (2016)
9. Paglia, C., Paderi, M., Mosca, C., Antonietti, S.: The recycling of a concrete with known properties to reproduce a durable material for the civil engineering infrastructure. In: International Meet on Civil, Structural and Environmental Engineering, 23–26 Mai, Munich, Germany, online, Civilmeet (2022)

10. Giner Cordero, E., Paglia, C.: The recovery of municipal solid waste inceneration bottom slags for mortars. Physicochemical characterization and analysis of the heavy metals leaching. In: The 2nd international conference smart materials & structures, Smart materials 2024, 22–24 July, Berlin, Germany (2024)
11. European technical committee, CEN/TC 51/WG 14, Hydraulic binders for road bases, Conveyor Miguel Ángel Sanjuán, committee member C. Paglia (2024)
12. Delatte, N.: Simplified design of roller compacted concrete composite pavements. Transp. Res. Rec. **1896**(1), 57–65 (2004)
13. ACI 327R-14, Guide to roller compacted concrete pavements, 2015
14. Rilem technical committee. Chair Christian Paglia, Deputy Chair Corey Zollinger: Rolled compacted concrete for pavement applications. https://www.rilem.net/groupe/rcc-rolled-compacted-concrete-for-pavement-applications-454
15. Paglia, C.: Recycled Concrete with Mixed Granulate Material: Still a Future Perspective? COJ Tech. Sci. Res. **3**(5). COJTS. 000575 (2022)
16. Paglia, C., Guerini, L.: Introduction to the durability and damage of asphalt mixes. In: The Second World Symposium on Materials Sciences and Engineering, Nov 8th–10th, 2023, Singapore (2023)
17. Paglia, C., Mosca, C., Antonetti, S., Paderi, M., Corredig, G.: The reclaimed asphalt pavement addition to produce concrete pavements, The 13th International conference on concrete pavements, Minneapolis, Minnesota, 25–29 August USA (2024)
18. Horat A., Self-compacting concrete with recycled concrete aggregates, Bachelor Thesis F23, USI, SUPSI, OST, Supervisors S. Stürwald, E. Illoret-Fritschi and C. Paglia, 10.06.2024
19. Sadeghi, J.M., Zakeri, J.A., Najar, M.E.M.: Developing track ballast characteristic guideline in order to evaluate its performance. Int. J. Railway **9**(2), 27–35 (2016)
20. Paglia, C., Mosca, C.: The conservation state of a funicular railway infrastructure from the beginning of the XX century. In: The 76th Rilem Annual Week and International Conference on Regeneration and Conservation of Structures 3–9 September, Kyoto, Japan (2022)
21. Koh, T., Moon, S.-W., Jung, H., Jeong, Y., Pyo, S.J.S.: A Feasibility study on the application of basic oxygen furnace (BOF). Steel Slag Railway Ballast Mater. **10**(2), 284 (2018)
22. Jia, W., Markine, V.L., Jing, G.: Analysis of furnace slag in railway sub-ballast based on experimental tests and DEM simulations. Constr. Build. Mater. **288** (2021)
23. Paglia, C.: The active and passive protection of reinforced concrete structures. In: Academia International Webinar on Materials Science & Engineering, Materials Spectrum, August 15–1 6, 2022 (2022)
24. Hussain, A., Hussaini, S.K.K.: Use of steel slag as railway ballast: a review. Transp. Geotech. **35** (2022)
25. Youventharan, D., Ramandhansyah, P.J., Jeevithan, K.M., Rokiah, O., Mohd Arif, S., Yaacob, H.: Durability performance of concrete debris and bottom ash as an alternative track ballast material. In: IOP Conference Series: Earth and Environmental Science, vol. 682. no. 1 (2021)
26. Sol-Sanchez, M., Thom, N.H., Moreno-Navarro, F., Rubio-Gamez, M.C., Airey, G.D.: A study into the use of crumb rubber in railway ballast. Constr. Build. Mater. **75**, 19–24 (2015)
27. D'Angelo, G.: Bitumen stabilized ballast: a novel track-bed solution towards a more sustainable railway, University of Nottingham (2018)
28. Standard EN 13108–1: 2016. Bituminous mixtures - Material specifications - Part 1: Asphalt Concrete
29. Ferro, E., Ajayi, O., Le Pen, L., Zervos, A., Powrie, W.: Settlement response of fibre reinforced railway ballast (2016)
30. UIC, Circular practices in the railway and ways forward REUSE Project final Report. International Union of Railways (2021). https://uic.org/projects/article/reuse
31. Guo, Y., Xie, J., Fan, Z., Markine, V., Connolly, D.P., Jing, G.: Railway ballast material selection and evaluation: a review. Constr. Build. Mater. **344**, 128218 (2022)

32. Ferdous, W., et al.: Recycling of landfill wastes (tires, plastics and glass) in construction – A review on global waste generation, performance, application and future opportunities. Resources, Conservation and Recycling, 173 (December 2020): 105745. Elsevier B.V (2021). https://doi.org/10.1016/j.resconrec.2021.105745
33. Paglia, C., Paderi, M.: Durability and Recycling of Rubber Concrete. J mate poly sci **4**(2), 1–5 (2024)
34. Jarjour, J., Meguid, M.: On the use of recycled waste materials in ballasted railway infrastructures: a review. In: Conference: Geosaskatoon 2023: Bridging Infrastructures and Resources, Saskatoon, Saskatchewan, October 2023, Canada (2023)
35. Boschmann, C., Kathler, E.U., Angst, A.M., Aguilar, B.: Elsener, Data Article, A novel approach to systematically collect critical chloride contents in concrete in an open access data base, Data in brief 27 (2019)
36. Wang, R., Zhang, Q., Li, Y.: Deterioration of concrete under the coupling effects of freeze–thaw cycles and other actions: a review. Constr. Build. Mater. **319**, 126045 (2022)
37. Paglia, C., Antonietti, S., Mosca, C.: The degradation of concrete bridge elements. In: The 5th International Conference on Building Materials and Materials Engineering, Barcellona, Spain, September, 2021 (2021)
38. Paglia, C.: The deterioration asymmetry of engineering structures. Case Stud. Constr. Mater. **16**, e00980 (2022)
39. Antonietti, S., Mosca, C., Paglia, C.: Concrete recycling aggregates to produce durable concrete. In: 12th ACI/RILEM International Conference on Cementitious Materials and Alternative Binders for Sustainable Concrete, ICCM2024, June 23 to June 26, 2024, Toulouse, France
40. Paglia, C.S., Wombacher, F., Boehni, H.: The influence of alkali-free and alkaline shotcrete accelerators within cement systems: hydration, microstructure and strength development. ACI Mater. J. **101**(5), 353–357 (2004)
41. Bellopede, R., Brusco, F., Oreste, P., Pepino, M.: Main aspects of tunnel rock recycling. Am. J. Environ. Sci. **7**, 338–347 (2011)
42. Paglia, C., Frigeri, G., Chollet, A.: The use of tunnel demolition rocks to produce shotcrete for a railway infrastructure. In: The first International Online Conference on Infrastructures-Rilem, 7–9 June, IOCI (2022)

Dynamic Characteristics Analysis of High-Speed Trains Considering Wheel Polygonal Wear Under Track Irregularity Excitation

Xin Wang, Hongzhang Yu, Lin Zhou, Junyi Mu, Imdad Ullah Khan, and Chunrong Hua(✉)

School of Mechanical Engineering, Southwest Jiaotong University, Chengdu 610031, Sichuan, China
hcrong@swjtu.cn

Abstract. Wheel polygon wear (WPW) is a common form of wheel non-circularity in high-speed trains. As the train speed increases, the WPW and track irregularity excitation become coupled, which not only excites the structural deformation of the flexible wheelset, but also aggravates the high-frequency components of the axle box's vibration response. This paper develops a rigid-flexible coupling dynamic model using the Craig-Chang reduction method, and analyzes the coupling effect of track irregularity excitation and WPW on the axle box vibration response at 300 km/h. The study also investigates the influence of WPW wear order and depth on the amplitude of the response harmonic component, the amplitude of the wheelset modal frequency, and the safety evaluation metrics under the coupling effects. The results show that the coupling effect significantly increases the vibration level of the axle box, but reduces the amplitude of the wheel bending, torsional deformation modal frequency and the WPW excitation frequency. This study provides theoretical and technical support for the dynamic optimization of high-speed train wheel groups.

Keywords: Track irregularity · Wheel polygonal wear · Rigid-flexible coupling; Dynamic characteristics

1 Introduction

Wheelsets are critical components for train safety and significantly affect vehicle operation [1]. With the increase in train speed, wheel polygon wear (WPW) exacerbates, leading to high-frequency excitations that cause vehicle vibration and noise [2]. The coupling of track irregularity excitation and WPW excitation intensifies the wheel-rail interaction, leading to fatigue in critical components and threatening the operation safety of the train [3]. Therefore, studying the influence of the coupling effect of track irregularity and WPW excitation on the vibration response of the axle box at high speeds is crucial for ensuring safe train operations.

Vehicle vibration response under WPW excitation is often investigated with numerical simulation approaches. To simulate the dynamic response of the wheel-rail under

WPW excitation at 100 km/h, Liu et al. [4] built a vehicle-track coupling dynamic model that utilized wheel-rail flexibility into account. They discovered that the out of circularity time-varying rate exhibits a strong linear correlation with the wheel-rail vertical force and the axle box vertical acceleration when the excitation frequency is near the wheel-rail system's natural frequency. Liao et al. [5] investigated how different wear orders and amplitudes affected the axle box system's dynamic performance. They found that high-order WPW (17th–21st order) had a greater impact on the axle box system than low-order (1st–5th order) wear, and that it caused high-frequency vertical vibration. Many studies have examined the impact of low-order WPW parameters on the dynamic performance of vehicles. However, as running speed increases, both the order and depth of WPW intensify, leading to coupling effects caused by track unevenness during actual train operations.

Guan et al. [6] analyzed the wheelset flexibility's impact on the dynamic response of the axle box bearing under wheel-rail coupling excitation. They examined different vehicle speeds (50–300 km/h) and WPW amplitudes and revealed that WPW excitation and wheel flexibility significantly influenced the axle box bearing's dynamic response, exhibiting a complex coupling effect. Wu et al. [7] examined the dynamic response and fatigue under wheel flat and WPW excitation, showing that frequency coupling induced a specific modal resonance in the gearbox. Liao et al. [5] investigated the contact load between the bearing roller and outer raceway under WPW and wheel-rail excitation at 250 km/h and displayed that coupling action increased the contact load and introduced high-frequency excitation components. Existing studies primarily focus on the influence of WPW excitation under medium and low-speed conditions. Although some studies have considered the coupling effect of track unevenness and WPW excitation, however, under high-speed conditions, WPW parameters affect the vibration amplitude of the harmonic components in the axle box vibration response and the bending-torsion modal frequency of the wheelset.

This study establishes a rigid-flexible wheel model using Craig-Chang reduction method, considering the main vibration mode of the wheel structure, which reflects the dynamic characteristics of the wheel structure. Wheel group flexibility's impact on vehicle dynamics is examined, as is the impact of track unevenness on the vibration response caused on by axle and hub wear. At a speed of 300 km/h, the effects of track unevenness and axle and hub coupling on the axle box's dynamic performance and the train's operational safety are also methodically investigated.

2 Development of Dynamic Model of Rigid-Flexible Coupling Vehicle System

WPW generates high-frequency impact loads between wheels and rails which excites the modal resonance in both the vehicle and wheel-rail system. This paper develops a flexible wheelset model by integrating Abaqus and Simpack software using the Craig-Chang reduction method, considering the main mode of the wheelset substructure. The high-frequency vibrations of the wheel-rail system under coupling excitation are then precisely simulated by building a rigid-flexible coupling dynamics model for high-speed trains.

2.1 Wheelset Flexibility

Based on the modal orthogonality and expansion theorems, the vibration response of the flexible component is calculated, which requires various finite element units and nodes. The substructure, with internal degrees of freedom eliminated, is analyzed based on the stiffness and mass of the retained node degrees of freedom [8]. In this paper, the Craig-Chang reduction method is employed, retaining unconstrained degrees of freedom and introduces generalized degrees of freedom q^a to enhance the internal modal characteristics of the wheelset structure. The displacements of the internal and generalized degrees of freedom of the wheelset system are expressed as:

$$\Delta u^E = (K^{EE})^{-1}(\Delta P^E - Ku^{ER}\Delta u^R) + (\phi^B)^T q^a \tag{1}$$

$$\Delta u^B = (K^{EE})^{-1} K^{ER} \delta u^R) + (\phi^B)^T q^a \tag{2}$$

where K^{EE} and K^{ER} are the stiffness submatrices for the internal degrees of freedom and between internal and retained degrees of freedom, respectively. K is the overall stiffness matrix, ΔP^E is the variation in external forces acting on the internal degrees of freedom, Δu^R is the displacement variation of the retained degrees of freedom, $\varnothing^B$ is the modal matrix of the substructure, and δu^R represents the virtual displacement vector of the retained degrees of freedom.

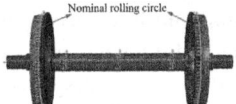

Fig. 1. Wheelset master node selection

Table 1. Free modes of the wheelset

Modal	Frequency (Hz)	Shaking type	Modal	Frequency (Hz)	Shaking type
7	68.64		16-19	440.18	
8-9	84.81		20-21	530.34	
10-11	150.47		22	650.97	
12	269.21		23	740.57	
13-14	307.31		24	912.57	
15	410.27		25	1110.76	

A finite element model of a high-speed train wheelset is developed using Abaqus, featuring an LMA profile and meshed with hexahedral solid elements. The model consists of 212 main nodes chosen to reduce the wheelset's degrees of freedom, as illustrated in Fig. 1. The Craig-Chang reduction method is employed to compute the modes of the wheelset model in its free state, as detailed in Table 1. Comparing the natural frequencies of the wheelset substructure model with values from the literature [9] reveals an average error of 4.14%, demonstrating the model's accuracy and reliability. The analysis considers the first 35 modes of the wheelset, with a cutoff frequency of 2023 Hz, encompassing the WPW passing frequency range addressed in this study.

2.2 Establishment of Vehicle-Track Rigid-Flexible Coupling Model

A high-speed train model is created using SIMPACK software, consisting of one car body, two bogies, four axle boxes, and four wheelsets, along with both primary and secondary suspensions. This model incorporates the intricate coupling of nonlinear contact geometry, nonlinear suspension forces, and nonlinear creep forces, featuring 62 degrees of freedom. The rigid wheelset has been substituted with a flexible one using the flexible body interface (FEMBS) module, as illustrated in Fig. 2.

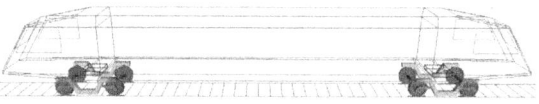

Fig. 2. Vehicle model considering wheelset flexibility

To validate the established model, the train speed is set to 300 km/h, with a sampling frequency of 5 kHz, running on a straight track using the measured Beijing-Tianjin track spectrum. The simulated spectrum of the vertical acceleration of the axle box is then compared to the measured vibration data from the literature [10], as depicted in Fig. 3.

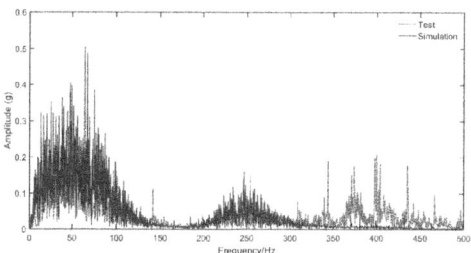

Fig. 3. Axle box vibration response spectrum

From the comparative analysis, the simulation results align well with the measured data, showing the obvious peaks at 38.2 Hz (main frequency of P2 force vibration), 67.4 Hz, and 248.4 Hz (nearby frequencies induced by the track unevenness). The measured data exhibits an increased vibration amplitude at frequencies of 300 Hz and above, due to the bending mode of the rails and the long-wave wear. The comparison results of the axle box vibration acceleration frequency domain confirm the model's accuracy and reliability for the simulation of vehicle vibration response under WPW.

2.3 Comparative Analysis of Rigid Wheelset and Flexible Wheelset Models

The interaction between the flexible structure and the vibrations of the entire vehicle system influences the train's dynamic performance. To examine the impact of wheelset flexibility on vehicle dynamics, a comparative analysis was performed using both rigid and flexible wheelsets based on the established model.

Figure 4 presents the time-domain diagram of the axle box vibration acceleration. The root mean square (RMS) values of vertical and lateral vibration acceleration for the rigid model are 12.48 m/s^2 and 5.21 m/s^2, respectively, which are 8.6% and 71.4% greater than those of the coupled model.

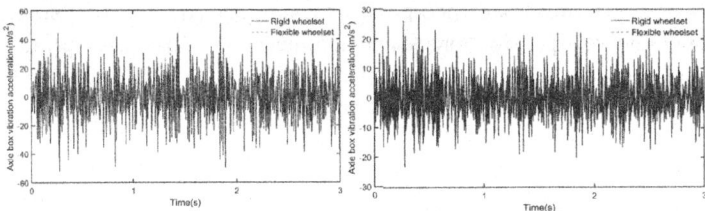

Fig. 4. Time domain response of axle box vibration acceleration under wheel polygon wear: (a) vertical; (b)lateral.

Figure 5 illustrates the time-frequency diagram of the axle box's vertical vibration acceleration. The primary frequencies are 67.4 Hz and 246.79 Hz, but the vibration energy within this frequency range is lower in the coupled model compared to the rigid model. The elastic deformation of the wheelset also leads the coupled model to exhibit additional vibration responses around 740 Hz and 910 Hz. Figure 6 presents the time-frequency diagram of the lateral vibration acceleration, where the main frequency aligns with that of the vertical vibration. However, the coupled model reveals significant vibration energy peaks at approximately 650 Hz, 740 Hz, and 1100 Hz. These findings suggest that under identical service conditions, the coupled model has a lower main frequency vibration amplitude than the rigid model, as the flexibility of the wheelset helps absorb and dissipate vibration and impact forces. However, the elastic deformation of the wheelset introduces additional natural modal frequencies, resulting in extra vibration response peaks in the coupled model at higher frequencies such as 650 Hz.

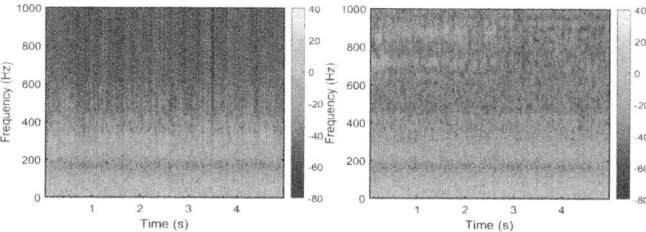

Fig. 5. Time-frequency diagram of vertical vibration acceleration of axle box. (a) rigid wheelset; (b) flexible wheelset.

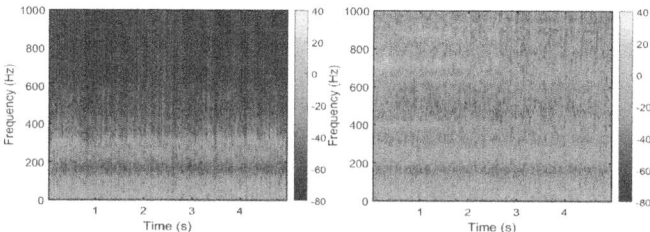

Fig. 6. Time-frequency diagram of axle box lateral vibration acceleration. (a) rigid wheelset; (b) flexible wheelset.

3 Analysis of Vibration Response Caused by WPW Under Track Irregularity

According to field test data, the high-speed train speed generally operates around 300 km/h with wheels exhibiting high-order WPW in the 15th-25th orders, while occasionally above the 30th order, with a wear depths ranging from 0.01 to 0.1 mm [11]. Consequently, the vibration characteristics of the axle box system brought on by these WPW situations are the focus of the work. This excitation frequency generated by WPW can be obtained by formula (3):

$$f_N = \frac{Nv}{\pi D} \tag{3}$$

where N is the WPW order, D is the wheel rolling circle diameter, and v is the train running speed.

3.1 Effect of Track Irregularity Excitation on Axle Box Vibration Response

Simulations were conducted at a train speed of 300 km/h to assess the impact of track irregularity excitation on axle box vibration response caused by wheel polygonal wear (WPW) under two scenarios: one with and one without track irregularity excitation. A 20th order harmonic WPW with a wave depth of 0.03 mm was applied to all wheels.

The axle box vibration response's time domain diagram is displayed in Fig. 7. With track unevenness, the RMS of vertical and lateral vibration accelerations increased by

0.12% and 16.76%, respectively, to 146.65 m/s^2 and 281.13 m/s^2, compared to the case without track unevenness. Figure 8 displays the response spectrum. Compared to literature [7], the main frequency and amplitude of the response without track unevenness remain nearly identical. With track unevenness, the vertical and lateral amplitudes at the main frequency (617 Hz) decrease by 13.92% and 15.57%, respectively, and at the 2nd harmonic frequency (1234 Hz), by 29.13% and 26.4%. Amplitudes near 307 Hz, 740 Hz, and 920 Hz appear in both conditions, aligning with the 2nd and 3rd order wheelset bending modes and the wheel torsional mode; however, these amplitudes decrease in the presence of track unevenness. To sum up, the coupling effect of WPW and track excitation reduces the vibration amplitude of the main and second-order frequency components, while increasing the overall vehicle vibrations in the high-frequency range.

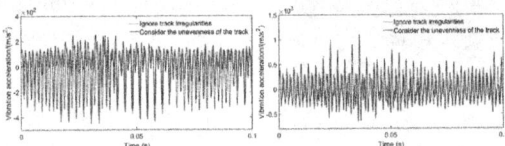

Fig. 7. Axle box vibration acceleration time domain response. (a) vertical; (b) lateral

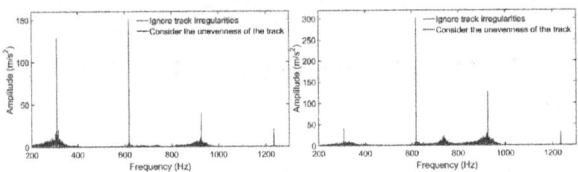

Fig. 8. Axle box vibration acceleration spectrum. (a) vertical; (b) lateral

3.2 Analysis of Vibration Characteristics of WPW Under Changing Wear Parameters

Two groups of excitations are applied to the wheels to further examine the impact of WPW order and depth on axle box vibration response under track irregularity excitation. The first group involves 15th to 30th order harmonic WPW with a wave depth of 0.05 mm, while the second group uses 20th order harmonic WPW with a wave depth of 0.01 mm to 0.1 mm.

Figure 9 illustrates how the axle box vibration acceleration frequency response varies with changes in wear order. WPW of the wheel excites the 23rd (f_{w23}) and 24th (f_{w24}) wheelset modes, causing the wheel to experience bending and torsion. The primary frequencies of the vertical and lateral vibration responses match the excitation frequencies of the WPW.

Figure 10(a) displays the change in the amplitude of the harmonic component of the response as the wear order varies. As the WPW order increases, the 1X amplitude increases first. When the order is 24, the vertical and lateral 1X amplitudes both have

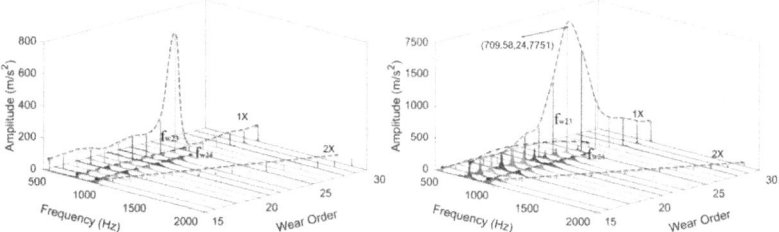

Fig. 9. Frequency response of vibration acceleration changes with wear order. (a) vertical; (b) lateral.

peaks, and finally stabilize at 103 m/s² and 330 m/s², respectively. The 2X amplitude exhibits slow nonlinear growth, with a significant increase at order 24. Under these WPW conditions, the excitation frequency of the 24th order WPW is 740.16 Hz, which is close to f_{w23}, triggers the resonance in the wheel-rail system.

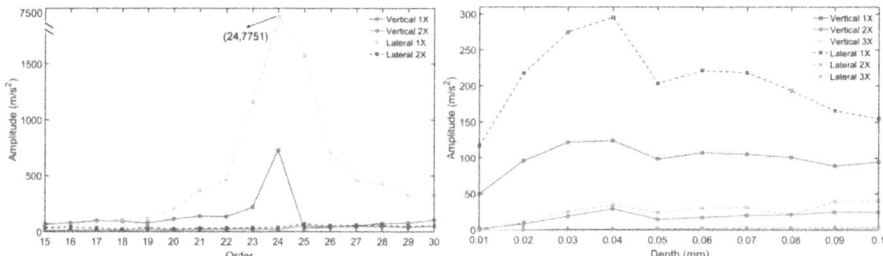

Fig. 10. The change law of the amplitude of the response harmonic component. (a) order change; (b) amplitude change.

Figure 10(b) illustrates the change of the harmonic component of the response as the wear depth increases. As the wear depth increases, the 1X amplitude increases initially. When the wear depth is 0.04 mm, the vertical and lateral 1X amplitudes both reach peaks, and eventually stabilize at 94 m/s² and 154 m/s², respectively. The 2X amplitude changes in the same way as the 1X, with steady values of 25 m/s² and 41 m/s², respectively. The 3X amplitude shows a slow nonlinear growth, with maximum values of 3.57 m/s² and 4.35 m/s², respectively.

Figure 11 illustrates the frequency response of the axle box vibration acceleration as the wear depth varies. The vertical vibration main frequency remains at f_{20}, while the lateral vibration main frequency changes from f_{20} to f_{w23} when the wear depth is 0.05 mm.

As the wear depth increases, the amplitudes of f_{w23} and f_{w24} grow unevenly. The standard deviations of the f_{w23} and f_{w24} amplitude growth rates of the vertical vibration response are 1.14 and 0.91, respectively, and 1.66 and 0.81 in the lateral direction. When the wear depth is 0.05 mm, the amplitudes increase significantly, with vertical and lateral growth rates of 306.4% and 377.6%, respectively. As wear depth increases, the amplitude of the wheel's bending and torsional modal frequencies is amplified, shifting the primary

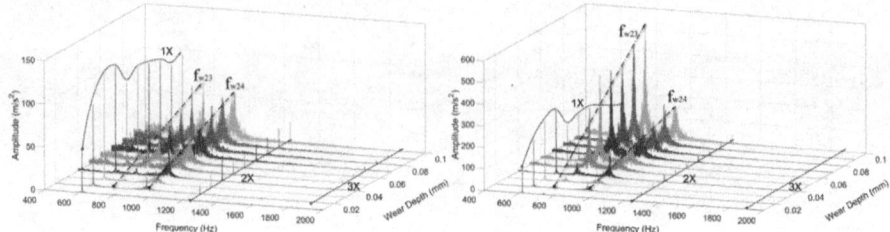

Fig. 11. Frequency response of vibration acceleration changes with wear depth. (a) vertical; (b) lateral.

frequency of the axle box vibration response from being dominated solely by the WPW excitation frequency to a combined effect of the WPW excitation frequency and the wheelset bending modal frequency.

The impact of variations in WPW parameters under coupled wheel-rail excitations on safety evaluation metrics was assessed using the wheel load reduction rate ($\Delta P/P_0$) and the derailment coefficient (Q/P).

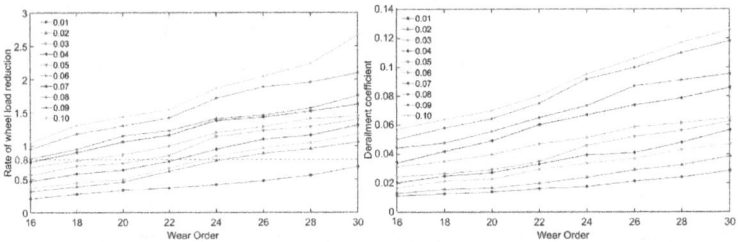

Fig. 12. Variation patterns of safety evaluation metrics with WPW parameters: (a) wheel load reduction rate; (b) derailment coefficient.

Figure 12 illustrates the variation patterns of ($\Delta P/P_0$) and Q/P with changes in WPW parameters. The safety limit for wheel polygonal wear decreases with increasing order. The average growth rates of ($\Delta P/P_0$) and Q/P are 14.81% and 13.98%, respectively, with increasing order, while the average growth rates with increasing amplitude are 18.73% and 19.91%. Thus, it is concluded that the amplitude of WPW has a more significant impact on vehicle safety performance metrics.

4 Conclusion

In this study, the Craig-Chang reduction method is applied to develop a flexible wheelset model that incorporates the primary mode of the wheelset substructure. A vehicle-track rigid-flexible coupling dynamic model is then created to investigate the influence of wheelset flexibility on vehicle dynamics, with a focus on the coupling effect of track irregularities and WPW. The study systematically analyzes how changes in wear parameters impact the axle box vibration response under coupling. The main conclusions are as follows:

- Under coupling excitation, the vehicle vibration intensifies, widening the vibration energy band near the WPW-induced main frequency. However, the harmonic component amplitude decreases as the coupling effect lowers the vibration amplitude near the wheelset's bending and torsional modal frequencies.
- The axle box vibration acceleration is primarily affected by WPW. As the WPW order increases, the main frequency of the vibration response remains consistent with the WPW excitation frequency. When the excitation frequency approaches f_{w23}, resonance in the wheel-rail system is triggered. Furthermore, as the WPW order increases, the main frequency of the vertical vibration response continues to correspond to the WPW excitation frequency. However, when the depth reaches 0.05 mm, the main frequency of the axle box's lateral vibration shifts from the WPW excitation frequency to f_{w23}. The amplitude of WPW has a more significant impact on vehicle safety performance metrics.

References

1. Peng, D., Liu, Z., Wang, H., Qin, Y., Jia, L.: A novel deeper one-dimensional CNN with residual learning for fault diagnosis of wheelset bearings in high-speed trains. IEEE Access **7**, 10278–10293 (2019)
2. Song, Z., Guo, L., Hu, X., Cheng, D., Li, Q.: Research on causes and countermeasures of wheel polygon wear based on kik-piotrowiski contact model. Veh. Syst. Dyn. **62**(4), 955–975 (2024)
3. Ma, Q., Liu, Y., Yang, S., Liao, Y., Wang, B.: A coupling model of high-speed train-axle box bearing and the vibration characteristics of bearing with defects under wheel rail excitation. Machines **10**(11), 1024 (2022)
4. Liu, X., Song, Y.: Research on the Influence of Wheel Polygons on Vehicle Dynamics Performance (2023)
5. Liao, X., Yi, C., Zhang, Y., Chen, Z., Ou, F., Lin, J.: A simulation investigation on the effect of wheel-polygonal wear on dynamic vibration characteristics of the axle-box system. Eng. Failure Anal. **139** (2022)
6. Guan, T., Deng, X., Wang, J.: Dynamic response of axle box bearing for high-speed train considering wheelset flexibility and polygonal wear. Sci. Rep. **13**(1), 1–18 (2023)
7. Wu, H., Wei, J., Wu, P., Zhang, F., Liu, Y.: Dynamic response analysis of high-speed train gearboxes excited by wheel out-of-round: experiment and simulation. Veh. Syst. Dyn. (2024)
8. Weng, S., Zhu, H., Xia, Y., Li, J., Tian, W.: A review on dynamic substructuring methods for model updating and damage detection of large-scale structures. Adv. Struct. Eng. **23**(3), 584–600 (2020)
9. Cong, R.: Influence of Train Wheelsets on Dynamic Performance [D] (2014)
10. Chu, M., Yang, L., Wang, Z., Jiang, Z., Wang, Q., Mo, J.: Study on the effects of curve geometry parameters and braking conditions on the dynamic characteristics of a train axle-box bearing. J. Mech. Eng. (2023)
11. Song, Z., et al.: Research on causes and countermeasures of wheel polygon wear based on kik-piotrowiski contact model. Veh. Syst. Dyn.Dyn. **62**(4), 955–975 (2024)

Modeling Interdependencies in Intelligent Traffic Systems and Sustainable Urban Development Using Graph Neural Networks

Qian Cao[1,3], Jing Li[2], and Paolo Trucco[1(✉)]

[1] Department of Management, Economics and Industrial Engineering, Politecnico di Milano, Milan, Italy
qian.cao@gsom.polimi.it, paolo.trucco@polimi.it
[2] School of Economics and Management, Tsinghua University, Beijing, China
jing_li@tsinghua.edu.cn
[3] Mogo Co.,, Beijing, China

Abstract. Urbanization and the proliferation of vehicles have intensified traffic congestion, leading to environmental degradation and impeding sustainable urban development. Intelligent Traffic Systems (ITS) have emerged as pivotal solutions to manage traffic efficiently. Recent advancements in Graph Neural Networks offer unprecedented capabilities in modeling complex spatial-temporal relationships inherent in traffic networks. This paper presents a comprehensive approach, Sustainable Urban Traffic Graph Neural Network (SUT-GNN), which integrates GNN into ITS to enhance traffic prediction accuracy, optimize traffic flow, and support sustainable city development. By incorporating environmental impact assessments and urban planning considerations, our model not only predicts traffic patterns but also aligns with sustainability objectives. Experimental results on real-world traffic datasets demonstrate the superiority of our approach over traditional methods, marking a significant step towards smarter and greener cities.

Keywords: Sustainable Urban Development · Graph Neural Networks · Traffic Prediction

1 Introduction

The rapid growth of urban populations has led to increased demand for transportation infrastructure, resulting in severe traffic congestion and environmental issues such as air pollution and elevated carbon emissions [2,4,13,15]. These challenges hinder the sustainable development of cities, affecting economic growth and the quality of life for urban dwellers. Many intelligent traffic systems (ITS) solutions [1,12] have been developed to address these challenges by utilizing advanced technologies to optimize traffic flow, enhance road safety, and reduce environmental impacts. However, the dynamic and complex nature of urban traffic networks presents significant modeling challenges. Traditional traffic prediction models often struggle to capture the intricate spatial-temporal dependencies and non-linear relationships within traffic data [9,14,16]. Graph Neural Networks (GNNs) have emerged as powerful tools for handling graph-structured data, making them well-suited for modeling road networks where nodes represent intersections and edges represent roads [7]. GNNs can effectively capture the spatial dependencies between different parts of the network while integrating temporal dynamics [19,20].

By leveraging GNNs within ITS, we can enhance the accuracy of traffic predictions, enabling more efficient traffic management strategies that contribute to sustainable urban development. Incorporating environmental factors and urban planning considerations into these models further aligns traffic management with sustainability goals.

Our main contribution can be summarized as follows:

- We propose a novel GNN-based model SUT-GNN that integrates environmental impact assessments and urban planning considerations into traffic prediction. The integration of graph convolution and temporal modeling enables the model to learn complex spatial-temporal patterns in traffic data.
- By capturing the interdependencies between ITS and SUD, the model provides insights into how traffic patterns and sustainability factors influence each other.
- Demonstrates the effectiveness of the proposed model through extensive experiments on real-world traffic datasets.

2 Related Work

2.1 Intelligent Traffic Systems

ITS encompass a range of technologies aimed at improving transportation efficiency and safety. Early ITS focused on signal control and traffic monitoring [10], but recent advancements have incorporated data analytics and machine learning to enhance decision-making processes [1,3,12]. Despite these improvements, challenges remain in accurately predicting traffic patterns due to the complex nature of urban networks.

2.2 Graph Neural Networks in Traffic Prediction

GNNs have gained attention for their ability to model graph-structured data. Techniques such as Graph Convolutional Networks (GCNs) [8] and Graph Attention Networks (GATs) [17] have been applied to traffic prediction with promising results. Studies have shown that GNN-based models outperform traditional methods by effectively capturing spatial-temporal dependencies [5,11,18-20]. However, existing GNNs for traffic prediction often overlook environmental factors and urban planning considerations crucial for sustainable development. Integrating these aspects into GNN frameworks remains an open research area.

3 Methodology

3.1 Problem Formulation

Let $G = (V, E)$ represent the traffic network graph, where V is the set of nodes (e.g., intersections, road segments) and E is the set of edges (e.g., roads). Each node $v \in V$ has associated features that may vary over time, such as traffic speed, volume, and sustainability factors. Our objective is to predict future traffic conditions by capturing both spatial and temporal dependencies while integrating sustainability considerations. Formally, we aim to learn a function f that predicts the future traffic state $\mathbf{Y}_{t+1}$ given historical traffic data $\mathbf{X}_{t-T+1:t}$, sustainability features $\mathbf{S}_{t-T+1:t}$, and the graph structure G:

$$\mathbf{Y}_{t+1} = f\left(\mathbf{X}_{t-T+1:t}, \mathbf{S}_{t-T+1:t}, G\right) \tag{1}$$

where $\mathbf{X}_{t-T+1:t} \in \mathbb{R}^{T \times N \times F}$ is the historical traffic data over T time steps, with $N = |V|$ nodes and F features per node, and $\mathbf{S}_{t-T+1:t} \in \mathbb{R}^{T \times N \times S}$ is the historical sustainability data, with S sustainability features per v.

3.2 Model Architecture

SUT-GNN integrates spatial and temporal modeling with sustainability feature integration. The architecture comprises four components: graph construction and preprocessing, spatial-temporal graph layer, sustainability feature integration module, and output layer. The pseudo-code is shown in Algorithm 1.

Graph Construction and Preprocessing. We construct the G based on the urban road network, where V represent intersections or road segments and E represent the connectivity between nodes. The adjacency matrix $\mathbf{A} \in \mathbb{R}^{N \times N}$ is defined as:

$$\mathbf{A}_{i,j} = \begin{cases} 1, & \text{if there is a direct road from node } i \text{ to node } j \\ 0, & \text{otherwise} \end{cases} \tag{2}$$

We normalize $\mathbf{A}$ using the symmetric normalized Laplacian:

$$\hat{\mathbf{A}} = \mathbf{D}^{-1/2} \mathbf{A} \mathbf{D}^{-1/2} \tag{3}$$

where $\mathbf{D}$ is the degree matrix with $\mathbf{D}_{ii} = \sum_j \mathbf{A}_{ij}$.

Spatial-Temporal Graph Layer. The SUT-GNN captures spatial dependencies using graph convolution and temporal dependencies using gated recurrent units (GRUs) [18]. At each time step t, the spatial features are updated using graph convolution:

$$\mathbf{H}_t^{(l+1)} = \sigma\left(\hat{\mathbf{A}} \mathbf{H}_t^{(l)} \mathbf{W}_s^{(l)}\right) \tag{4}$$

where $\mathbf{H}_t^{(l)} \in \mathbb{R}^{N \times D^{(l)}}$ is the hidden representation at layer l and time t. $\mathbf{W}_s^{(l)} \in \mathbb{R}^{D^{(l)} \times D^{(l+1)}}$ is the spatial weight matrix. $\sigma(\cdot)$ is an activation function, such as ReLU.

To model temporal dependencies, we use a GRU for each node v:

$$\mathbf{h}_v^{(l)} = \text{GRU}\left(\mathbf{h}_v^{(l-1)}, \mathbf{H}_{v,t}^{(l)}\right), \tag{5}$$

where $\mathbf{h}_v^{(l)} \in \mathbb{R}^{D^{(l)}}$ is the hidden state of v at layer l and $\mathbf{H}_{v,t}^{(l)}$ is the feature vector of v at time t. The GRU updates are defined as:

$$\begin{aligned}
\mathbf{z}_v &= \sigma\left(\mathbf{W}_z \mathbf{H}_{v,t}^{(l)} + \mathbf{U}_z \mathbf{h}_v^{(l-1)} + \mathbf{b}_z\right) \\
\mathbf{r}_v &= \sigma\left(\mathbf{W}_r \mathbf{H}_{v,t}^{(l)} + \mathbf{U}_r \mathbf{h}_v^{(l-1)} + \mathbf{b}_r\right) \\
\tilde{\mathbf{h}}_v &= \tanh\left(\mathbf{W}_h \mathbf{H}_{v,t}^{(l)} + \mathbf{U}_h \left(\mathbf{r}_v \odot \mathbf{h}_v^{(l-1)}\right) + \mathbf{b}_h\right) \\
\mathbf{h}_v^{(l)} &= (1 - \mathbf{z}_v) \odot \mathbf{h}_v^{(l-1)} + \mathbf{z}_v \odot \tilde{\mathbf{h}}_v
\end{aligned} \tag{6}$$

where $\mathbf{z}_v$ is the update gate vector. $\mathbf{r}_v$ is the reset gate vector. $\odot$ denotes element-wise multiplication. $\mathbf{W}_z, \mathbf{W}_r, \mathbf{W}_h, \mathbf{U}_z, \mathbf{U}_r, \mathbf{U}_h$ are weight matrices. $\mathbf{b}_z, \mathbf{b}_r, \mathbf{b}_h$ are bias vectors.

Sustainability Feature Integration Module. We integrate sustainability features into the model to capture their influence on traffic dynamics. we use an attention mechanism to weigh sustainability features:

$$\begin{aligned}
\mathbf{a}_t &= \text{softmax}\left(\mathbf{W}_a \mathbf{F}_t + \mathbf{b}_a\right) \\
\mathbf{H}_t^{(0)} &= \mathbf{a}_t \odot \mathbf{F}_t
\end{aligned}$$

where $\mathbf{W}_a$ and $\mathbf{b}_a$ are learnable parameters and $\mathbf{a}_t \in \mathbb{R}^{N \times (F+S)}$ assigns attention weights to features.

Output Layer. The final output layer generates the predicted traffic state for each node: $\mathbf{Y}_{t+1} = \mathbf{H}_t^{(L)} \mathbf{W}_o + \mathbf{b}_o$, where $\mathbf{H}_t^{(L)}$ is the hidden representation from the last layer L, $\mathbf{W}_o \in \mathbb{R}^{D^{(L)} \times O}$ maps to the output dimension O, and $\mathbf{b}_o$ is the bias vector.

Algorithm 1. Training Procedure of SUT-GNN

Initialize: Model parameters Θ
1 **for** *each epoch* **do**
2   **for** *each time window $t - T + 1$ to t* **do**
        Input : $\mathbf{X}_{t-T+1:t}$, $\mathbf{S}_{t-T+1:t}$, $\mathbf{A}$
3   **Feature Fusion**: Combine $\mathbf{X}_t$ and $\mathbf{S}_t$ to get $\mathbf{F}_t$;
4   Compute attention weights $\mathbf{a}_t$ and initial features $\mathbf{H}_t^{(0)}$;
5   **Spatial Convolution**: Update spatial features using

$$\mathbf{H}_t^{(l+1)} = \sigma\left(\hat{\mathbf{A}} \mathbf{H}_t^{(l)} \mathbf{W}_s^{(l)}\right)$$

        Temporal Modeling: Update temporal features using GRU;
    Output : Predicted output $\mathbf{Y}_{t+1}$
6   **Compute Loss**:

$$\mathcal{L} = \text{MSE} + \lambda \|\Theta\|_2^2$$

    Backpropagation: Update Θ using gradient descent;
7 **return** *Trained model*
---

4 Experiments

4.1 Dataset and Preprocessing

To evaluate the effectiveness of SUT-GNN, we utilize a comprehensive dataset comprising traffic data ($\mathbf{X}_{t-T+1:t}$) and sustainability data ($\mathbf{S}_{t-T+1:t}$):

- Traffic Data. Collected from loop detectors and traffic cameras in a major city area (Hengyang, the second largest city of Hunan Province, China.) over one year, including vehicle speeds, flow rates, road delay, queue length, time occupancy and space occupancy levels at 60-minute intervals.
- Environmental Data. Air quality measurements (e.g., CO_2, NO_x, PM2.5 levels) from environmental monitoring stations located throughout the city.
- Public Transportation Data. Real-time schedules, ridership numbers, and route information for buses and metro lines.
- Urban Planning Data. Information on zoning laws, land use, population density, and infrastructure projects obtained from municipal records.
- External Factors. Data on weather conditions (temperature, humidity, precipitation), special events, and reported traffic incidents.

4.2 Baseline Models

We compare our proposed model with the following baselines: HA, a naive model predicting traffic conditions based on historical averages over the same time periods. ARIMA, a traditional statistical model for time series prediction. LSTM [6], a long short-term memory deep learning network capturing temporal dependencies in sequential data. GCN [8], a graph convolutional network without temporal modeling. ST-GCN [18], a spatial-temporal GCN without sustainability features. DCRNN [11], a diffusion convolutional recurrent neural network integrates diffusion processes to model spatial-temporal dynamics.

4.3 Evaluation Metrics

We follow the classic works mentioned above, then assess model performance using the following metrics:

Mean Absolute Error (MAE). MAE measures the average magnitude of errors between the predicted and actual values, without considering their direction. It is calculated as MAE $= \frac{1}{N} \sum_{v=1}^{N} |y_v - \hat{y}_v|$, where y_v is the true value, $\hat{y}_v$ is the predicted value, and N is the total number of observations.

Root Mean Squared Error (RMSE). RMSE is another common metric for model evaluation, placing a higher penalty on larger errors compared to MAE. It is particularly useful when larger errors are less desirable. RMSE is defined as: RMSE $= \sqrt{\frac{1}{N} \sum_{v=1}^{N} (y_v - \hat{y}_v)^2}$.

Mean Absolute Percentage Error (MAPE). MAPE expresses prediction errors as a percentage, allowing for scale-independent comparison across different datasets or domains. It is calculated as MAPE $= \frac{100\%}{N} \sum_{v=1}^{N} \left| \frac{y_v - \hat{y}_v}{y_v} \right|$

R-squared (R^2) Score. The R^2 score, or coefficient of determination, indicates the proportion of the variance in the dependent variable that is predictable from the independent variables. It is computed as $R^2 = 1 - \frac{\sum_{v=1}^{N}(y_v - \hat{y}_v)^2}{\sum_{v=1}^{N}(y_v - \bar{y})^2}$, where $\bar{y}$ is the mean of the observed values. An R^2 score of 1 indicates perfect prediction, while a score of 0 indicates that the model performs no better than simply predicting the mean of the observed data.

4.4 Environment Setup

The models are implemented using PyTorch v2.4.1 and trained on 4 * NVIDIA Tesla V100 GPU. We split the dataset into 70% training, 15% validation, and 15% testing sets. Hyperparameters are tuned using grid search.

4.5 Results

Quantitative Performance. The performance of SUT-GNN and baseline models over the test set is summarized in Table 1. Our proposed model outperformed all baseline models across all evaluation metrics. Notably, incorporating sustainability features into the SUT-GNN architecture resulted in a significant improvement, with a 18% reduction in MAE compared to the ST-GCN. Additionally, the R^2 score of 0.93 indicates that our model is capable of explaining a substantial proportion of the variance, thus demonstrating its superior predictive ability in comparison to other models.

Table 1. Model Performance Comparison

Model	Core Idea	MAE	RMSE	MAPE (%)	R^2 Score
HA	historical average	7.84	10.50	18.2	0.68
ARIMA	time series	6.95	9.12	15.4	0.75
LSTM	long-term dependencies	5.60	7.80	12.0	0.82
GCN	graph convolutions	5.45	7.55	11.7	0.83
DCRNN	diffusion processes	4.70	6.50	10.2	0.88
ST-GCN	spatial-temporal	4.55	6.30	9.8	0.89
SUT-GNN	sustainability integration	**3.85**	**5.40**	**8.5**	**0.93**

Temporal Performance Analysis. The performance of models across different time periods is crucial for urban traffic prediction, particularly during peak hours when traffic flow can fluctuate significantly. By analyzing the changes in MAE of various models during the morning peak (7 AM–9 AM), evening peak (5 PM–7 PM), and off-peak hours (10 AM–4 PM), we gain a comprehensive understanding of each model's adaptability and stability.

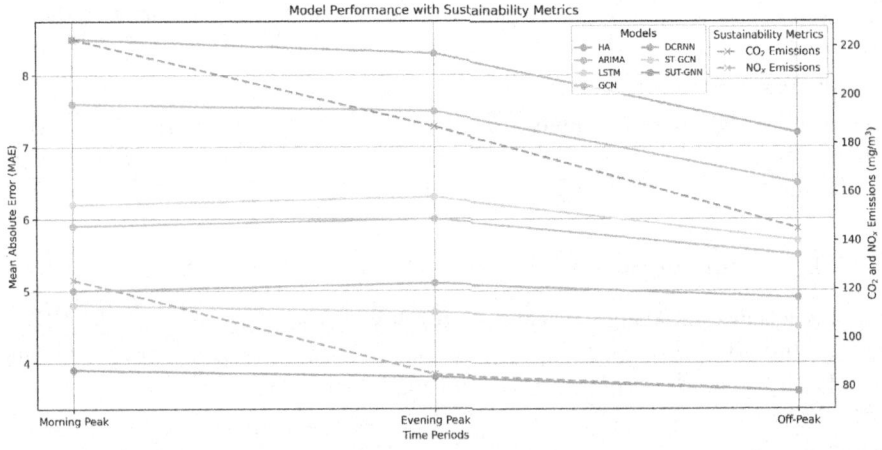

Fig. 1. Model Performance with Sustainability Metrics

The experimental results, as depicted in Fig. 1, demonstrate that SUT-GNN consistently outperforms all baseline models across all time segments. Notably, during the morning and evening peak hours-characterized by heightened traffic volatility-SUT-GNN exhibits a significantly lower MAE compared to other models. This finding highlights the model's capability to effectively leverage the historical data from the previous T time steps, allowing it to capture short-term traffic patterns and trends with remarkable accuracy. With sustainability metrics, CO_2 and NO_x emissions during the morning peak hours are noticeably higher compared to the evening peak. This disparity likely stems from the higher concentration of vehicles during the morning commute.

The superior performance of SUT-GNN during these critical time periods underscores its potential for real-time applications in urban traffic management. The model's design, which integrates sustainability features and spatial-temporal dependencies, enhances its predictive power, making it particularly suited for navigating the complexities of urban traffic dynamics. These insights emphasize the importance of adopting advanced modeling techniques like SUT-GNN in efforts to develop smarter, more resilient urban transportation systems.

Ablation Study. To evaluate the contribution of each key component within the SUT-GNN model, an ablation study was conducted by systematically removing sustainability features, temporal modeling, and spatial modeling. The modified models included SUT-GNN without Sustainability Features (w/o S), SUT-GNN without Temporal Modeling (w/o T), and SUT-GNN without Spatial Modeling (w/o G). The results of this study are presented in Table 2.

Table 2. Ablation Study Results

Model	MAE	RMSE	MAPE (%)	R^2 Score
w/o S	4.55	6.25	9.8	0.89
w/o T	5.15	7.05	11.1	0.85
w/o G	5.35	7.30	11.6	0.83
SUT-GNN	**3.85**	**5.40**	**8.4**	**0.93**

The results indicate that removing each component significantly impacts model performance. w/o S showed a notable increase in errors, with a 18% increase in MAE compared to the full SUT-GNN model, underscoring the critical role of sustainability features in capturing interdependencies between environmental factors and traffic patterns. w/o T, which excludes temporal modeling, exhibited even larger errors, confirming the necessity of incorporating historical data (over the time window $t - T + 1 : t$) to model traffic dynamics accurately. w/o G, which omits spatial dependencies, performed the worst, with the highest MAE and RMSE values, emphasizing the importance of the graph structure in effectively modeling spatial interactions across the road network. These findings highlight that the integration of all three components—sustainability, temporal, and spatial features—is crucial for achieving optimal predictive performance in traffic forecasting.

5 Conclusion

With the swift progress in deep learning, graph neural networks have gained prominence as a powerful tool in intelligent traffic systems. Nevertheless, the majority of current studies have primarily focused on traffic prediction, often neglecting other essential aspects like sustainable urban development. In this paper, we present a novel approach, Sustainable Urban Traffic Graph Neural Network (SUT-GNN), to modeling urban traffic by integrating sustainability factors into a spatial-temporal graph neural network framework. This method captures the complex interdependencies between traffic dynamics and sustainable urban development, providing more accurate predictions and supporting environmentally friendly traffic management strategies. By providing a tool that considers environmental impacts, this work contributes to the advancement of smarter and greener cities.

Limitation and Future Work. While the model shows promising results, limitations include the reliance on the availability and quality of sustainability data. In regions where such data is scarce or unreliable, the model's effectiveness may be reduced. And our method necessitates further rigorous mathematical analysis and theoretical grounding.

Acknowledgments. The authors sincerely appreciate the anonymous reviewers for their thorough evaluation of this article and for offering valuable suggestions that significantly enhanced its quality. This research was supported in part by the Institute for Industrial Innovation and Finance (IIIF), Tsinghua University.

Disclosure of Interests. The authors have no competing interests to declare that are relevant to the content of this article.

References

1. Akhtar, M., Moridpour, S.: A review of traffic congestion prediction using artificial intelligence. J. Adv. Transp. **2021**(1), 8878011 (2021)
2. Babaei, A., Khedmati, M., Jokar, M.R.A., Tirkolaee, E.B.: Sustainable transportation planning considering traffic congestion and uncertain conditions. Expert Syst. Appl. **227**, 119792 (2023)
3. Bastarianto, F.F., Hancock, T.O., Choudhury, C.F., Manley, E.: Agent-based models in urban transportation: review, challenges, and opportunities. Eur. Transp. Res. Rev. **15**(1), 19 (2023)
4. Chen, Y., Zhang, H., Wang, F.Y.: Society-centered and DAO-powered sustainability in transportation 5.0: an intelligent vehicles perspective. IEEE Trans. Intell. Veh. **8**(4), 2635–2638 (2023)
5. Guo, S., Lin, Y., Feng, N., Song, C., Wan, H.: Attention based spatial-temporal graph convolutional networks for traffic flow forecasting. In: Proceedings of the AAAI Conference on Artificial Intelligence, vol. 33, pp. 922–929 (2019)
6. Hochreiter, S., Schmidhuber, J.: Long short-term memory. Neural Comput. **9**(8), 1735–1780 (1997)

7. Ju, W., et al.: A comprehensive survey on deep graph representation learning. Neural Netw. 106207 (2024)
8. Kipf, T.N., Welling, M.: Semi-supervised classification with graph convolutional networks. arXiv preprint arXiv:1609.02907 (2016)
9. Kong, J., Fan, X., Jin, X., Lin, S., Zuo, M.: A variational bayesian inference-based En-decoder framework for traffic flow prediction. IEEE Trans. Intell. Transp. Syst. **25**(3), 2966–2975 (2023)
10. Li, J., Yu, C., Shen, Z., Su, Z., Ma, W.: A survey on urban traffic control under mixed traffic environment with connected automated vehicles. Transp. Res. Part C: Emerg. Technol. **154**, 104258 (2023)
11. Li, Y., Yu, R., Shahabi, C., Liu, Y.: Diffusion convolutional recurrent neural network: data-driven traffic forecasting. arXiv preprint arXiv:1707.01926 (2017)
12. Mall, P.K., et al.: Fuzzynet-based modelling smart traffic system in smart cities using deep learning models. In: Handbook of Research on Data-Driven Mathematical Modeling in Smart Cities, pp. 76–95. IGI Global (2023)
13. Musa, A.A., Malami, S.I., Alanazi, F., Ounaies, W., Alshammari, M., Haruna, S.I.: Sustainable traffic management for smart cities using internet-of-things-oriented intelligent transportation systems (its): Challenges and recommendations. Sustainability **15**(13), 9859 (2023)
14. Othman, B., De Nunzio, G., Di Domenico, D., Canudas-de Wit, C.: Analysis of the impact of variable speed limits on environmental sustainability and traffic performance in urban networks. IEEE Trans. Intell. Transp. Syst. **23**(11), 21766–21776 (2022)
15. Rehman, F.U., Islam, M.M., Miao, Q.: Environmental sustainability via green transportation: a case of the top 10 energy transition nations. Transp. Policy **137**, 32–44 (2023)
16. Singh, S., et al.: A novel framework to avoid traffic congestion and air pollution for sustainable development of smart cities. Sustain. Energy Technol. Assess. **56**, 103125 (2023)
17. Veličković, P., Cucurull, G., Casanova, A., Romero, A., Lio, P., Bengio, Y.: Graph attention networks. arXiv preprint arXiv:1710.10903 (2017)
18. Yan, S., Xiong, Y., Lin, D.: Spatial temporal graph convolutional networks for skeleton-based action recognition. In: Proceedings of the AAAI Conference on Artificial Intelligence, vol. 32 (2018)
19. Zhang, W., et al.: Irregular traffic time series forecasting based on asynchronous spatio-temporal graph convolutional networks. In: Proceedings of the 30th ACM SIGKDD Conference on Knowledge Discovery and Data Mining, pp. 4302–4313 (2024)
20. Zheng, C., et al.: Spatio-temporal joint graph convolutional networks for traffic forecasting. IEEE Trans. Knowl. Data Eng. **36**(1), 372–385 (2023)

Towards a Green Future: Case Analysis and Technological Prospects for Sustainable Transportation Decarbonization

Shu Liu, Shanshan Shi(✉), and Chen Fang

State Grid Shanghai Municipal Electric Power Company Electric Power Research Institute, Shanghai, China
sss3397@163.com

Abstract. The transportation industry is a significant contributor to global greenhouse gas emissions, accounting for roughly 24% of global energy-related emissions. Various strategies have been employed worldwide to reduce these emissions, including the adoption of public transportation systems, shared mobility solutions, electrified transportation, and intelligent transportation applications. This paper explores case studies within these categories, analyzing the measures taken and evaluating their effectiveness in reducing carbon emissions. The findings highlight the importance of technological integration, policy support, and innovative approaches in achieving significant emission reductions and moving towards a sustainable transportation future.

Keywords: Sustainable Transportation · Carbon Reduction · Electric Vehicles

1 Introduction

Global climate change has become one of the most pressing environmental issues of the 21st century. Climate change is primarily driven by the emission of greenhouse gases, leading to global temperature rise, sea-level rise, and frequent extreme weather events. According to the Intergovernmental Panel on Climate Change (IPCC), human endeavors have led to a rise in global mean temperatures by around 1.2 °C since the latter half of the 19th century, and this upward momentum is intensifying. Carbon dioxide (CO_2) is the primary component of greenhouse gases, accounting for about 76%. To combat climate change, the international community has formulated the Paris Agreement, seeking to confine the increase in global temperatures to significantly below 2 °C and making efforts to cap it at 1.5 °C.

The transportation industry is a major contributor to worldwide carbon emissions. As reported by the International Energy Agency (IEA), this sector is responsible for about 24% of the energy-related CO_2 emissions on a global scale. Among these, road transportation contributes the most, followed by aviation and shipping. Since transportation vehicles use fossil fuels (such as gasoline and diesel) that produce a large amount of carbon dioxide when burned, reducing carbon emissions in the transportation sector has become a vital step toward reaching global climate goals.

Sustainable transportation and carbon reduction have been widely studied due to their crucial role in mitigating climate change. Existing research summarizes various solutions for sustainable transportation carbon reduction. Reference [1] systematically reviews electric vehicles, hydrogen fuel cells, and public transport optimization, emphasizing the importance of technological integration and policy support. It concludes that effective policies and technological innovations are crucial for carbon reduction and calls for more interdisciplinary research. Reference [2] investigates the function of green technologies and sustainable energy sources in reducing CO_2 emissions in the EU transport sector, suggesting a mix of technologies and policies for significant emission reductions, but highlights the need for long-term impact assessments. Reference [3] evaluates the carbon reduction potential of passenger transport systems in a medium-sized Colombian city through scenario simulations, particularly focusing on electric vehicles and public transport optimization, and stresses that sound urban planning and technological application can substantially reduce emissions.

However, there is limited research on combining different low-carbon technologies into a single system to maximize carbon emissions reductions, and insufficient research on the economic costs, employment impacts, and public acceptance of sustainable transportation solutions.

This paper aims to explore various approaches to sustainable transportation carbon reduction, classify these approaches, and forecast future technological advancements. Specifically, the analysis of four types of sustainable transportation decarbonization is placed in the second section. The third section introduces carbon reduction assessment and policy support in detail. Technological outlook is provided and discussed in the fourth section. Finally, the last section summarizes the findings and suggests avenues for future research.

2 Case Classification and Analysis

The case studies in sustainable transportation carbon reduction are categorized into four main types to highlight the diverse approaches and technologies used to achieve emission reductions. These categories include Public Transportation Systems, Shared Mobility, Electrified Transportation, and Intelligent Transportation Applications. Each category demonstrates unique strategies and their impact on reducing carbon emissions in urban environments.

2.1 Public Transportation Systems

The London Underground has implemented several carbon reduction initiatives, including the use of renewable energy sources like wind and solar power. These efforts are part of a broader strategy to achieve a zero-carbon railway by 2030. The system's electricity needs, which amount to 1.6 TWh annually [4], are gradually being met through Power Purchase Agreements (PPAs) with renewable energy suppliers. This shift not only reduces carbon emissions but also supports the development of new renewable energy projects in the UK [5].

Shenzhen, China, has transitioned its entire public bus fleet to electric vehicles, significantly cutting down on transportation emissions. The fleet, which comprises over 16,000 electric buses, has led to a yearly decrease of around 1.3 million tons of CO_2 emissions [6]. This transition is part of Shenzhen's broader efforts to combat air pollution and promote sustainable urban transportation, showcasing the city's commitment to environmental sustainability.

Copenhagen's bike-sharing system, effectively reduces transportation-related carbon emissions by promoting cycling. The system's extensive network of bike lanes and easy access to shared bicycles encourage residents to opt for cycling over driving. This initiative significantly decreases daily car trips and reduces urban carbon emissions by about 10,000 tons annually. Copenhagen's approach serves as a model for integrating cycling into urban transport planning to enhance sustainability [7] (Fig. 1).

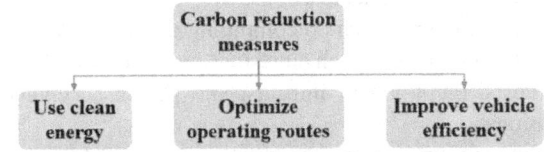

Fig. 1. Public transportation systems' carbon reduction measures

These cases illustrate successful strategies for reducing carbon emissions in public transportation through the adoption of clean energy, the promotion of electric vehicles, and the integration of cycling infrastructure.

2.2 Shared Mobility

New York City's Citi Bike is a textbook example of how bike-sharing programs can effectively reduce carbon emissions. By providing an accessible alternative to private cars for short trips, Citi Bike has replaced an estimated 30,000 car trips per month, significantly lowering greenhouse gas emissions [8]. The program increases the efficiency of bicycles by making them available to multiple users throughout the day, maximizing their utility and contributing to a more sustainable transportation system.

Companies like Lime and Bird offer electric scooter sharing in various cities, providing a convenient and eco-friendly alternative to car trips. Studies have shown that a significant portion of e-scooter trips replace car trips, leading to substantial reductions in CO_2 emissions. For instance, research in Portland, Oregon, found that 34% of e-scooter users would have used a car if the scooters were not available, demonstrating the effectiveness of electric scooter sharing in promoting sustainable urban mobility [9] (Fig. 2).

By analyzing these shared mobility programs, it is evident that they have a substantial impact on decreasing carbon emissions through decreased reliance on private cars and improved efficiency of transportation resources.

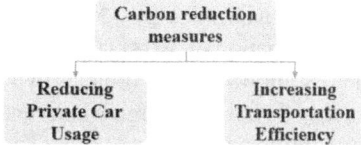

Fig. 2. Shared mobility's carbon reduction measures

2.3 Electrified Transportation

Norway leads the world in electric vehicle (EV) adoption, with EVs making up nearly 70% of new car sales [10]. The Norwegian government has implemented various incentive policies to promote EV usage, such as tax exemptions, free parking, and toll-free roads. These measures have significantly increased EV adoption, resulting in an estimated annual reduction of 500,000 tons of CO_2 emissions by 2019.

California's ZEV program aims to have 1.5 million zero-emission vehicles, on the road by 2025. The state has introduced mandates for automakers to produce and sell a certain percentage of zero-emission vehicles, along with providing purchase subsidies and tax incentives. This program has led to a substantial decrease in transportation-related emissions, with a reported reduction of 1.3 million tons of CO_2 in 2019, according to the California Air Resources Board (CARB) [11].

China has aggressively promoted new energy vehicles (NEVs) through an amalgamation of financial incentives, tax breaks, and the establishment of a comprehensive charging network. The government's dual credit policy incentivizes automakers to produce more NEVs. As a result, China has seen a significant increase in NEV adoption, leading to a reduction of approximately 5 million tons of CO_2 emissions in 2020, based on data from the China Association of Automobile Manufacturers [12] (Fig. 3).

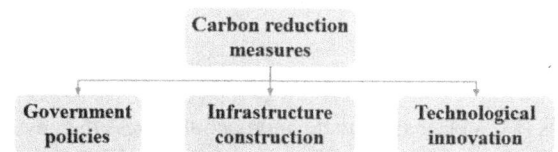

Fig. 3. Electrified transportation's carbon reduction measures

These cases demonstrate the significant effects of electrified transportation in achieving carbon emission reduction goals through government policy support, infrastructure construction and technological innovation.

2.4 Intelligent Transportation Applications

The Automated Traffic Surveillance and Control (ATSAC) system in Los Angeles uses artificial intelligence to manage and adjust traffic signals dynamically in real-time. Through real-time modification of signal timings in response to the existing traffic flow, ATSAC has significantly improved traffic flow and reduced congestion. Studies have shown that this system reduces travel times by 12% and fuel consumption by 13%, leading to an estimated annual reduction of 30,000 tons of CO_2 emissions [12].

The V2G program in the UK allows electric vehicles to interact with the power grid, facilitating bi-directional energy flow. EVs charge when renewable energy is abundant and discharge back to the grid during peak demand times. This integration helps balance grid loads, reduces the need for additional fossil fuel-based power generation, and supports grid stability. Research indicates that V2G could save approximately £3.5 billion per year in grid infrastructure costs and significantly reduce CO_2 emissions by utilizing renewable energy more effectively [13] (Fig. 4).

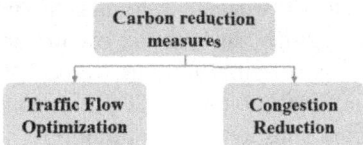

Fig. 4. Intelligent transportation applications' carbon reduction measures

These cases illustrate the potential of intelligent transportation systems to enhance traffic efficiency and significantly reduce carbon emissions through advanced AI and communication technologies.

3 Carbon Reduction Assessment

These cases demonstrate the significant effects of electrified transportation in achieving carbon emission reduction goals through government policy support, infrastructure construction and technological innovation.

3.1 Lifecycle Analysis

Life cycle analysis (LCA) is a systematic approach to assess the environmental impact of a product, process or service throughout its life cycle. LCA covers all stages from the extraction of raw materials, production, use to final treatment (such as recycling or disposal). By quantitatively evaluating resource consumption and pollution emissions at each stage, LCA can provide comprehensive environmental impact data. The process is shown in the following figure (Fig. 5).

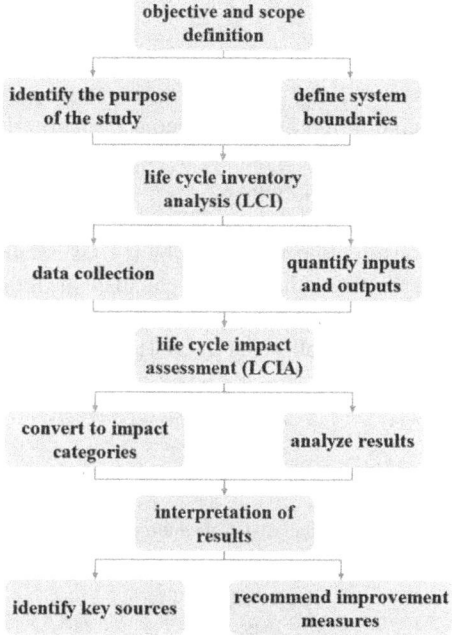

Fig. 5. LCA procedure

In LCA, the purpose of the study needs to be clarified, such as evaluating the carbon reduction effect of EVs relative to internal combustion engine vehicles (ICEVs), and the system boundaries need to be defined, including life cycle stages such as production, use, maintenance and recycling. Next, a Life Cycle Inventory (LCI) analysis is carried out to gather and measure the inputs (like energy, raw materials) and outputs (such as emissions, waste) for every phase. For example, when evaluating electric vehicles, it is necessary to consider battery production, power sources, and power consumption during vehicle production and use. Subsequently, there is the Life Cycle Impact Assessment (LCIA), which translates the information gathered during the inventory analysis into specific environmental impact categories, including potential for climate change, acidification, and eutrophication. The assessment of carbon reduction in transportation focuses on global warming potential. Finally, the results of the impact assessment are analyzed and interpreted to identify key sources of environmental impacts and emission reduction potential. For example, an LCA may show that electric vehicles have lower carbon emissions in the use phase, but higher carbon emissions in the battery production phase.

Through the LCA method, we can systematically and comprehensively evaluate the carbon emissions of sustainable transportation technologies and identify the carbon reduction potential at each stage, so as to formulate more effective carbon reduction strategies and promote the development of sustainable transportation.

3.2 Emission Inventories

Emission inventories are comprehensive databases that track and quantify the sources and amounts of greenhouse gases (GHGs) emitted by various transportation activities. These inventories provide detailed information on emissions from different modes of transport, such as cars, trucks, airplanes, and ships, helping to identify the major contributors to GHG emissions within the transportation sector.

The process of creating emission inventories involves collecting data from various sources, including fuel consumption records, vehicle usage statistics, and direct measurements of emissions. This data is then used to calculate the total emissions of specific greenhouse gases, such as CO_2, CH_4, and N_2O, for a given time period and geographic area. The process is shown in the following figure (Fig. 6).

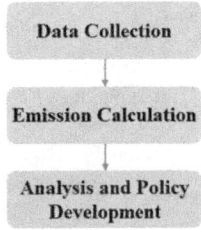

Fig. 6. Emission inventories' procedure

First, collect comprehensive data on transportation activities, such as fuel consumption records, vehicle registration data, travel activity surveys, and direct emission measurements, covering various modes of transportation such as cars, trucks, buses, trains, airplanes, and ships, and different fuel types (such as gasoline, diesel, and electricity). Then, use these data and emission factors to calculate the total greenhouse gas emissions of the transportation sector and generate a detailed emission inventory. Finally, analyze these data, identify the main sources of emissions, formulate and adjust policy measures, such as promoting low-emission vehicles, improving fuel efficiency, and optimizing public transportation systems, and monitor progress through regular updates of the inventory to ensure that sustainable transportation goals are achieved.

3.3 Simulation Models

Simulation model is a method that uses computer technology to construct mathematical models of transportation systems to predict the impact of different transportation scenarios on carbon emissions. These models can simulate changes in various transportation modes and policy measures, assess their potential impact on carbon emissions, and help formulate and optimize transportation carbon reduction strategies (Fig. 7).

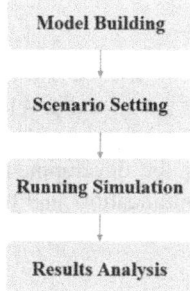

Fig. 7. Simulation Models' procedure

First, define the system boundaries by determining the simulation scope, including modes of transportation (e.g., road, rail, air) and geographic areas (e.g., cities, regions, countries). Collect relevant data on transportation activities, such as traffic flow, vehicle types, fuel usage, and emission factors. Establish a baseline scenario of the current transportation system to serve as a reference for evaluating other scenarios. Create alternative scenarios involving different policies or technologies, such as increasing the number of electric vehicles, improving public transport, or implementing carbon taxes, and simulate their impact on carbon emissions. Use computational models to run these scenarios and calculate carbon emissions for each. Finally, compare the emission results of baseline and alternative scenarios to assess the effectiveness of various measures, and conduct a sensitivity study to assess the influence of different input variables on the outcomes of the simulation, ensuring the robustness and reliability of the models.

3.4 Field Measurements

Field measurement is a method of assessing the environmental impact of vehicles by directly collecting emission and fuel consumption data from actual vehicles. Compared with laboratory tests or model simulations, field measurement can provide more realistic and accurate data, reflecting the performance of vehicles under actual operating conditions.

Field measurement methods collect emissions and fuel consumption data from actual vehicles under real operating conditions by using devices such as portable emission measurement systems (PEMS), on-board diagnostic systems (OBD) and flow meters. These devices can record multiple emissions including CO_2, CO, nitrogen oxides (NO_X) and particulate matter (PM), along with the vehicle's fuel use in real time. By testing under various driving conditions and road types, field measurement methods can provide highly accurate and widely applicable environmental impact data.

Although field measurement methods have the advantages of high accuracy, dynamic monitoring and diversity, they also face challenges such as high cost, large amount of data processing and the impact of environmental variables. This method is widely used in policy making, technology evaluation and urban planning, providing scientific basis and data support for governments, automakers and urban traffic management, thereby helping to achieve sustainable transportation and carbon reduction goals.

4 Technology Outlook

With the continuous advancement of science and technology, sustainable transportation systems will usher in more innovation and development opportunities. This section explores the potential of emerging technologies in reducing carbon emissions in the transportation sector, emphasizing the development of autonomous vehicles (AVs), Vehicle-to-Everything (V2X) communication, and big data analytics, alongside policy recommendations to support their adoption and integration.

4.1 New Technology Development

AVs can optimize driving patterns and reduce fuel consumption through efficient traffic management. V2X technology facilitates communication between vehicles and infrastructure, enhancing traffic efficiency and minimizing periods of idleness. Big data analytics can analyze traffic patterns, optimize routes, and predict maintenance needs, enhancing overall efficiency.

4.2 Potential Carbon Reduction

These technologies have the potential to significantly reduce carbon emissions by optimizing various aspects of transportation. Autonomous driving can reduce fuel consumption by up to 20% by minimizing stop-and-go driving. V2X technology can cut emissions by optimizing traffic flow, reducing congestion, and improving fuel efficiency. Additionally, big data analytics can enhance overall efficiency by predicting and preventing traffic congestion, leading to further reductions in carbon emissions.

4.3 Policy Recommendations

Governments should invest in V2X infrastructure and smart traffic systems to support these technologies. Providing tax incentives and subsidies will encourage the adoption of AVs and V2X technologies. Additionally, developing regulations and standards is crucial for the safe and efficient deployment of these technologies. Public-private partnerships should be encouraged to foster collaboration between governments, tech companies, and automotive manufacturers, accelerating innovation and implementation.

5 Conclusion

In conclusion, achieving sustainable transportation requires a combination of policy measures, technological advancements, and infrastructure investments. The successful examples discussed in this paper highlight the critical role of interdisciplinary collaboration and innovation in driving down transportation-related carbon emissions. Future research should focus on integrating various low-carbon technologies into cohesive systems, assessing their economic and social impacts, and enhancing public acceptance to maximize the benefits of sustainable transportation initiatives.

Acknowledgement. This work was sponsored by the Technology Project of State Grid Corporation of China (1400-2024400301A-1-1-ZN).

References

1. Reis, J., Costa, J., Marques, P., et al.: Sustainable transport: a systematic literature review. In: International Conference on Flexible Automation and Intelligent Manufacturing, pp. 898–908. Springer, Cham (2024)
2. Kwilinski, A., Lyulyov, O., Pimonenko, T.: Reducing transport sector CO2 emissions patterns: environmental technologies and renewable energy. J. Open Innovation: Technol. Market, Complexity **10**(1), 100217 (2024)
3. Montoya-Torres, J., Akizu-Gardoki, O., Alejandre, C., et al.: Towards sustainable passenger transport: Carbon emission reduction scenarios for a medium-sized city. J. Clean. Prod. **418**, 138149 (2023)
4. Greater London Authority: Mayor's plan to power Tube network with 100% renewable energy. Press release. http://www.london.gov.uk/press-releases/mayoral/plans-to-power-tfl-network-with-green-energy. Accessed 10 Jul 2024
5. Burges Salmon: Burges Salmon advises Transport for London on first steps to powering Tube on 100% renewable energy. Press release. http://www.burges-salmon.com/news-and-insight/press-releases/burges-salmon-advises-transport-for-london-on-first-steps-to-powering-tube-on-100-renewable-energy. Accessed 10 Jul 2024
6. World Bank: Electrification of Public Transport: A Case Study of the Shenzhen Bus Group. http://documents.worldbank.org. Accessed 10 Jul 2024
7. Copenhagen Carbon Neutral Cities Alliance: Copenhagen – CNCA. http://carbonneutralcities.org/cities/copenhagen/. Accessed 10 Jul 2024
8. NYC DOT: NYC DOT, Lyft Unveil New York City's First Electrified Citi Bike Charging Stations. http://www.nyc.gov/press-releases/mayoral/plans-to-power-tfl-network-with-green-energy. Accessed 10 Jul 2024
9. Portland Bureau of Transportation: E-Scooter Findings Report. http://www.portlandoregon.gov/transportation/article/709719. Accessed 10 Jul 2024
10. European Environment Agency: Electric vehicles in Europe. http://www.eea.europa.eu/themes/transport/electric-vehicles/electric-vehicles-in-europe. Accessed 10 Jul 2024
11. California Air Resources Board (CARB): California's Advanced Clean Cars Midterm Review. http://ww2.arb.ca.gov/resources/documents/advanced-clean-cars-midterm-review. Accessed 10 Jul 2024
12. International Energy Agency (IEA): Global EV Outlook 2020. http://www.iea.org/reports/global-ev-outlook-2020. Accessed 10 Jul 2024
13. American Association of State Highway and Transportation Officials: Los Angeles Automated Traffic Surveillance and Control (ATSAC) System. http://www.transportation.org/atsac. Accessed 10 Jul 2024
14. Energy Advances: Can vehicle-to-grid facilitate the transition to low carbon energy systems? http://pubs.rsc.org. Accessed 10 Jul 2024
15. Hauschild, M.Z.: Introduction to LCA methodology. In: Life Cycle Assessment: Theory and Practice, pp. 59–66 (2018)
16. Bender, M.A., Farach-Colton, M.: The LCA problem revisited. In: LATIN 2000: Theoretical Informatics: 4th Latin American Symposium. Punta del Este, Uruguay, April 10–14, 2000 Proceedings, pp. 88–94. Springer, Berlin Heidelberg (2000)
17. Li, M., Liu, H., Geng, G., et al.: Anthropogenic emission inventories in China: a review. Natl. Sci. Rev. **4**(6), 834–866 (2017)

18. Peters, G.P.: From production-based to consumption-based national emission inventories. Ecol. Econ. **65**(1), 13–23 (2008)
19. Liu, Z., Qiu, Z.: A systematic review of transportation carbon emissions based on CiteSpace. Environ. Sci. Pollut. Res. **30**(19), 54362–54384 (2023)
20. Tong, Y., Li, H., Pang, L.: China's tourism transportation carbon emissions: dynamic mechanisms and multi-regulatory strategies simulation, pp. 1–27. Environment, Development and Sustainability (2024)

Author Index

A
Abdolbaghi, Ghazal 74
Ahmadov, Emil 162
Alvarez, Arjel 311
Atalay, Ali Serdar 118
Aziziaghdam, Elif Toy 118

B
Bairy, Akhila 106
Bao, Hongqing 282
Baudru, Julien 3
Bersini, Hugues 3
Brocchini, Lorenzo 260

C
Cao, Qian 406
Cao, Yaoguang 83
Chen, Liang-Kuang 50
Chen, Mingjie 83
Chen, Yiyang 271
Comi, Antonio 173

D
de Guzman, Mark 311
Dinar, Yousuf 38

E
Ergün, Salih 118

F
Fang, Chen 416
Farahnakian, Farshad 220
Fränzle, Martin 106
Fukuda, Daisuke 360

G
Gao, Kun 63
Gertz, Carsten 38

Gilb, Tom 324
Gong, Weifeng 83
Guang, Haoran 83

H
Heikkonen, Jukka 220
Herkiloğlu, Oğuzhan 118
Hua, Chunrong 396
Hürten, Christian 188

I
Imbugwa, Gerald B. 324

J
Jarofka, Maximilian 188
Jiang, Luqing 295

K
Kanak, Alper 118
Karcı, Ahu Ece Hartavi 118
Khan, Imdad Ullah 396
Kim, Dong-Ju 98
Kim, Hyo-Jin 98
Kracht, Frédéric E. 188
Kuo, Chia-Chiun 50

L
Lee, Chang-Yeop 98
Li, Bin 295, 340
Li, Jing 406
Liu, Shu 416
Long, Chao 375
Luo, Xia 282

M
Maaß, Jacqueline Bianca 38
Mazzara, Manuel 324
Mu, Junyi 396

N
Nakamura, Kaori 360
Nevalainen, Paavo 220

P
Paglia, Christian 384
Papagni, Guglielmo 25
Pratelli, Antonio 260

Q
Qian, YongSheng 200
Qian, Yongsheng 246
Queck, Elena 38
Quinto, Ronnel C. 311

R
Reuss, Hans-Christian 63
Riemer, Thomas 63
Rindone, Corrado 173
Rosario, Roberto D. 311
Rubenis, Aivars 375
Russo, Francesco 173
Rzayev, Ramin 162

S
Schramm, Dieter 188
Schrammel, Johann 25
Shi, Shanshan 416
Shi, Yi 83
Song, Bo 200
Su, Qiming 282
Su, Shun 360
Suh, Young-Joo 98
Sun, Hu 295, 340
Susilo, Yusak 360
Sutanto, Jason 188

T
Trucco, Paolo 406
Tscheligi, Manfred 25

V
Vähämäki, Tanja 220
Vincenti, Michele 137

W
Wang, Hongjie 282
Wang, Xin 396
Wang, Xuan 246
Wei, Xu 200, 246
Weinrich, Ulrike 63
Wen, Zhiyong 147
Weng, Xiaoxiong 147

X
Xiao, Shi 295, 340
Xie, Bangquan 147
Xu, Biao 375

Y
Yang, Peijin 340
Yang, Shichun 83
Yazdizadeh, Alireza 74
Yu, Hongzhang 396

Z
Zafari, Setareh 25
Zeng, JunWei 200
Zeng, Junwei 246
Zhang, Zhiqiang 340
Zhao, Jun 209
Zhao, Yajie 295
Zheng, Bowen 83
Zheng, Jingwen 295
Zhong, Zhihong 295, 340
Zhou, Lin 396

Made in the USA
Monee, IL
03 May 2026